DISRUPTING THE DIGITAL HUMANITIES

BEFORE YOU START TO READ THIS BOOK, take this moment to think about making a donation to punctum books, an independent non-profit press,

@ https://punctumbooks.com/support/

If you're reading the e-book, you can click on the image below to go directly to our donations site. Any amount, no matter the size, is appreciated and will help us to keep our ship of fools afloat. Contributions from dedicated readers will also help us to keep our commons open and to cultivate new work that can't find a welcoming port elsewhere. Our adventure is not possible without your support.

Vive la Open Access.

Fig. 1. Hieronymus Bosch, *Ship of Fools* (1490–1500)

First published in 2018 by punctum books, Earth, Milky Way.
https://punctumbooks.com

ISBN-13: 978-1-947447-71-4 (print)
ISBN-13: 978-1-947447-72-1 (ePDF)

LCCN: 2018948914
Library of Congress Cataloging Data is available from the Library of Congress

Book design: Vincent W.J. van Gerven Oei

HIC SVNT MONSTRA

DISRUPTING THE DIGITAL HUMANITIES

EDITED BY
DOROTHY KIM &
JESSE STOMMEL

Difference Is Our Operating System

Cathy N. Davidson

The following preface is revised from a blog post originally pub-lished at HASTAC (Humanities, Arts, Science, and Technology Alliance and Collaboratory, hastac.org) on August 3, 2011.[1] This collection takes Cathy Davidson's words as a challenge, a point of departure, an opening to new questions.

It was our HASTAC Scholars Director Fiona Barnett who, in 2011 coined our pithy HASTAC motto: "Difference is not our deficit; it's our operating system." Fiona has a talent for getting to the heart of the matter.

What Fiona's defining quote states so well is that, if you begin from the conviction that everything that powers you — your op-erating system — is grounded in asking the question of "how are we different?" then the answers themselves won't look ordinary. That is our starting place: "What makes this network unique?"

The same is true of any enterprise, including this volume, *Dis-rupting the Digital Humanities*. There may or may not be right answers but there certainly are limiting questions. Depending

---

1   See Cathy N. Davidson, "'Difference Is Our Operating System' — Fiona Bar-nett," *HASTAC,* 3 August 2011.

on what questions you ask, the answers are not just different but restrictive. They can be blinders that obscure other options and better solutions.

Questioning our questions — are they open enough? What are we missing? — is key. What about this particular way of doing things is intrinsically and determinedly different than other ways of doing things? What are we leaving out? What are the range of differences by which we are defining ourselves? What are we for — and against? What new ways of seeing the world are we espousing? What blurring do we see between the binaries that have shaped academe — thinking vs. doing, imagining vs. building, form vs. content, academic vs. non-academic, K-12 vs. higher ed, formal vs. informal learning, technology vs. humanities. Such binaries need to be blurred, if not entirely erased, across all disciplines, departments, and fields as well as across all of the identity categories you can ever imagine and in all combinations you've ever seen or heard of if we are to arrive at a DH defined by difference, not by simple automation and replication of the past.

These are deeply, profoundly epistemological questions: questions about how we know the world — and how we know it differently depending on what we bring to it. Within those epistemological questions, there are three interconnected areas of interest that also help to shape the ways that, if we are disruptive enough, we can learn better together:

1. *New Media*: How does new media afford us different ways of interacting, motivating, connecting, and networking than do previous forms of communication and publication? How can we envision and build new and different communities and connections as we design new media? How can we safeguard those communities from not just disruption but invasion, theft, insecurity, and manipulation while still promoting openness? How can we develop new tools, not for the sake of new tools, but because they offer the possibilities of retooling, of powering different, important, creative new ways of thinking and acting in the world?

2. *Critical Thinking*: How can we make sure that, in our excitement to create new tools, new communities, new interactions, we always ask ourselves the why? How? Who is included? Who is excluded? What is the cost — human and material? What is the benefit — human and material? What different contribution does this make? How is it different? Does that difference matter? How do critical thinking, creative thinking, and computational thinking all support and contest one another and help us arrive at a more equitable world?

3. *Participatory or Blended Learning*: In or out of academe, in or out of school, in the classroom and in the streets, in the classroom and globally distributed across networks, HASTAC has been dedicated to maximizing the affordances of new media to model new and different ways of learning together. When we began in 2002, we didn't know that we would that we would be dubbed, by the National Science Foundation no less, the "world's first and oldest academic social network" — older than Facebook or Twitter, older even than MySpace, and older than the oldest academic science network nanoHUB. What we knew was that, given the affordance of openness, we wanted to support a community in which every participant felt both safe and free to participate openly. Anyone can register. Once you register, anyone can contribute, so long as it is relevant to HASTAC's broad mission and respectful of other community members. That doesn't sound so hard but it turns out those principles — a respectful, free, diverse, welcoming, open community that does not exploit its users' data for any commercial purpose — is an anomaly.

Anomaly — by definition, that which deviates from the standard, normal and expected — is a *good thing*.

Anomaly — difference — is a value to be supported, embraced, cherished, and rewarded. It is a core value in *Disrupting the Digital Humanities*.

If all we do is produce and reproduce the same hierarchical, traditional, rote forms of learning and measuring the content of what we learn, then we have failed our principle that "difference is our operating system." All the possibilities of interactive learning, of "collaboration by difference," a methodology that selects difference as a key and defining principle, must be embodied by what we do. And that means in projects, not just in ideas, that are carried out in real world settings.

"Difference is not our deficit; it's our operating system": I am honored that the editors of this volume have asked me to return to this 2011 blog post announcing the core value of difference and have made it a kind of celebratory banner for this volume. Let's wave that flag high! When you have difference as a mandate, you can make a difference in the world.

HASTAC has grown from a group of about twenty of us from the worlds of academe, the arts, and technology to a network over 15,500 members strong. As a technology and organization, it certainly does not run itself (any more than this volume magically put itself together). However, as a community of ideas, it exists and thrives, growing every day, precisely because it offers a platform for openness and difference and does not dictate, in advance, how individuals and groups use that platform to advance their ideas. Rather, we seek to communicate, model, and help anyone who wishes to contribute to explore what possibilities are here. Even better, we encourage the community to go for it! The administration is centralized but the content is contributed by users, day in and day out, and going strong.

I believe that the editors of this volume asked me to contribute this Preface because *Disrupting the Digital Humanities* begins from a parallel conviction and commitment. The volume bookmarks a similar place of difference. In the "Introduction," the editors describe the contributors as a "motley crew": historically, that means a diverse and even antic group made of people and things of different colors, a "we" defined not by similarity but a commitment to a community of difference.

"We" can only be as different as those of us committed to changing the rules, changing the game, changing the boundaries, changing the limits, changing the questions and therefore the answers. *Differently*.

Bibliography

Davidson, Cathy N. "'Difference Is Our Operating System' — Fiona Barnett." HASTAC. 3 August 2011. https://www.hastac.org/blogs/cathy-davidson/2011/08/03/difference-our-operating-system-fiona-barnett.

Contents

§Jeremiad

§Labor

§Networks

§Play

§Structure

§Conclusion

Disrupting the Digital Humanities: An Introduction

Dorothy Kim & Jesse Stommel

Much of this introduction was written before the world went to shit. The chapters here, no matter how recently written, can't keep at bay a world being actively undone. We find ourselves wondering why and how this work even matters. What has the digital humanities community done collectively for #BlackLivesMatter? What place is there for pedagogy in a world where education has been so systematically devalued, where students worry that even their classroom isn't safe from an ICE raid? How do we rally when so many are complicit? Scholarship can only vaguely hope to keep up. And so these are not really the questions of this volume. But they should be. As a field like digital humanities squabbles, the world around it is laid to waste. Academic turf wars have no place in a world of mass-shootings, fear-mongering, xenophobia, and white supremacy. Demanding fellow scholars do a literature review before speaking their mind has no place in a world of AR-15 assault rifles and weaponized algorithms. When something as basic as going to the bathroom lacks dignity for so many, we have no use for double-blind peer review.

All too often, defining a discipline becomes more an exercise of exclusion than inclusion.

Disrupting the Digital Humanities seeks to rethink how we map disciplinary terrain by directly confronting the gatekeeping impulse of many other so-called field-defining collections. What is most beautiful about the work of the digital humanities is exactly the fact that it *can't* be tidily anthologized — that it's messy and pushes in uncomfortable ways. The desire to neatly define the digital humanities (to filter the DH-y from the DH) is a way of excluding the radically diverse work that actually constitutes the field. Ultimately, it's exactly the fringes, the outliers, that make the digitial humanities both lovely *and* rigorous.

Many scholars originally were drawn to the digital humanities because we felt like outcasts, because we had been marginalized within the academic community. We gathered together because our work collectively disrupted the hegemony and insularity of the "traditional" humanities. Our work was collaborative, took risks, flattened hierarchies, shared resources, and created new and risky paradigms for humanities work. As attentions have turned increasingly toward the digital humanities, many of us have found ourselves more and more disillusioned. Much of that risk-taking, collaborative, community-supported, and open-to-all practice has been elided for a digital humanities creation-and-inclusion narrative that has made a turn towards traditional scholarship with a digital hand, government or institutionally funded database projects and tools, and a turn away from critical analysis of its own embedded practices, especially in relation to multilingualism, race, class, gender, sexuality, disability, and global praxis. This is not a new critique, and we have no desire to duplicate other efforts.[1] As Jessica Marie Johnson writes, "the humanities has a justice imperative that it has not

---

[1]   It has been most recently discussed in relation to digital humanities entanglement with the stakes of the neoliberal university. See David Allington, Sarah Brouillette, and David Golumbia's "Neoliberal Tools (and Archives): A Political History of Digital Humanities," *LA Review of Books,* 1 May 2015. See also Jacque Wernimont's "Whence Feminism? Assessing Feminist Interventions in Digital Literary Archives," *Digital Humanities Quarterly* 7, no. 1 (2013), as well as work by Martha Nell Smith, Alex Juhasz, #tranformdh, Adeline Koh, Deb Verhoeven, #thistweetcalledmyback, and many others.

quite fulfilled as a mission (even as individuals continue to work and push that)."[2] Our contributors point to a myriad of places where this work has productively begun. There is nothing novel about asking these questions. There is nothing novel about being professionally marginalized for asking these questions.

This collection does not constitute yet another reservoir for the new digital humanities canon. We are not positing how the "big tent" has expanded and how that canon may or may not look. We are arguing, instead, for a digital humanities that is irrevocably destabilized from the outset. Matthew K. Gold and Lauren F. Klein's recent introduction to *Debates in the Digital Humanities* (2016) articulates a vision of the DH canon:

> We posit the book as a reflection of the current, site-specific conditions of the field. In the multivalent shape of its arguments, progressing across a range of platforms and environments, *Debates in the Digital Humanities 2016* offers a vision of DH as an expanded field — a vision of new possibilities, differently structured.[3]

While this is important work, our aim is less about surveying, assembling, or re-assembling the field's structures or conversations as it is about creating points of entry to dialogue. In the words of James Weldon Johnson (NAACP, 1930s), our collection is about "creative disorder,"[4] about moving the margins to the center. Building a truly communal space for the digital humanities requires that we approach that space with a commitment to: 1) creating open and non-hierarchical dialogues; 2) championing non-traditional work that might not otherwise be recognized through conventional scholarly channels; 3) amplifying

2 Melissa Dinsman, "The Digital in the Humanities: An Interview with Jessica Marie Johnson," *LA Review of Books,* 23 July 2016.

3 Lauren F. Klein and Matthew K. Gold, "Digital Humanities: The Expanded Field," in *Debates in the Digital Humanities,* eds. Matthew K. Gold and Lauren F. Klein (2016).

4 Derek Bell, *And We Are Not Saved: The Elusive Quest for Racial Justice* (New York: Basic Books, 1989), 70.

marginalized voices; 4) advocating for students and learners; and 5) sharing generously to support the work of our peers. *Disrupting the Digital Humanities* is more than a mere time capsule, more than just disciplinary navel-gazing. Our aim in gathering this material is to construct something that uses all of the talk about what the digital humanities is and isn't as a jumping off point for a much deeper inquiry about education as social justice, the future of higher education, and what it is to be radically and diversely human in the digital age.

Disrupting the Digital Humanities offers a rowdy assemblage of works brought together, published open-access and in print. We have commissioned new chapters and are also republishing pieces that have stirred conversation elsewhere. However, we would not assume to anthologize the best of disrupted or disruptive DH. Rather, our goal is to bring to the surface voices that aren't adequately heard in mainstream discussions of the digital humanities.

Our contributors are a motley crew prodding at the constraints of conventional academic prose. Contributors work within a diverse array of digital humanities subfields, including postcolonial, queer, critical race, disability, radical librarianship, feminist digital humanities, adjunct DH, public humanities, and digital pedagogy. The goal is to make more space for broader perspectives in the digital humanities, to bring otherwise marginalized voices (or bits of voices no matter how small) to the fore. The collection includes critique, manifestos, art, poetry, play, listicles, and other forms.

dis·rup·tion
/dis'rəpSH(ə)n/

In feminist critical race theory, black, indigenous, and women of color (BIWOC) bodies disrupt the narratives of mainstream white feminism by having voices, by creating counternarratives, by calling out the frameworks of the hegemonic center. Thus we take for this volume the productive term "disruption" in the same vein, to decenter the digital humanities. We reimagine DH as not the seamless products of the neoliberal academy, non-profit "philanthropists," fascist government, and the military industrial complex, but as the work of people, labor, and voices at the margins creating friction and fantasy, mapping edges and new locations, playing slanted and in glitches with distributed resources and global communities.

In tech circles, the word "disruption" has come to mean something altogether different and more insidious.[5] The tech industry is saturated with a rhetoric that imagines "disruptive innovation" as a system of creative disruption when in fact it is more often influenced by profit, efficiency, and the faux-revolution of technology bound up in the trappings of commerce. "Disruptive innovation" has co-opted the term "disruption" as a far more sanitizing mechanism whereby increasing the efficiency and spreadability of the capitalist status quo becomes a so-called "revolution." And so, we here reclaim the word "disruption" in order to rehabilitate it and to return its originally intended critical heft.

Though we will not offer an exhaustive history of the use of the term "disruption," we do wish to point to an extensive bibliography that grapples with this term and theory in critical race studies.[6] In a 1989 article, Richard Delgado writes, "the stories

5   See Audrey Watters' piece, "The Myth and Millenialism of Disruptive Innovation," in this volume for a full discussion of this term in tech circles.

6   See Catharine A. MacKinnon, *Only Words* (Cambridge: Harvard University Press, 1993); Mari J. Matsuda, Charles R. Lawrence II, Richard Delgado, and Kimberlé Crenshaw, *Words That Wound: Critical Race Theory, Assaultive Speech, and the First Amendment* (Boulder: Westview Press, 1993); Richard Delgado, ed., *Critical Race Theory: The Cutting Edge* (Philadelphia: Temple University Press, 1995); Gloria Ladson-Billings and William F. Tate IV, "Toward a Critical Race Theory of Education," *Teachers College Record* 97 (1995): 47–68; Kimberlé Crenshaw, Neil Gotanda, Gary Peller, and Kendall Thomas, *Critical Race Theory: The Key Writings That Formed the Movement* (New York: The New Press, 1995); Richard Delgado, *When Equality Ends: Stories about Race and Resistance* (Boulder: Westview Press, 1999); Gloria Ladson-Billings, "Just What is Critical Race Theory and What's It Doing in a Nice Field Like Education," in *Race Is…Race Isn't: Critical Race Theory and Qualitative Studies in Education,* eds. Lawrence Parker, Donna Deyhele, and Sofia Villenas, 7–39 (Boulder: Westview Press, 1999); Gloria Ladson-

of outgroups aim to subvert that ingroup reality."[7] He continues, "stories and counterstories, to be effective, must be or must appear to be noncoercive. They invite the reader to suspend judgment, listen for their point or message, and then decide what measure of truth they contain. They are insinuative, not frontal; they offer a respite from linear, coercive discourse."[8] Delgado's idea of counternarrative very much resonates with the purpose of our collection. As Charles R. Lawrence III explains in his 1990 discussion of civil rights protest and speech, "we are aware that the struggle for racial equality has relied heavily on the persuasion of peaceful protest protected by the first amendment, but experience also teaches us that our petitions often go unanswered until they disrupt business as usual and require the self-interested attention of those persons in power."[9] Social justice and equity are as urgent now, and much of the "business as usual" in need of disrupting is currently fortified by the tech industry.

"Disruption," as a critical term is not something that tech circles have invented, but rather have erased or coopted from the work of scholars on race. The terrain of our collection highlights

---

Billings, "Racialized Discourses and Ethnic Epistemologies," in *Handbook of Qualitative Research,* 2nd edn., eds. N. Denzin and Y. Lincoln (Thousand Oaks: Sage, 2000); Richard Delgado and Jean Stefancic, eds., *Critical Race Theory: An Introduction,* 1st edn. (New York: New York University Press, 2001; republished 2012), 144; Lani Guinier and Gerald Torres, *The Miner's Canary: Enlisting Race, Resisting Power, Transforming Democracy* (Cambridge: Harvard University Press, 2002); Laurence Parker and Marvin Lynn, "What's Race Got to Do With It? Critical Race Theory's Conflicts with and Connections to Qualitative Research Methodology and Epistemology," *Qualitative Inquiry* 8 (February 2002): 7–22; and Sharon M. Chubbuck, "Whiteness Enacted, Whiteness Disrupted: The Complexity of Personal Congruence," *American Educational Research Journal* 41, no. 2 (2004): 301–33.

7   Richard Delgado, "Storytelling for Oppositionists and Others: A Plea for Narrative," *Michigan Law Review* 87, no. 8: Legal Storytelling (1989): 2411–41, at 2413.

8   Ibid, 2415.

9   Charles R. Lawrence III, "If He Hollers Let Him Go: Regulating Racist Speech on Campus," *Duke Law Journal* (1990): 431–83, at 466–67.

the critical importance of subjectivity and autoethnography in the work of resisting oppressive systems:

> Critical race theory writing embraces an experimentally grounded, oppositionally expressed, and transformatively aspirational concern with race and other socially constructed hierarchies. [...] The narrative voice, the teller, is important to critical race theory in a way not understandable by those whose voices are tacitly deemed legitimate and authoritative. The voice exposes, tells and retells, signals resistance and caring, and reiterates the most fearsome power — the power of commitment to change.[10]

Counternarratives from the margins, as Garrett Albert Duncan writes, "provide potent counterpoints to challenge existing narratives."[11] Counternarratives or counterstories[12] in this

---

10 "Who's Afraid of Critical Race Theory," in *The Derrick Bell Reader, Critical America,* eds. Richard Delgado and Jean Stefancic, 78–84 (New York: New York University Press 2005), at 80.

11 Garret Albert Duncan, "Critical Race Ethnography in Education: Narrative, Inequality and the Problem of Epistemology," *Race, Ethnicity and Education* 8, no. 1 (2005): 93–114. The article can also be found in Adrinne D. Dixson and Celia K. Rousseau, eds., *Critical Race Theory in Education: All God's Children Got a Song,* 191–212 (New York: Routledge, 2006), at 200.

12 Sabina Vaught writes: "Counterstorytelling is used to challenge grand narratives of Whiteness and its self-characterization as the norm (Ladson-Billings 2000). [...] Counterstorytelling was adopted as a CRT methodology for a number of purposes: first, in the hope 'that well-told stories describing the reality of black and brown lives can help readers bridge the gap between their world and those of others' (Delgado and Stefancic 2001, 41); second, it provides people of color a means by which to 'name their own reality' (Choe 1999; Hermes 1999; Ladson-Billings 1999); third, counterstories can disrupt and challenge the totalizing, erasing discourse of dominant White society in transformative and liberatory ways (Parker and Lynn 2002)." See Sabina Vaught, *Racism, Public Schooling, and the Entrenchment of White Supremacy: A Critical Race Ethnography* (Albany: State University of New York Press, 2011), 19. Vaught cites Ladson-Billings, "Racialized Discourses and Ethnic Epistemologies"; Delgado and Stefancic, *Critical Race Theory: An Introduction*; Lena Domyung Choe, "Negotiating Borders of Consciousness in the Pursuit of Education: Identity Politics and Gender of Second Generation Korean American Women," in *Race Is...Race Isn't,* 205–30;

volume narrate away from the center of the digital humanities which has been consistently and frequently imagined as a white, male, able-bodied, cisgendered, heternormative space.[13]

This assemblage takes the critical term "disruption" and plays with these theoretical strands to produce multiple-voiced counternarratives about the hegemonic center of digital humanities. Together these different, non-straight approaches, disrupt the formation of DH and its definitions of itself by slanting the outlook, mixing and blending hierarchical frameworks, shaking up the terms and frames. One can't substantively "include" everyone without disrupting hierarchies — without transforming the field's critical lens and practice. Different geographies, different languages, and different modes of scholarship demand new frames. Put simply, the digital humanities must reimagine itself, and its boundaries, in order to make way for a more radically inclusive and activist community.

An origin counternarrative

This collection was inspired by two open conversations at the Modern Language Association's 2015 and 2016 conferences. Many of the chapters evolved from short position papers re-

---

Mary Hermes, "Research Methods as a Situated Response: Toward a First Nation's Methodology," in *Race Is...Race Isn't,* 83–100; Ladson-Billings, "Just What Is Critical Race Theory"; and Parker and Lynn, "What's Race Got to Do with It?"

13 "Counterstories purposely disrupt stock stories by telling personal accounts that contradict the stock stories. [...] These are stories that people of color of all economic backgrounds, and of all educational and professional attainments, tell." See Sherry Marx, "Critical Race Theory," in *The SAGE Encyclopedia of Qualitative Research Methods,* eds. Lisa M. Given, 164–68 (Thousand Oaks: SAGE Publications, 2008), 166.

However, we also take the criticism of counternarratives as potentially a form of "empathic fallacy" that particularly is done for "white sympathies" (see Vaught, *Racism, Public Schooling, and the Entrenchment of White Supremacy,* 19). See also Richard Delgado, *The Coming Race War?: And Other Apocalyptic Tales of America after Affirmative Action and Welfare* (New York: New York University Press, 1996) and Delgado and Stefancic, *Critical Race Theory.*

leased in advance of these conversations. Selections from these chapters were published on disruptingdh.com and also spurred conversation on social media and across the open web. "Disrupting the Digital Humanities" was, according to MLA Commons, the most tweeted session at MLA 2015 in Vancouver. It was again one of the most tweeted sessions at MLA 2016.

The conversation at these sessions included panelists, voices from virtual contributors, and extensive audience interaction. At both sessions, panelists offered brief opening remarks with a facilitator "leading" discussion, but papers were published openly in advance, effectively "flipping" the conference presentation. The time spent together during the session was used mostly for dialogue and debate between panelists and between the panel and audience. At MLA 2015, Dorothy Kim absented herself from the panel to make way for the voices of #thistweetcalledmyback, a group of black, indigenous, and women of color activists working in digital spaces.[14] We broadened this further during MLA 2016 (in light of the Presidential theme, "Literature and Its Publics: Past, Present, and Future") by hosting a live Twitter chat (on #digped and #disruptingDH) leading up to and during the session. Our aim was to question the boundaries of the digital humanities as an academic discipline and to redirect our work outward towards an ecosystem of publics.

To launch the conversation at our first panel, Sean Michael Morris asked a series of questions, which have helped frame the work of this collection:

---

14 Academia and DH particularly must ask why these public writers, citizens, activists did not opt to contribute their discussions and work to this volume. Is academic work already too attuned to the "empathic fallacy" of white academic audiences (academia's demographics makes this particularly clear)? We are, thus, missing a vital counterstory and counternarrative to the mainstream, academic, white, hegemonic discussions of the digital humanities. It is an absence in our collection that narrates how the digital humanities, in the end, fails to allow adequate space for these voices to tell their stories. See Lauren Chief Elk et al., "This Tweet Called My Back," *Model View Culture*, 13 December 2014.

- What are the best ways for us to practice radical inclusion? How do we amplify marginalized voices, and what are the complications associated with trying to do so?
- Attempts have been made *ad nauseam* to define the digitial humanities. Can we come to some sense of DH that encapsulates the field and the work without delimiting it?
- Who is left out of the DH conversation? When and how have they been left out? Or perhaps more to the point, how has digital humanities been inscribed in such a way as to omit their work?
- Where is the conversation about digitial humanities really taking place? Is it in conference rooms like this one? If so, who is guiding that conversation? And if not, where are the richest conversations happening, and who is leading them?

And, in her position paper for our MLA 2016 panel (revised for this collection), Annemarie Perez offered one answer to this last question by describing her encounter with the digitial humanities at MLA 2012: "the rooms, crowded to bursting were visibly, notably white spaces. This was a bit jarring, but what was even more so was that no one was talking about this. No one was asking where the brown people were. The absence of racialized bodies was un-noted." She felt, as many others have (and still do), like the shapes of the rooms and who could comfortably occupy them had been determined (and delimited) in advance of her arriving.

We have no interest in duplicating the institutional structures of prestige and privilege that have already led to certain voices being left out of this conversation. We decidedly did not use a traditional CFP or peer-review process to choose panelists or chapters for the collection. Rather, we imagined this project as a grassroots effort from the start, less about championing the legitimacy of individual projects and more about questioning who decides what counts (and how). About reimagining legitimacy as less a product of gatekeeping and more a product of community building. On the other hand, this collection remains troublingly academic. And, after this process, we recognize the charge to continue to find ways to breach the gap between scholarly networks, academic publics, and extra-academic publics. To highlight and make central the "undercommons" and "maroon spaces" that Johnson has discussed.[15] But these are not just our gaps to breach; academia must also (and *first*) find ways to step aside so these publics can speak and we can listen. These publics must feel safe, compensated, credited in order for dialogue to emerge.

We have organized this book in a non-linear and overlapping set of keywords: Etymology, Play, Structure, Labor, Identity, Networks, Jeremiad. The pieces in this collection can be identified with at least one but often are identified with more than one of these keywords. In this way, the organization of this volume does not have a rigid linear structure. For example, Meg Worley's piece "The Rhetoric of Disruption" is an example of Etymology as she digs deeply into the history of the word "disruption" itself while simultaneously also being a narrative about DH's identity, and a subtle jeremiad about the impossibility of a homogenous DH community. She proposes, "that we camouflage ourselves with disruptive coloration. Let us play up our contrasts in order to simultaneously disguise and preserve the unity of the whole." Likewise, the Mongrel Coalition Against Gringpo's two poetic pieces show the contours of identity, jeremiads, and structures that give out a "GOLD STAR FOR FEELING 'MOVED' BY CLAUDIA RANKINE'S CITIZEN BUT BEING MOVED TO DO NOTHING IN AN ACTUALITY THAT MIGHT IMPACT YOU." We are delighted to include these pieces in the collection because they point so loudly to the problems of whiteness in academic and digital spaces: "GOLD STAR FOR PROTECTING YOUR NETWORK AND REFUSING TO CALL OUT YOUR RACIST BUDS OR PUBLICLY SUPPORT THOSE WHO DO. PROPS TO WHITENE$$!" Their work is also an example of a collective activist group, protesting ingeniously and playing with the limits of a digital platform to call out this whiteness.

---

15 See Dinsman, "The Digital in the Humanities: An Interview with Jessica Marie Johnson."

The volume also includes a discussion of MOOCs (a topic the digital humanities has proclaimed taboo) but through a postcolonial and global lens. As Maha Bali writes in "The 'Unbearable' Exclusion of the Digital":

> In spite of good intentions (and sometimes blatant pretensions of altruism and respect), platforms like the once wildly popular xMOOCs only exposed and intensified fake universality of design and practices in transnational higher education. Inclusion, we suggested, cannot be achieved by imposing or assuming local values as universal, representing others as tokens, refusing to look beyond those who are already in, denying the hegemony of power, or using stories of those who have bought in to suggest inclusion of everyone from everyone.

Their piece also traverses our keywords from structure to labor, identity, networks, and jeremiads.

Chris Bourg's piece on "The Library is Never Neutral" plainly explains one of the foundational slices in this collection:

> A fundamental tenet that undergirds this article, and frankly undergirds much of the work I have done in and for libraries, is the simple assertion that libraries are not now nor have they ever been merely neutral repositories of information. […] [W]e live in a society that still suffers from racism, sexism, ableism, transphobia and other forms of bias and inequity; but libraries also fail to achieve any mythical state of neutrality because we contribute to bias and inequality in scholarship, and publishing, and information access.

The digital humanities is not neutral and this is not a statement of passive reflection. The digital humanities is not neutral because — in its current big-tentish, expanding-terrainish configuration — it still does not (and sometimes refuses to) consistently, rigorously, methodically, theoretically bring the perspectives of the margins in relation to race, gender, disability, sexuality, etc.

into the center of its default discourse. Our data, our algorithms, our databases, our tools, our methodologies, our objects, our networks, our writing, our funding streams, our conferences are never neutral.

Disrupting the introduction

We want to end by acknowledging how hard it is to write an introduction to a book that aims to push productively at the edges of a field when that field is in turmoil. Our hope is that the brand of "disruption" we're offering here is not taken merely as critique but offers paths to formulate questions that facilitate a different path and view. We know this may not spur the collective move forward in the way we'd both hope for, but we will (and we hope the field will) stumble forward multiply nonetheless, finding increasing (not diminishing) points of entry to dialogue.

Our aim is not to agree with every word of every piece in this collection. Our aim is to push upon thinking (our own, our contributors, our readers) more than allow it to congeal into something like agreement. But all of the pieces here are motivated by a desire to make the digital humanities more open, more inclusive, more generous. There is anger among these pieces, and sadness and frustration, but also hope. Of the kind Paulo Freire advocates for, a hope that demands struggle, action, activism. This collection is about dreams and possibilities. Audre Lorde writes on the importance of poetry:

> Possibility is neither forever nor instant. It is not easy to sustain belief in its efficacy. [...] [W]e must constantly encourage ourselves and each other to attempt the heretical actions our dreams imply and some of our old ideas disparage. In the forefront of our move toward change, there is only our poetry to hint at possibility made real.[16]

---

16 Audre Lorde, *Sister Outsider: Essays and Speeches* (Berkeley: Crossing Press, 1984; reprint 2007), 38; available online at "'Poetry is Not a Luxury,' by Audre Lorde," *On Being with Krista Tippett* (blog), 23 July 2015.

Particularly in a moment of increasing fascism, deadly racism, virulent antifeminism, and violent transphobia, it is important to staunchly continue in the belief that dissent can make space for hope and community. Play, disruption, and the articulation of dreams and possibilities are not "a luxury" but an essential part of what it means to be human in a digital world. The digital humanities should never be so cloyingly academic as to turn its nose from the rigors of this kind of work — a very specific kind of rigor that might at times seem anathema to academia. A rigor that values dialogue over peer review, poetry over data, community over citation, asking honest questions over demands for evidence.

We are decidedly not defining DH. It is not for us to define. Not for any single voice or collection or discipline to define.

The goal of our collection is to highlight gaps, fissures, and points of productive contact. It is not a history. It is not a representative anthology. It is not even an intervention in any direct way. It is "creative disorder" interested in letting a rhizomatic counternarrative of the digital humanities speak, breathe, play. It offers no linear reading but asks its readers to forge their own narrative from our shifting assemblage. We are publishing an assortment of outliers and pieces that productively open (rather than police) the boundaries of DH. Together, they work to short circuit the worst tendencies of the increasingly corporate university that would have us constantly in competition with each other for limited resources. Rather, the work asks us to, as Jesse has said, "make friends as an act of radical political resistance."

Our aim is to leave no DH stone unturned and to revel in what we discover and what we can weave together from so many parts.

We end this introduction by invoking bell hooks from her book *Teaching to Transgress*: "The classroom, with all its limitations, remains a location of possibility. In that field of possibility we have the opportunity to labor for freedom, to demand of ourselves and our comrades an openness of mind and heart that allows us to face reality even as we collectively imagine ways to move beyond boundaries, to transgress."[17] We end this introduction by asking our readers to transgress, to resist, to hope, to protest, to play slant, to create communities, to demand change. Together. This is what matters.

17 bell hooks, *Teaching to Transgress: Education as the Practice of Freedom* (New York: Routledge, 1994), 207.

Bibliography

Allington, David, Sarah Brouillette, and David Golumbia. "Neoliberal Tools (and Archives): A Political History of Digital Humanities." *LA Review of Books*. 1 May 2015. https://lareviewofbooks.org/article/neoliberal-tools-archives-political-history-digital-humanities/.

Bell, Derek. *And We Are Not Saved: The Elusive Quest for Racial Justice*. New York: Basic Books, 1989.

Chief Elk, Lauren, et al. "This Tweet Called My Back." *Model View Culture*. 13 December 2014. https://modelviewculture.com/pieces/thistweetcalledmyback.

Choe, Lena Domyung. "Negotiating Borders of Consciousness in the Pursuit of Education: Identity Politics and Gender of Second Generation Korean American Women," *In Race Is… Race Isn't: Critical Race Theory and Qualitative Studies in Education,* edited by Lawrence Parker, Donna Deyhele, and Sofia Villenas, 205–30. Boulder: Westview Press, 1999.

Chubbuck, Sharon M. "Whiteness Enacted, Whiteness Disrupted: The Complexity of Personal Congruence." *American Educational Research Journal* 41, no. 2 (2004): 301–33. http://www.jstor.org/stable/3699368.

Crenshaw, Kimberlé, Neil Gotanda, Gary Peller, and Kendall Thomas. *Critical Race Theory: The Key Writings That Formed the Movement*. New York: The New Press, 1995.

Delgado, Richard. "Storytelling for Oppositionists and Others: A Plea for Narrative." *Michigan Law Review* 87, no. 8: Legal Storytelling (1989): 2411–41.

———, ed. *Critical Race Theory: The Cutting Edge.* Philadelphia: Temple University Press, 1995.

Delgado, Richard. *The Coming Race War?: And Other Apocalyptic Tales of America after Affirmative Action and Welfare*. New York: New York University Press, 1996.

———. *When Equality Ends: Stories about Race and Resistance*. Boulder: Westview Press, 1999.

———— and Jean Stefancic, eds. *Critical Race Theory: An Introduction.* 1st edition. New York: New York University Press, 2001; republished 2012.

Dinsman, Melissa. "The Digital in the Humanities: An Interview with Jessica Marie Johnson." *LA Review of Books.* 23 July 2016. https://lareviewofbooks.org/article/digital-humanities-interview-jessica-marie-johnson/.

Dixson, Adrienne D., and Celia K. Rousseau, eds. *Critical Race Theory in Education: All God's Children Got a Song,* 191–212. New York: Routledge, 2006.

Duncan, Garret Albert. "Critical Race Ethnography in Education: Narrative, Inequality and the Problem of Epistemology." *Race, Ethnicity and Education* 8, no. 1 (2005): 93–114. DOI: 10.1080/1361332052000341015.

Guinier, Lani, and Gerald Torres. *The Miner's Canary: Enlisting Race, Resisting Power, Transforming Democracy.* Cambridge: Harvard University Press, 2002.

Hermes, Mary. "Research Methods as a Situated Response: Toward a First Nation's Methodology." In *Race Is...Race Isn't: Critical Race Theory and Qualitative Studies in Education,* edited by Lawrence Parker, Donna Deyhele, and Sofia Villenas, 83–100. Boulder: Westview Press, 1999.

hooks, bell. *Teaching to Transgress: Education as the Practice of Freedom.* New York: Routledge, 1994.

Klein, Lauren F., and Matthew K. Gold. "Digital Humanities: The Expanded Field." In *Debates in the Digital Humanities,* edited by Matthew K. Gold and Lauren F. Klein. 2016. http://dhdebates.gc.cuny.edu/debates/text/51.

Ladson-Billings, Gloria, and William F. Tate IV. "Toward a Critical Race Theory of Education." *Teachers College Record* 97 (1995): 47–68.

————. "Just What is Critical Race Theory and What's It Doing in a Nice Field Like Education." In *Race Is... Race Isn't: Critical Race Theory and Qualitative Studies in Education,* edited by Lawrence Parker, Donna Deyhele, and Sofia Villenas, 7–39. Boulder: Westview Press, 1999.

————. "Racialized Discourses and Ethnic Epistemologies." In *Handbook of Qualitative Research.* 2nd edition, edited by N. Denzin and Y. Lincoln. Thousand Oaks: SAGE, 2000.

Lawrence III, Charles R. "If He Hollers Let Him Go: Regulating Racist Speech on Campus." *Duke Law Journal* (1990): 431–83.

Lorde, Audre. *Sister Outsider: Essays and Speeches by Audre Lorde.* Berkeley: Crossing Press, 1984; reprint 2007.

MacKinnon, Catharine A. *Only Words.* Cambridge: Harvard University Press, 1993.

Marx, Sherry. "Critical Race Theory." In *The SAGE Encyclopedia of Qualitative Research Methods,* vol. 1 Vol. 2, M–Z Index, edited by Lisa M. Given, 164–68. Thousand Oaks: SAGE Publications, 2008.

Matsuda, Mari J., Charles R. Lawrence II, Richard Delgado, and Kimberlé Crenshaw. *Words That Wound: Critical Race Theory, Assaultive Speech, and the First Amendment.* Boulder: Westview Press, 1993.

Parker, Laurence, and Marvin Lynn. "What's Race Got to Do With It? Critical Race Theory's Conflicts With and Connections to Qualitative Research Methodology and Epistemology." *Qualitative Inquiry* 8 (February 2002): 7–22.

"'Poetry is Not a Luxury,' by Audre Lorde." 23 July 2015. *On Being with Krista Tippett* (blog). http://www.onbeing.org/program/words-shimmer/feature/poetry-not-luxury-audre-lorde/318.

Vaught, Sabina E. Racism, *Public Schooling, and the Entrenchment of White Supremacy: A Critical Race Ethnography.* Albany: State University of New York Press, 2011.

Wernimont, Jacque. "Whence Feminism? Assessing Feminist Interventions in Digital Literary Archives." *Digital Humanities Quarterly* 7, no. 1 (2013). http://www.digitalhumanities.org/dhq/vol/7/1/000156/000156.html.

"Who's Afraid of Critical Race Theory." In T*he Derrick Bell Reader, Critical America,* edited by Richard Delgado and

Jean Stefancic, 78–84. New York: New York University Press 2005. http://www.jstor.org/stable/j.ctt9qg47z.12.

1

A Letter to the Humanities: DH Will Not Save You

Adeline Koh[1]

I am often asked about the digital humanities and how it can update, make relevant, and provide funding for many a beleaguered humanities department. Some faculty at underfunded institutions imagine DH is going to revitalize their discipline — it's going to magically interest undergraduates, give faculty research funding, and exponentially increase enrollment.

Well, the reality is this: what has until recently been commonly understood as *real* "Digital Humanities" is already belated and is *not* going to save humanities departments from ever bigger budget cuts and potential dissolution.

Yes, of course, everyone will tell you that there are multiple debates over what actually defines Digital Humanities as a field, whether it *is* a field or not, yadda yadda yadda. But the projects which have until very recently dominated the federal digital humanities grants — the NEH grants, the ACLS grants, among others — *are* by default, the definition of the field, or the "best" the field has to offer. This means that until very recently and with few exceptions, the list of awardees rarely includes digital work

---

1   Originally published as Adeline Koh, "A Letter to the Humanities: DH Will Not Save You," *Hybrid Pedagogy,* 19 April 2015.

that focuses more on culture than computation, projects that focus on digital pedagogy, or digital recovery efforts for works by people of color.[2]

If you look through the projects that have been funded in the last decade you're going to see a lot of repeated themes. Heck, even when you look at the roster for who is being invited to give DH talks and what they are talking about, you see many of the same names and the same topics. You're going to see a lot of emphasis on tools. A lot of emphasis on big data analysis. A lot of emphasis on computation, and the power of computation. What *aren't* you going to see as much of? Emphasis on *why computing,* the *conditions under which computing is manufactured,* a *cultural analysis* of the ideologies of computing. Why is that?

Because "digital humanities" is currently defined in many existing works as coming out of a field previously known as "humanities computing."[3] This field is cast as the primary antecedent for what is now called the digital humanities, immortalized by the publication of the *Blackwell Companion to Digital Humanities,* in which the term switched from "humanities computing" to "Digital Humanities," the use of DH in forming the Alliance of Digital Humanities Organizations as an umbrella global organization, and the development and naming of the NEH ODH branch.[4] "Humanities computing" projects have primarily focused on digitization of canonical texts, text encoding and markup, the creation of tools to facilitate humanities research, and more recently, "big data" and ways to study it, such as "topic modeling."[5] Uniformly, advocates of DH as humanities comput-

2   "Announcing 17 Digital Humanities Start-Up Grant Awards," N*ational Endowment for the Humanities,* 24 March 2015. See also Amy E. Earhart, "Can Information Be Unfettered? Race and the New Digital Humanities Canon," in *Debates in the Digital Humanities,* ed. Matthew K. Gold, 309–18 (Minneapolis: University of Minnesota Press, 2012).

3   See Matthew Kirschenbaum, "What is Digital Humanities and What's It Doing in English Departments?" in *Debates in the Digital Humanities,* 3–11, at 3.

4   Ibid., 3–6.

5   For example, see Ed Folsom and Kenneth M. Price, eds., *The Walt Whitman Archive,* http://www.whitmanarchive.org.

ing have argued that DH is, in the words of Matt Kirschenbaum, "more akin to a common methodological outlook than an investment in any one specific set of texts or even technologies."[6]

This focus on methodology is important, because throughout the majority of humanities computing projects, the social, political, and economic underpinnings, effects, and consequences of methodology are rarely examined. Too many in this field prize method without excavating the theoretical underpinnings and social consequences of method. In other words, humanities computing has focused on using computational tools to further humanities research, and not to study the effects of computation *as a humanities question*.

But "digital humanities" in the guise of "humanities computing," "big data," "topic modelling," "object oriented ontology" is *not* going to save the humanities from the chopping block. It's only going to push the humanities further over the precipice. Because these methods alone make up a field which is simply a handmaiden to STEM. Think about this: Why would you turn to a pseudo-STEM field that uses STEM methods to answer your questions, rather than to STEM directly? Indeed, when I brought up "critical making"—what some consider to be the perfect marriage of "yack" and "hack"[7]—with my engineer spouse, he commented, "Isn't engineering already 'critical making'?" The editorial preface to an article on critical making by Matt Ratto describes critical making as "processes of material and conceptual exploration and creation of novel understandings by the makers themselves."[8] After mulling over my husband's remark, I realized that engineering is indeed already practicing critical making as its DH practitioners often prescribe it—arguably better than they are. But in relation to the humanities, engineering does not integrally inspect critical identity categories, access and

6   Kirschenbaum, "What is Digital Humanities," 4.

7   Natalia Cecire, "When DH Was in Vogue; or, THATCamp Theory," *Works Cited* (blog), 19 October 2011.

8   Matt Ratto, "Critical Making," in *Open Design Now: Why Design Cannot Remain Exclusive,* eds. Bas van Abel et al. (Amsterdam: BIS Publishers, 2011).

privilege in the process of making, issues that designate what the humanities considers to be "critical."

Another thing: if you want to start a DH program to save your probably much underfunded humanities department from extinction, trying to practice DH the way resource-rich, research-oriented institutions do might be prohibitively expensive. Big data analysis, 3D printing, tool-building: these are expensive endeavors to undertake, even on a small scale. Because of their mission and resources, the majority of non-wealthy, non-R1 institutions are going to concentrate on smaller-scale projects involving undergraduate students. These are not normally the sorts of projects that receive federal funding for DH.

So this is what I want to say: If you want to save humanities departments, champion the new wave of digital humanities: one which has humanistic questions at its core. Because the humanities, centrally, is the study of how people process and document human cultures and ideas, and is fundamentally about asking critical questions of the methods used to document and process. And because these questions can and should be dealt with by people in departments who care about research with undergraduates, by people without the resources to develop the latest and greatest cutting edge digital humanities tool (which, quite frankly, will be enveloped by commercial industries in the blink of an eye).

So instead of pouring more money into tool building or the latest and greatest 3D printer, let's not limit the history of the digital humanities to humanities computing as a single origin point. Let's consider "sister fields" to the digital humanities as actually *foundational* to the digital humanities.[9] Consider work with undergraduates and digital pedagogy (Rebecca Frost Davis, Kathryn Tomasek, Katherine D. Harris, Angel David Nieves, Janet Simons, Jesse Stommel, Sean Michael Morris)[10] as foundational

---

9   Adeline Koh, Twitter post, 11 April 2015, 2:03 p.m.

10   Their Twitter handles are as follows: Rebecca Frost Davis (@FrostDavis), Kathryn Tomasek (@KathrynTomasek), Katherine D. Harris (@triproftri), Angel David Nieves (@angeldnieves), Janet Simons (@janettsimons), and Jesse Stommel (@Jessifer). See also Jesse Stommel, "Articles by Author,"

to the field. Consider the work of scholars who engage media studies as foundational — especially as they deeply engage with questions of race and ethnicity, gender and sexuality, ability and the digital (Lisa Nakamura, Anna Everett, Alondra Nelson, Tara McPherson, Elizabeth Losh, Alexandra Juhasz, Wendy Chun, Cathy Davidson, Fiona Barnett, David Theo Goldberg, David Golumbia, Martha Nell Smith, Cheryl E. Ball, Edmond Chang, Anastasia Salter, Carly Kocurek, Jessie Daniels, Amy Earhart, Anne Cong-Huyen, Alexis Lothian, Radhika Gajjala, Carol Stabile, Nishant Shah, Michelle Moravec, Monica Mercado, Simone Browne, Moya Bailey, Brittney Cooper & the Crunk Collective, etc.).[11] Consider Sandra Harding and the postcolonial and feminist work of Science and Technology studies foundational to the field.[12] Consider HASTAC, FemTechNet, and FemBot foundational initiatives, none of whom have ever received NEH funding for their operations, but have been instrumental to the

---

Hybrid Pedagogy, http://www.digitalpedagogylab.com/hybridped/author/jessifer.

11 Twitter handles include: Lisa Nakamura (@lnakamur), Alondra Nelson (@alondra), Tara McPherson (@tmcphers), Elizabeth Losh (@lizlosh), Wendy Chun (@whkchun), Cathy Davidson (@CathyNDavidson), Fiona Barnett (@fiona_barnett), David Golumbia (@dgolumbia), Martha Nell Smith (@MarthaNellSmith), Cheryl E. Ball (@s2ceball), Edmond Chang (@edmondchang), Anastasia Salter (@AnaSalter), Carly Kocurek (@sparklebliss), Jessie Daniels (@JessieNYC), Amy Earhart (@amyeetx), Anne Cong-Huyen (@anitaconchita), Alexis Lothian (@alothian), Radhika Gajjala (@cyberdivalivesl), Carol Stabile, (@castabile), Michelle Moravec (@ProfessMoravec), Monica Mercado (@monicalmercado), Simone Browne (@wewatchwatchers), Moya Bailey (@moyazb), and Brittney Cooper (@ProfessorCrunk). See also the academic profiles and webpages for the following: Anna Everett (http://www.filmandmedia.ucsb.edu/people/faculty/everett/everett.html), Alexandra Juhasz (http://pzacad.pitzer.edu/~ajuhasz/), David Theo Goldberg (http://www.faculty.uci.edu/profile.cfm?faculty_id=4716), Nishant Shah (http://www.leuphana.de/universitaet/personen/nishant-shah.html), and Crunk Feminist Collective (http://www.crunkfeministcollective.com).

12 Sandra Harding, ed., *The Postcolonial Science and Technology Studies Reader* (Durham: Duke University Press, 2011).

recent shift in federal digital humanities awards towards the "H" in DH rather than the "D."[13]

The insistent focus on computing and methodology in the humanities without incisive, introspective examination of their social implications is devaluing the humanities. We shouldn't be pouring federal money into building tools without making the ideological structure of the process explicit and their social effects and presuppositions open to inspection; we shouldn't be funding the digitization of canonical (read: white, often male) authors without the simultaneous digitization of works by people of color, especially women of color. To do both is to betray some of the most important lessons which the humanities has learned with the rise of women, gender and sexuality studies, race, ethnic and postcolonial studies, and disability studies.

Instead, let's reconsider what "core" digital humanities means. Let's redefine what we mean by the "best," most critical, and seminal digital humanities research. Let's open digital humanities research to people who don't have the time and resources to learn a programming language like R,[14] but are happy to use Wordle as an entry into literary texts as data. Let's consider pedagogy central to DH. Let's consider class, race, ethnicity, gender, sexuality, ability, nationality primary to and constitutional of the digital humanities, not simply the "diversity box" of political correctness. Let's remember the fringe fields and movements who did this in the past, but did not receive widespread support and funding, as part of the central history of DH. Only when we completely reconfigure and re-center the *humanities* in DH will we be able to talk about using the field to "save" humanities departments from extinction.

13 *HASTAC*, https://www.hastac.org; *FemTechNet*, http://femtechnet.newschool. edu; and *FemBot*, http://fembotcollective.org.

14 *The R Project for Statistical Computing*, https://www.r-project.org.

Bibliography

"Announcing 17 Digital Humanities Start-Up Grant Awards."
National Endowment for the Humanities. 24 March 2015.
http://www.neh.gov/divisions/odh/grant-news/announcing-
17-digital-humanities-start-grant-awards-march-2015.

Bailey, Moya (@moyazb). Twitter feed. https://twitter.com/
moyazb.

Ball, Cheryl E (@s2ceball). Twitter feed. https://twitter.com/
s2ceball.

Barnett, Fiona (@fiona_barnett). Twitter feed. https://twitter.
com/fiona_barnett.

Browne, Simone (@wewatchwatchers). Twitter feed. https://
twitter.com/wewatchwatchers.

Cecire, Natalia. "When DH Was in Vogue; or, THATCamp
Theory." *Works Cited* (blog). 19 October 2011. http://
nataliacecire.blogspot.com/2011/10/when-dh-was-in-vogue-
or-thatcamp-theory.html.

Chang, Edmond (@edmondchang). Twitter feed. https://
twitter.com/edmondchang.

Chun, Wendy (@whkchun). Twitter feed. https://twitter.com/
whkchun.

Cong-Huyen, Anne (@anitaconchita). Twitter feed. https://
twitter.com/anitaconchita.

Cooper, Brittney (@ProfessorCrunk), Twitter feed. https://
twitter.com/ProfessorCrunk.

Crunk Feminist Collective. http://www.crunkfeministcollective.
com.

Daniels, Jessie (@JessieNYC). Twitter feed. https://twitter.com/
JessieNYC.

Davidson, Cathy (@cathyndavidson). Twitter feed. https://
twitter.com/cathyndavidson.

Earhart, Amy (@amyeetx). Twitter feed. https://twitter.com/
amyeetx.

———. "Can Information Be Unfettered? Race and the
New Digital Humanities Canon." In *Debates in the
Digital Humanities,* edited by Matthew K. Gold, 309–18.

Minneapolis: University of Minnesota Press, 2012. http://
dhdebates.gc.cuny.edu/debates/text/16.

Everett, Anna. "People." *University of California Santa Barbara,
Film and Media Studies.* http://www.filmandmedia.ucsb.
edu/people/faculty/everett/everett.html.

FemBot. http://fembotcollective.org.

FemTechNet. http://femtechnet.newschool.edu.

Folsom, Ed, and Kenneth M. Price, eds. *The Walt Whitman
Archive.* http://www.whitmanarchive.org.

Frost Davis, Rebecca (@FrostDavis). Twitter feed. https://
twitter.com/FrostDavis.

Gajjala, Radhika (@cyberdivalivesl). Twitter feed. https://
twitter.com/cyberdivalivesl.

Goldberg, David Theo. "Faculty Profile." *University of
California, Irvine.* http://www.faculty.uci.edu/profile.
cfm?faculty_id=4716.

Golumbia, David (@dgolumbia). Twitter feed. https://twitter.
com/dgolumbia.

Harding, Sandra, ed. *The Postcolonial Science and Technology
Studies Reader.* Durham: Duke University Press, 2011.

Harris, Katherine D (@triproftri). Twitter feed. https://twitter.
com/triproftri.

HASTAC. https://www.hastac.org.

Juhasz, Alexandra. "Faculty Profile." *Pitzer College.* http://
pzacad.pitzer.edu/~ajuhasz.

Kirschenbaum, Matthew. "What is Digital Humanities and
What's It Doing in English Departments?" In *Debates in
the Digital Humanities,* edited by Matthew K. Gold, 3–11.
Minneapolis: University of Minnesota Press, 2012.

Kocurek, Carly (@sparklebliss). Twitter feed. https://twitter.
com/sparklebliss.

Koh, Adeline. "A Letter to the Humanities: DH Will Not
Save You." *Hybrid Pedagogy.* April 19, 2015. https://
hybridpedagogy.org/a-letter-to-the-humanities-dh-will-
not-save-you/.

————. (@adelinekoh). Twitter post. 11 April 2015, 2:03 p.m.
https://twitter.com/adelinekoh/status/586952634579288064.

Losh, Elizabeth (@lizlosh). Twitter feed. http://www.twitter. com/lizlosh.

Lothian, Alexis (@alothian). Twitter feed. https://twitter.com/ alothian.

McPherson, Tara (@tmcphers). Twitter feed. http://twitter. com/tmcphers.

Mercado, Monica (@monicalmercado). Twitter feed. https:// twitter.com/monicalmercado.

Moravec, Michelle (@ProfessMoravec). Twitter feed. https:// twitter.com/ProfessMoravec.

Nakamura, Lisa (@lnakamur). Twitter feed. https://twitter. com/lnakamur.

Nelson, Alondra (@alondra). Twitter feed. https://twitter.com/ alondra.

Nieves, Angel David (@angeldnieves). Twitter feed. https:// twitter.com/angeldnieves.

The R Project for Statistical Computing. https://www.r-project. org.

Ratto, Matt. "Critical Making." In *Open Design Now: Why Design Cannot Remain Exclusive,* edited by Bas van Abel et al. Amsterdam: BIS Publishers, 2011. http://opendesignnow. org/index.php/article/critical-making-matt-ratto.

Salter, Anastasia (@AnaSalter). Twitter feed. https://twitter. com/AnaSalter.

Shah, Nishant. "Personen." *Universität Lüneburg.* http://www. leuphana.de/universitaet/personen/nishant-shah.html.

Simons, Janet (@janettsimons). Twitter feed. https://twitter. com/janettsimons.

Smith, Martha Nell (@MarthaNellSmith). Twitter feed. https:// twitter.com/MarthaNellSmith.

Stabile, Carol (@castabile). Twitter feed. https://twitter.com/ castabile.

Stommel, Jesse (@Jessifer). Twitter feed. https://twitter.com/ Jessifer.

———. "Articles by Author." *Hybrid Pedagogy.* http://www. digitalpedagogylab.com/hybridped/author/jessifer.

Tomasek, Kathryn (@KathrynTomasek). Twitter feed. https://
twitter.com/KathrynTomasek.

2

The Myth and the Millennialism of "Disruptive Innovation"

Audrey Watters[1]

> *Turning and turning in the widening gyre*
> *The falcon cannot hear the falconer;*
> *Things fall apart; the centre cannot hold;*
> *Mere anarchy is loosed upon the world,*
> *The blood-dimmed tide is loosed, and everywhere*
> *The ceremony of innocence is drowned;*
> *The best lack all conviction, while the worst*
> *Are full of passionate intensity.*

> *Surely some revelation is at hand;*
> *Surely the Second Coming is at hand.*
> *The Second Coming! Hardly are those words out*
> *When a vast image out of Spiritus Mundi*
> *Troubles my sight: a waste of desert sand;*
> *A shape with lion body and the head of a man,*
> *A gaze blank and pitiless as the sun,*
> *Is moving its slow thighs, while all about it*
> *Wind shadows of the indignant desert birds.*

---

1   Originally published as Audrey Watters, "The Myth and the Millennialism of 'Disruptive Innovation,'" *Hack Education,* May 24, 2013.

> *The darkness drops again but now I know*
> *That twenty centuries of stony sleep*
> *Were vexed to nightmare by a rocking cradle,*
> *And what rough beast, its hour come round at last,*
> *Slouches towards Bethlehem to be born?*
> — William Butler Yeats, "The Second Coming"[2]

Folklorists often balk at the common usage of the word "myth" to mean "lie." A myth, by their disciplinary definition, is quite the opposite. A myth is a culture's sacred story. It involves supernatural or supreme beings — gods. It explains origins and destinies. A myth is the Truth.

So when I say that "disruptive innovation" is one of the great myths of the contemporary business world, particularly of the tech industry, I don't mean by "myth" that Clayton Christensen's explanation of changes to markets and business models and technologies is a falsehood. (I have an MA in Folklore, not an MBA — so that's part of it, for sure.) Rather, my assigning "myth" to "disruptive innovation" is meant to highlight the ways in which this narrative has been widely accepted as unassailably true. No doubt (as a Harvard professor) Christensen has faced very little skepticism or criticism about his theory concerning the transformation of industries — why, it's as if *The Innovator's Dilemma* were some sort of sacred text.[3]

Helping to enhance its mythic status, the storytelling around "disruptive innovation" has taken on another broader and looser dimension as well, as the term is now frequently invoked in

---

2 William Butler Yeats, "The Second Coming," in *The Collected Poems of W. B. Yeats* (Hertfordshire: Wordsworth Poetry Library, 2000), 158. See also "The Second Coming," *Poetry Foundation,* http://www.poetryfoundation. org/poems-and-poets/poems/detail/43290.

3 Clayton M. Christensen, T*he Innovator's Dilemma: When New Technologies Cause Great Firms to Fail* (New York: Harvard Business School Press, 1997). See also Erwin Danneels, "Disruptive Technology Reconsidered: A Critique and Research Agenda," *Journal of Product Innovation and Management* 21, no. 4 (2004): 246–58.

many quarters to mean things quite different from Christensen's original arguments in *The Innovator's Dilemma.*

In this vein, almost every new app, every new startup, every new tech — if you believe the myth-making-as-marketing at least — becomes a disruptive innovation: limo-summoning iPhone apps (e.g., Uber), photo-sharing iPhone apps (e.g., Path), email on your iPhone (e.g., Mailbox), online payments (e.g., PayPal), electric vehicles (e.g., Tesla), cloud computing (e.g., Amazon Web Services), 3D printers (e.g., Makerbot), video-based lectures (e.g., Khan Academy), social search (e.g., Facebook Graph Search), the entire internet, etc., *ad nauseam.*[4]

The millennialism of disruptive innovation

The companies above might very well be innovative — in their technologies and their business models. That's beside the point if you're looking for disruption. Per Christensen's framework, these could also be "sustaining innovations" — that is, products and services that strengthen the position (and the profits) of incumbent organizations.[5]

But that's not the mythology embraced by the tech industry, which despite its increasing economic and political power, continues to see itself as an upstart rather than an incumbent.

And as a self-appointed and self-described disruptor, the tech industry seems to have latched on to the most millennial elements of Christensen's theories — that is, the predictions about the destruction of the old and the ascension of the new at the hands of technology: The death of the music industry. The death of newspapers. The death of print. The death of Hollywood. The death of books. The death of the Web. The death of RSS. The death of Microsoft.[6] All predicted to be killed — suddenly or

---

4   Matthew Yglesias, "Stop 'Disrupting' Everything: How a Once-Useful Concept Turned into a Meaningless Buzzword," *Slate,* 1 May 2013.

5   Christensen, *The Innovator's Dilemma,* xv.

6   See the following articles, respectively: Michael DeGusta, "The REAL Death of the Music Industry," *Business Insider,* 18 February 2011; Don Irvine, "New Study Predicts the Death of Newspapers in Five Years," *Accuracy in Media,*

gradually or in the library with a candlestick — by some sort of "disruptive innovation."

The structure to this sort of narrative is certainly a well-known and oft-told one in folklore — in tales of both a religious and secular sort. Doom. Suffering. Change. Then paradise.

People seemingly love to believe in the "end of the world as we know it" stories — for reasons that have to do with both the horrors of the now and the heaven of the future. Many cultures (and Silicon Valley is, despite its embrace of science and technology, no different here) tell a story that predicts some sort of cataclysmic event(s) that will bring about a radical cultural (economic, political) transformation and, eventually, some sort of paradise.

The Book of Revelations. "The Hollow Men." The Mayan Calendar. The Shakers. The Ghost Dance. Nuclear holocaust. Skynet. The Singularity.

I'll be the first to admit that the data in folklore professor Dan Wojcik's book *The End of the World As We Know It* is dated (um, he was my Master's Thesis advisor, circa 2000); he wrote the book in 1997 — oh! the same year that *The Innovator's Dilemma* was originally published![7] Wojcik's analysis of a sweeping societal belief in "the end of the world" was well-timed with the technological anxieties surrounding Y2K, making it an interesting and contrasting companion to Christensen's contention that we'll witness "the end" of certain organizations thanks to technological "innovation."

For his part Wojcik noted that, according to Nielsen, some 40% of Americans believed that there was nothing we could do to prevent nuclear holocaust. Sixty percent believed in Judg-

17 December 2011; Christopher Mims, "Predicting the Death of Print," *MIT Technology Review,* 23 August 2010; Michael White, Ronald Grover, and Andy Fixmer, "Jobs's Death Leaves Hollywood Without Trusted Tech Envoy," *Bloomberg Technology,* 7 October 2011; Leah Price, "Dead Again," *The New York Times,* 20 August 2012; Chris Anderson and Michael Wolff, "The Web Is Dead. Long Live the Internet," *Wired,* 17 August 2010; Steve Gillmor, "Rest in Peace, RSS," *TechCrunch,* 5 May 2009; and Paul Graham, "Microsoft Is Dead," blog post, April 2007.

7   Daniel Wojcik, *The End of the World as We Know It: Faith, Fatalism, and Apocalypse in America* (New York: New York University Press, 1997).

ment Day, 44% in the Battle of Armageddon, 44% in the Rapture.[8] He didn't say how many believed in Y2K. He didn't say how many believed in "disruptive innovation." He did not ask how many believed in "the singularity" and such.

I'd argue that despite its staid Harvard Business School origins, Christensen's "disruptive innovation" story taps into these same powerful narratives about the end-times — told, as always, by the chosen ones (be they Americans, Christians, Shakers, Heaven's Gate followers, survivalists, Java programmers, or "my generation"). Folks do seem drawn to these millennial stories, particularly when they help frame and justify our religious, moral, economic, political, cultural, social, technological worldview.

Adjustments to the disruptive innovation eschatology

Here are a couple of (education-related) end-times predictions from Clayton Christensen:

- In 15 years, half of US universities may be bankrupt.[9]
- By the year 2019 half of all classes for grades K–12 will be taught online.[10]
- Disruptive innovation will be, as TechCrunch (among other acolytes) is happy to profess, the end of school as we know it.[11]

Such is its inevitability, so the story goes, that new players can enter the education market and, even though their product is

---

8   Ibid., 7–8. Wojcik cites statistics from George Gallup, Jr., and Jim Castelli, *The People's Religion: American Faith in the Nineties* (New York: Macmillan, 1989), 4, as well as a survey from *U.S. News and World Report,* 19 December 1994, 64.

9   Mark Suster, "In Fifteen Years From Now Half of US Universities May Be in Bankruptcy: My Surprise Discussion with @ClayChristensen," *Both Sides,* 3 March 2013.

10 Courtney Boyd Myers, "Clayton Christensen: Why Online Education Is Ready for Disruption, Now," *The Next Web,* 13 November 2011.

11 Gregory Ferenstein, "How California's Online Education Pilot Will End College As We Know It," *TechCrunch,* 15 January 2013.

of lower quality[12] and appeals to those who are not currently "customers," oust the incumbent organizations. (Incumbents, in this case, are publicly funded, brick-and-mortar schools.) As Christensen and his co-authors argued in *Disrupting Class* in 2008, "disruption is a necessary and overdue chapter in our public schools."[13]

But like many millennialist prophets are wont to do when their end-times predictions don't quite unfold the way they originally envisioned, Clayton Christensen and his disciples at the Clayton Christensen Institute (which was recently renamed from the Innosight Institute) have just tweaked their forecast about (public) education's future. Five years post-*Disrupting Class,* "disrupting class" will look a bit different, they now say.

In May 2013, the organization released a new white paper, detailing a new path for transformation that winds a new future between the disruptive and sustaining innovations: they call them "hybrid innovations."[14]

"A hybrid is a combination of the new, disruptive technology with the old technology and represents a sustaining innovation relative to the old technology."[15]

It's an interesting revision (a refinement, really) of the organization's predictions in *Disrupting Class,* the book which first applied "disruptive innovation" to education technology and that argued online learning would be a way to "modularize the system and thereby customize learning."[16] (In other buzzwords, to "unbundle" and "personalize" education.)[17]

---

12 Clay Shirky, "Napster, Udacity, and the Academy," 12 November 2012.

13 Clayton M. Christensen, Michael B. Horn, and Curtis W. Johnson, *Disrupting Class: How Disruptive Innovation Will Change the Way the World Learns,* 2nd edn. (New York: McGraw-Hill, 2011), v.

14 Clayton M. Christensen, Michael B. Horn, and Heather Staker, "Is K–12 Blended Learning Disruptive? An Introduction to the Theory of Hybrids," (Clayton Christensen Institute, 2013).

15 Ibid., 2.

16 Clayton M. Christensen and Michael B. Horn, "Online Learning for Student-Centered Innovation," *Deseret News,* 8 March 2011.

17 Michael Staton, "Unbundling Education, A Simple Framework," 5 February 2012. See also "Leveraging Intelligent Adaptive Learning to Personalize

Not so fast, the organization now says. Hybrid innovation. "Blended learning." A little bit online and a little bit offline. And while middle and high schools (and colleges, although that isn't the subject of this latest white paper) might offer opportunities for "rampant non-consumption," — that is, classically, an opportunity for "disruption" — "the future of elementary schools at this point is likely to be largely, but not exclusively, a sustaining innovation story for the classroom."[18] Computer hardware and software and internet-access in the classroom, as those of us who've been thinking about education technology for decades now keep saying, won't necessarily change "everything." (Go figure.)

Of course, even in *Disrupting Class,* the predictions of the ed-tech end-times were already oriented towards changing the business practices, not (necessarily) the pedagogy or the learning. And the promise of a thriving education technology eschatology were already muted in Christensen's earliest formulation by the "restrictions" placed upon the education sector — restrictions by virtue of education being a public and not a private institution, of education not being beholden to market forces quite the same way that the other examples that the mythology of "disruptive innovation" has utilized to explain itself.

"People did not create new disruptive business models in public education, however. Why not? Almost all disruptions take root among non-consumers. In education, there was little opportunity to do that. Public education is set up as a public utility, and state laws mandate attendance for virtually everyone. There was no large, untapped pool of non-consumers that new school models could target."[19]

Education," *DreamBox Learning* (Project Tomorrow, 2012), http://www2.dreambox.com/personalize-education-wp.

18 Christensen, Horn, and Staker, "Is κ–12 Blended Learning Disruptive?," 33.

19 Christensen, Horn, and Johnson, *Disrupting Class,* 60.

Agitating for the end times

This Christensen Institute's latest white paper on "hybrids" clarifies then that the future of education isn't necessarily (or utterly or easily) "disrupted." There are limits to the predictions, to the predictive models, to the business school approach to education change and such. There are, for example, lots of non-consumers of learning (a necessary piece of the "disruptive innovation" framework) if you're willing to frame education as something that happens outside the officially sanctioned, brick-and-mortar institutions. But it's not so easy to woo "non-consumers" if you're really just focused on the market and policy and practices of an otherwise compulsory schooling setting. (And the distinction between "consumers," "non-consumers," "students," and "learners" is important too, although all get lumped into a consumption framework by Christensen.)

Like so many millennialist entities faced with the harsh realities of faltering predictions, the Innosight Institute (now under its new name) offers a new prediction.

But, let's be clear, the organization doesn't just predict the future of education. The Clayton Christensen Institute does not just offer models — business models — for the future. It does not simply observe an always changing (education) technology market. It has not simply diagnosed the changes due to technological advancements. It has not simply prophesied or predicted what future outcomes might be.

It's written a best-selling book (or two) about disruptive innovation. It has actively lobbied governments for certain aspects of its agenda (its mythology?), becoming a vocal proponent for its particular vision of a disrupted and innovative future. The Clayton Christensen Institute is a member of ALEC, for example, a corporate lobbying organization whose education initiatives include writing and pushing for legislation that enables the outsourcing of education to for-profit, online education providers and that eases the restrictions of entry to the market of the very virtual schools.[20]

"Over time," the new white-paper reads, "as the disruptive models of blended learning improve, the new value propositions will be powerful enough to prevail over those of the traditional classroom."[21] And so, according to the Christensen mythology, despite any sort of hesitation about the hybridity of disruption now, disruption will prevail.

And so, indeed, it is written. And so, it is told.

---

20 *SourceWatch* entry for the "ALEC Education Task Force": http://www.source-watch.org/index.php/ALEC_Education_Task_Force. See also Audrey Watters, "Google Summit Answers Oregon Educators' Questions about Apps for Edu," *Hack Education,* 8 October 2010. I was contacted by a spokesperson from the Clayton Christensen Institute saying that "we have never been a member of ALEC" and asking for a correction to that effect in this article. ALEC itself does not publish the list of its members, so I cannot verify information that way. The institute is listed as a member on the ALEC *Exposed* site: http://www.alecexposed.org/wiki/ALEC_Exposed. It is also referenced by the public accountability group Little Sis as an education reform group endorsed by ALEC: http://littlesis.org/lists/188-education-reform-organiza-tions-endorsed-by-alec/members.

21 Christensen, Horn, and Staker, "Is K–12 Blended Learning Disruptive?," 41.

Bibliography

ALEC *Exposed.* http://www.alecexposed.org/wiki/ALEC_
Exposed.

Anderson, Chris, and Michael Wolff. "The Web Is Dead. Long
Live the Internet." *Wired.* 17 August 2010. http://www.wired.
com/2010/08/ff_webrip/.

Christensen, Clayton M. *The Innovator's Dilemma: When New
Technologies Cause Great Firms to Fail.* New York: Harvard
Business School Press, 1997.

———, Michael B. Horn, and Curtis W. Johnson. *Disrupting
Class: How Disruptive Innovation Will Change the Way the
World Learns.* 2nd edition. New York: McGraw-Hill, 2011.

——— and Michael B. Horn. "Online Learning for Student-
Centered Innovation." *Deseret News.* 8 March 2011. http://
www.deseretnews.com/article/700116326/Online-learning-
for-student-centered-innovation.html?pg=all.

———, Michael B. Horn, and Heather Staker. "Is K–12
Blended Learning Disruptive? An Introduction to the
Theory of Hybrids." Clayton Christensen Institute, 2013.
http://www.christenseninstitute.org/publications/hybrids.

Danneels, Erwin. "Disruptive Technology Reconsidered:
A Critique and Research Agenda." *Journal of Product
Innovation and Management* 21, no. 4 (2004): 246–58. DOI:
10.1111/j.0737-6782.2004.00076.x.

DeGusta, Michael. "The REAL Death of the Music
Industry." *Business Insider.* 18 February 2011. http://www.
businessinsider.com/these-charts-explain-the-real-death-of-
the-music-industry-2011–2.

DreamBox Learning. "Leveraging Intelligent Adaptive Learning
to Personalize Education." Project Tomorrow, 2012. http://
www2.dreambox.com/personalize-education-wp.

Ferenstein, Gregory. "How California's Online Education Pilot
Will End College As We Know It." *TechCrunch.* 15 January
2013. http://techcrunch.com/2013/01/15/how-californias-
new-online-education-pilot-will-end-college-as-we-know-
it.

Gallup, George, Jr., and Jim Castelli. *The People's Religion: American Faith in the Nineties.* New York: Macmillan, 1989.

Gillmor, Steve. "Rest in Peace, RSS." *TechCrunch.* 05 May 2009. https://techcrunch.com/2009/05/05/rest-in-peace-rss/.

Graham, Paul. "Microsoft Is Dead." Blog post. April 2007. http://www.paulgraham.com/microsoft.html.

Irvine, Don. "New Study Predicts the Death of Newspapers in Five Years." *Accuracy in Media.* 17 December 2011. http://www.aim.org/don-irvine-blog/new-study-predicts-the-death-of-newspapers-in-five-years/.

LittleSis. "Education Reform Organizations Endorsed by ALEC." http://littlesis.org/lists/188-education-reform-organizations-endorsed-by-alec/members.

Mims, Christopher. "Predicting the Death of Print." *MIT Technology Review.* 23 August 2010. https://www.technologyreview.com/s/420329/predicting-the-death-of-print/.

Myers, Courtney Boyd. "Clayton Christensen: Why Online Education Is Ready for Disruption, Now." *The Next Web.* 13 November 2011. http://thenextweb.com/insider/2011/11/13/clayton-christensen-why-online-education-is-ready-for-disruption-now.

Poetry Foundation. "The Second Coming." http://www.poetryfoundation.org/poems-and-poets/poems/detail/43290.

Price, Leah. "Dead Again." *The New York Times.* 20 August 2012. http://www.nytimes.com/2012/08/12/books/review/the-death-of-the-book-through-the-ages.html.

Shirky, Clay. "Napster, Udacity, and the Academy." 12 November 2012. http://www.shirky.com/weblog/2012/11/napster-udacity-and-the-academy.

SourceWatch. "ALEC Education Task Force." http://www.sourcewatch.org/index.php/ALEC_Education_Task_Force.

Staton, Michael "Unbundling Education, A Simple Framework." 5 February 2012. http://edumorphology.com/2012/02/unbundling-education-a-simple-framework.

Suster, Mark. "In Fifteen Years From Now Half of US
    Universities May Be in Bankruptcy: My Surprise Discussion
    with
    @ClayChristensen." *Both Sides.* 3 March 2013. https://
    bothsidesofthetable.com/in-15-years-from-now-half-
    of-us-universities-may-be-in-bankruptcy-my-surprise-
    discussion-with-979f93bd6874#.t5cvedlu4.

U.S. News and World Report. 19 December 1994.

Watters, Audrey. "Google Summit Answers Oregon Educators'
    Questions about Apps for Edu." *Hack Education.* 8 October
    2010. http://hackeducation.com/2010/10/08/google-
    summit-answers-oregon-educators-questions-about-apps-
    for-edu.

———. "The Myth and the Millennialism of 'Disruptive
    Innovation.'" *Hack Education.* 24 May 2013. http://
    hackeducation.com/2013/05/24/disruptive-innovation.

White, Michael, Ronald Grover, and Andy Fixmer. "Jobs's
    Death Leaves Hollywood Without Trusted Tech Envoy."
    Bloomberg Technology. 7 October 2011. http://www.
    bloomberg.com/news/articles/2011–10–06/steve-jobs-s-
    death-leaves-hollywood-without-trusted-silicon-valley-
    envoy.

Wojcik, Daniel. *The End of the World as We Know It: Faith,
    Fatalism, and Apocalypse in America.* New York: New York
    University Press, 1997.

Yeats, William Butler. *The Collected Poems of W. B. Yeats.*
    Hertfordshire: Wordsworth Poetry Library, 2000.

Yglesias, Matthew. "Stop 'Disrupting' Everything: How a Once-
    Useful Concept Turned into a Meaningless Buzzword."
    Slate. 1 May 2013. http://www.slate.com/articles/business/
    moneybox/2013/05/disrupting_disruption_a_once_useful_
    concept_has_become_a_lame_catchphrase.html.

3

The Rhetoric of Disruption:
What Are We Doing Here?

Meg Worley

I started out hating the title of this volume. Disruption has a special place in the mythology of my native land of Silicon Valley:
Every startup promises disruptive technologies that will change
the industry forever, and "Don't think outside the box — blow
up all the boxes!" is written in invisible ink on the business card
of every VC on Sand Hill Road. "Why do we want to *disrupt* the
digital humanities?" I muttered to myself. Why would we want
to borrow the rhetoric and methodology of Kleiner Perkins
(which was known as a hegemonic boys club long before Ellen
Pao sued them)? Is the Harvard MBA Program really the place to
turn for new models of innovation in humanistic inquiry? Can't
we effect change — deep, meaningful change — without adopting the language of one of the most inequitable neighborhoods
of late-stage capitalism? But in the course of trying to find a
place for my thoughts about community, conversation, and the
digital humanities under an umbrella that is labeled DISRUPT,
I have found my way to a new understanding of what we are
doing here. What follows is an anatomy, a taxonomy, even a genealogy of the term "disrupt" and a discussion of the questions
that precipitate out of each usage of the word. The emphasis on
questions rather than answers is intentional and, I think, impor-

tant, for if we are to diversify the digital humanities, there must be room for multiple solutions to every problem.

But first let me take a step back and explain my disenchantment with the currently ubiquitous form of disruption, Clayton Christensen's notion of *disruptive innovation*. Christensen is a professor of business administration at the Harvard School of Business, describing himself as the "World's Top Management Thinker."[1] His webpage defines disruption as "a process by which a product or service takes root initially in simple applications at the bottom of a market and then relentlessly moves up market, eventually displacing established competitors."[2] Karl Ulrich, Vice Dean of Innovation at the Wharton School of Business, situates the process in the realm of discourse, noting that the chief requirement is that incumbents are unable to respond.[3] In other words, the criterion of disruption is the silencing of competitors. This concept has been taken up eagerly by academics: The 2014 annual Educause meeting focused on the topic; Utah State University offers a prize in disruption case-study writing; institutions all across the country offer classes on disruptive innovation.

Christensen's model of disruption emphasizes competition between producers ("eventually displacing established competitors"), but in her *New Yorker* critique of his work, Harvard's Jill Lepore casts new light on the disruptive component. Drawing on the *New York Times*'s "2014 Innovation Report," Lepore summarizes disruptive innovation as making "cheaper and inferior alternatives" that create new markets and make old markets irrelevant.[4] The cheaper, lower-quality products may catch our immediate attention, but at least as important, and possibly more so, is the shift from displacing producers (Christensen's definition) to displacing consumers (the *Times*'s). In other words, disruption uses new products as the pretext for a change

1   *Clayton Christensen,* http:/www.claytonchristense.com.

2   "Disruptive Innovation," *Clayton Christensen,* http://www.claytonchristensen.com/key-concepts.

3   Karl Ulrich, "The Fallacy of 'Disruptive Innovation,'" *The Wall Street Journal,* 6 November 2014.

4   Jill Lepore, "What the Gospel of Innovation Gets Wrong," *The New Yorker,* 23 June 2014.

in business models, a change that frequently leaves users worse off. At the same time that it is creative, it is inherently destructive: It does away with better products and the market for them and caters instead to consumers who didn't need the product until they were constituted as a new market. If that is what we practice in the digital humanities, we should rethink our goals. I am uninterested in disrupting DH by doing cut-rate low-quality research in order to reach new audiences at the direct expense of old audiences (i.e., our scholarly peers).

Luckily, that is not the only model of disruption available to us. The word "disruption" (from *dis-,* "apart, asunder," and *rumpere,* "to burst or break") has gone through several phases of meaning since its adoption into the English language, and each of these phases has the potential to tell us something about the digital humanities and about ourselves. The earliest definition is the one that Christensen roots his concepts in: disruption as destruction and disintegration. The 19th century introduced a new definition: disruption as misbehavior. Both of these definitions are inadequate to our brief here, however. I propose that, instead, we are practicing what the biologist Hugh Cott termed "disruptive coloration": using high contrast and difference, counterintuitively, to emphasize unity and preserve the organism. The takeaway message from all this etymological microscopy is that we not just should but must scrutinize our rhetoric, for it sets the boundaries of what DH is in the world.

* * *

DISRUPT: verb, *intrans.* To burst asunder; to break into pieces, shatter; to disintegrate. 1657, R. Tomlinson, *Pharmaceutical Shop*: "Almonds may be agitated over a slow fire, till the Involucrum disrupt."[5]

I include a full definition here to draw attention to what I see as the most salient feature of the earliest form of disruption in English, namely, that it is an intransitive verb, like "disintegrate." It

---

5   *Oxford English Dictionary,* s.v. "disrupt," http://www.oed.com.

does not take a direct object; it is a process internal to the organism, rather than a process that is done to it by an external agent. The skin of the almond disrupts itself. It disintegrates, with no clearly implied actor. The intransitivity of disruption is interesting not only because it removes the question of an outside agent. It also emphasizes the coherence of the thing: until disruption, there is a unified object, an entity, an organism. This, then, is a critical question when we set about to disrupt the digital humanities: To what extent is DH a single entity?

One of the principles behind this volume is that not all the members of the DH community agree on that extent. We have only to look at Stephen Ramsay's definitions of DH1 and DH2 for a perfect illustration of this.[6] Ramsay explored a divide that most of us in the digital humanities have long been aware of, and he names the sides, using the regrettable metaphor of diabetes: Type 1 Digital Humanities is made of up of coders, and to his way of thinking it is a community gathered around a shared set of tools rather than shared objects of study. Type 2 Digital Humanities is definitively *not* part of this community, and Ramsay suggests that it does not form a community at all. For Ramsay, DH2 is best described as humanistic inquiry that in some way relates to the digital — it can be media studies, it can be digital art, it can be cultural criticism, it can be digital pedagogy. I don't agree with Ramsay's timeline: He sees DH2 as arising well after the coinage of the term "digital humanities" (which he dates to 2003), whereas Vannevar Bush, Ted Nelson, Katherine Hayles, and George Landow would surely argue otherwise. But when he describes the relationship between DH1 and DH2 as an "ideological war," it is not just a battle over who owns the digital humanities; it is an argument over the extent to which coders and critics are engaged in the same endeavor.

---

6   Stephen Ramsay, "DH Types One and Two," *Sitewide ATOM,* http://stephen-ramsay.us/2013/05/03/dh-one-and-two. Also important is Adeline Koh's "Niceness, Building, and Opening the Genealogy of the Digital Humanities: Beyond the Social Contract of Humanities Computing," *differences: A Journal of Feminist Cultural Studies* 25 (2014): 93–106.

Coders vs. critics (which we may also think of as tools vs. topics) is only one divide in the digital humanities. The terrain is striated with fissures, including gender, ability, race, and the assorted intersectionalities that are realized when fissures inevitably bisect each other. Paradoxically, these fissures stand out all the more starkly when the digital humanists on one side of the divide aren't even aware of the split. The first step toward disrupting the digital humanities is to recognize that DH is already (and perhaps always-already) a fractured community at best, and that it is sometimes hard to discern whether it is drawn together by shared inquiry (encompassing both tools and topics) or merely by competing for the same resources. I take it as a given that the self-proclaimed practitioners of digital humanities do not agree — not on what the fissures are, not on which fissure is most in need of attention, and not on whether DH is a single entity in the first place.

Given this portrait, the digital humanities does seem to be on the brink of dissolution and disintegration — disruption in the earliest historical sense of the word. But at the same time, it seems clear that few of us actually *want* the digital humanities to break apart into its constituent factions. DH1 may criticize DH2 for not being able to make anything (harking back to Ramsay's now-retracted claim that "If you're not making anything, you're not a digital humanist"[7]) but it has belatedly developed an interest in theory after all.[8] DH2 has argued from the start that critique is intertwined with creation, rather than giving up and creating a separate discipline. As Adeline Koh puts it in her critique of the implied social contract of the digital humanities, "yack is already present in hack,"[9] referring to the hack/yack distinction that sometimes distills DH1 vs. DH2. Meanwhile, digital human-

---

7   Stephen Ramsay, "On Building," *Sitewide ATOM,* http://stephenramsay.us/text/2011/01/11/on-building.

8   Matthew Kirschenbaum (@mkirschenbaum), Twitter post, "While we're acknowledging writing theory as making stuff, can we also acknowledge making stuff as doing theory?" 4 January 2013, 2:33 p.m.

9   Koh, "Niceness, Building, and Opening the Genealogy of the Digital Humanities," 99.

ists of color demand recognition and respect from white digital humanists, who are usually happy to give it — as long as they don't have to change anything. Admittedly, some (or most, or all, depending on whom you ask) of the glue that keeps the digital humanities together is resources. With institutions like the NEH Office of Digital Humanities and the DH centers (and jobs) that are springing up at universities around the world, remaining part of the DH community means access to grants, fellowships, and above all scholarly attention. But in the fervency of marginalized groups' fight for recognition, I see more than just a demand for resources: I see a demand for a place at the intellectual table where important claims are made. I see a discipline.

It is worth noting that while I have pointed out a few fissures here (DH1 vs DH2, race, and so on), these are not the only ones. There are surely even major divisions in the digital humanities that I am unaware of. One shared feature in most of the cases, however, is that one side of the divide feels slighted and disrespected, while the other — when it gives any thought to the divide at all — feels unfairly demonized. In other words, they all call out a situation where power (in whatever form it takes) is not distributed equitably. This is also the point at which we should remind ourselves that such inequities, and the conditions that they engender, are rarely attributable to individual action or an intent to oppress. As we know, these inequities are systemic and can coexist with an entire population of well-intentioned participants. None of this is news, but again, by articulating them as part of our rhetoric, we call them out as a crucial aspect of the digital humanities.

DISRUPTIVE DISCHARGE: noun. a sudden, increased flow of electrical current due to the complete failure of the insulating material under electrostatic stress.[10]

Disruptive discharge is what occurs when the electrical charge of a source is greater than the resistance of the insulation around

---

10 Modified from *Academic Press Dictionary of Science and Technology*, s.v. "Disruptive Discharge," ed. Christopher G. Morris (New York: Harcourt

it — so great that the insulating ability breaks down, or disintegrates, and the electricity is discharged. Lightning is a good example: The electricity stays in the storm cloud until it overwhelms the air's power to insulate, at which point a bolt of lightning strikes the ground (or another cloud). The insulation *disrupts* in the original sense of the word. Disruptive discharge was proven by Benjamin Franklin in his 1752 kite experiment,[11] and the term may have been coined by Michael Faraday for his presentation to the Royal Society in 1838.[12]

Disruptive discharge is an apt metaphor for the role of rhetoric in fractured discourse communities (and no matter how one views the existence of community in the digital humanities, DH certainly forms a discourse community). Crucially, electricity is governed, or even defined, by inequity: In the example of lightning, the inequity between the negative ions of the storm cloud and the positive ions of the ground overwhelms the insulating power of the air between them, resulting in a lightning bolt. If we replace the cloud and the ground with the two sides of any fissure in DH (whether DH1/DH2, gender, global north vs. global south, etc.), the inequities can result in disruptive discharges with the power to destroy. Only rhetoric — the ability and intent to persuade — insulates and prevents shocks, and it frequently breaks down.

If we are to control language's ability to both focus and diffuse violence, we must understand it, and that means being able to describe it. And while we have many adjectives to describe language that channels hostility, there is a paucity (if not a nullity) of terms to describe language that promotes peace. Nearly all of the most common adjectives for constructive social rhetoric are problematic, and that is symptomatic of the problem, proof of the incommensurability of the two sides of a social divide.

Brace Jovanovich Publishers, 1992), 661.

11 If, indeed, he ran that experiment at all — there is considerable doubt. See Alberto A. Martinez, *Science Secrets: The Truth About Darwin's Finches, Einstein's Wife, and Other Myths* (Pittsburg: University of Pittsburgh Press, 2011), 118–27.

12 Michael Faraday, "Experimental Researches in Electricity." *Philosophical Transactions of the Royal Society of London (1776–1886)* 122 (1832): 125–62.

- *Civility* is the word most frequently deployed, but scholars were critiquing the term for decades before the Steven Salaita/Phyllis Wise case turned it into a tinderbox word. As early as 1939, Norbert Elias observed that civility is the means by which we distinguish between Us and Them. Building on Jai Sen's equation of civility with oppression, Joan W. Scott points out in her recent anatomy of the term that "the watchword of [Western society's] colonizing movement is 'civility'"[13]. This connection of civility with colonialism carries us directly back to early Rome, where a *civis* was a citizen, fully endowed with legal rights, and a *colonia* was an outpost established to control a barbaric local population. The alternative to civility is barbarism (and it is worth noting that the term "barbaric" derives from "to stammer or speak badly"). In other words, "the dissident claims of minority groups" — claims that are by definition badly expressed — are automatically classified as unorthodox and therefore as uncivil.[14] This understanding of civility certainly accords with the way it is deployed in our post-Salaita era.
- *Courteous* is another possibility one hears mentioned in discussions of rhetoric, but in many ways it is as problematic as "civility." In his essay "From Civilitas to Civility," John Gillingham notes that the notion of civility arose in opposition to courtesy, understood in the Early Modern period as the code of behavior appropriate to a royal court. As Gillingham describes it (although he takes issue with the periodization), civility taught people — certain kinds of people — how to be virtuous members of society, whereas courtesy taught them to be subjects. In that sense, civility is something of an improvement. But the people interpellated into this system are still those who fit the hegemonic ideal in terms of race, class, religion, sexuality, and so forth. In other words, civility shift-

---

13 Norbert Elias, *The Civilizing Process,* vol. 2, trans. by Edmund Jephcott (New York: Pantheon Books 1982), 300.

14 Jai Sen, "The Power of Civility: Some Critical Reflections on Global Civil Society," *Development Dialog* 49 (2007): 51–67. See also Joan W. Scott, "The New Thought Police," *The Nation,* 15 April 2015.

ed the locus from court to city, but the organizational principle remained a seat of governmental hierarchy.

- The same holds true for a third option, *polite,* which traces its
  roots back to the Greek *polis.* Again, the city is the dominant
  paradigm for nonviolent interactions with others. Even for
  those living on farms and in convents, the city is a cornerstone of their construction as rhetorical subjects. Politeness
  is particularly connected with the 18th century and the public sphere, as documented — and critiqued — by Lawrence
  Klein, David Alvarez, J.G.A. Pocock, and others. The standards of politeness articulated by Joseph Addison, the Earl
  of Shaftesbury, and Adam Smith were an effort to contain
  social instability and dictate behavioral norms that extended
  beyond court to cover the middle and gentry classes. Pocock
  writes, "Commerce was the parent of politeness" — and anyone without a stake in the world of commerce was likewise
  excluded from the possibility of polite behavior.[15] Like civility, politeness was created in order to support existing power
  structures.[16]

All three of these frequently used terms — "civil," "courteous,"
and "polite" — define acceptable language as that which aligns
with and advances the interests of the locus of power. This holds
for *collegial* as well, a term that has been thoroughly examined
and excoriated by the academic media for some time.[17] These
words are all unacceptable because they privilege the needs of
the hegemon over, and sometimes at the expense of, those of the
margin. Other terms — *friendly, agreeable, pleasant* — are mis-

---

15 Ibid.
16 J.G.A. Pocock, "Cambridge Paradigms and Scotch Philosophers: A Study of
    Relations Between the Civic Humanists and the Civil Jurisprudential Inter-
    pretation of Eighteenth Century Social Thought," in *Wealth and Virtue: The
    Shaping of Political Economy in the Scottish Enlightenment,* eds. Istvan Hont
    and Michael Ignatieff, 235–52 (Cambridge: Cambridge University Press,
    1983), 241.
17 Thomas Woodman, *Politeness and Poetry in the Age of Pope* (Rutherford
    Fairleigh Dickinson University Press, 1989).

leading and/or out of line. *Nice* has the additional baggage of having originally meant "imbecilic" or "silly."

So what are we to call the language that we require in order to have productive conversations across ideological and other divides? We need to think about what we want our language to accomplish and move on from there. "Constructive" will do in a pinch, as perhaps might "language of respect" — although "respectful" has too much of the schoolroom about it and is altogether too Oliver Twist-y. I have no firm answers here. But I am convinced that until we can settle on adjectives that give a true description of the language of fruitful discussion rather than ones that reduplicate social inequities, the potential for disruptive discharge — rhetoric as destruction — will loom over all of our encounters.

DISRUPT: verb, *intrans.* To throw into disorder; to misbehave. 1994, *Diagnostic and Statistical Manual of Mental Disorders,* 4th Edition: "312.9: Disruptive Behavior Disorders, Not Otherwise Specified."[18]

I wasn't a troublemaker in grade school, but I lived in fear that my teacher would put an X next to the line on my report card that read "Disrupts class and distracts others." Back then, that was seen as a sign of plain old badness, but we know now that children disrupt class when their needs aren't being met. Almost by definition, then, disruption occurs on the margins. Thus it is no surprise that the first reference I can locate to disruption as misbehavior is in Tennyson's "Guinevere" (part of the first installment of *Idylls of the King,* published in 1859). It concerns the disruption of the Round Table by Modred, the ultimate outsider, and Tennyson goes so far as to attribute this disruption ultimately to Guinevere and her efforts toward self-fulfillment.[19]

18 One of the best examinations of collegiality can be found at: Dr. Crazy, "What Does It Mean to Be Collegial? What Constitutes Civility? And How Do We Promote These Things?" *Reassigned Time 2.0* (blog), 19 June 2012.

19 *Diagnostic and Statistical Manual of Mental Disorders,* 4th ed. (Washington, DC: American Psychiatric Association, 1994), 94.

The flower of (white, male, aristocratic) chivalry is brought down by a woman and a low-born bastard.

If we are disrupting the digital humanities, is it because needs are not being met? Certainly. Every essay in this volume is a witness to a lack — of attention, of respect, of resources — for work that is marginalized by dint of geography, ethnicity, methodology, and so forth. This book would not exist if its editors had not felt that important DH projects are being overlooked and critical methodologies neglected. But I'm not entirely comfortable with the misbehavior model, because it constitutes the disruptor as ill at best, and at worst a bad egg. The subalterns of DH (however they/we are defined, or define themselves) are in no way sick, nor are they pointlessly obstreperous. We play into the status quo if we accept the portraiture implied by "disruptive behavior."

Here I'd like to pursue a tangent that very much relates to the rhetoric of the margin and how it makes its needs known: the gerund. In case you've forgotten those seventh-grade grammar drills, a gerund is a verb used as a noun — "*Seeing* is believing." "Thanks for *asking*." "*Parting* is such sweet sorrow." Gerunds have always been with us, of course, but over the last few years, they have played a special role in the margin's talking back to the mainstream. I speak here of the terms "tone-policing," "slut-shaming," and "mansplaining." They are also used as conjugatable verbs (i.e., "He tone-policed him"), but they appear far more often as gerunds — and this is no coincidence. As gerunds, they assert that THIS IS A THING. The gerund form argues back against the frequently made claim that the offense was accidental, a mere slip of the tongue. No, the gerund insists: That is a common speech pattern that needs to be called out.

And the speech pattern called out by "tone-policing," "slut-shaming," and "mansplaining" is the covert critique of rhetoric masquerading as engagement with the speaker's ideas. A classic example of tone-policing is asking an interlocutor "Why must you be so hostile? It's impossible to have a conversation with someone with such anger." Slut-shaming, in the words of Duhaime's *Legal Dictionary*, is "The judgment objectifying and passed on the sexuality of woman […] that her choice of apparel is indicative of consent if not encouragement to sexual intimacy."[20] Finally, mansplaining is a patronizing and masculinist style of explanation, typified in Rebecca Solnit's classic essay "Men Explain Things to Me."[21] All three of these specifically concern rhetoric, although they may not at first appear to. Tone-policing, for example, is often understood as attacking the level of emotion in a discussion, while slut-shaming would seem to be about sexuality. But in both cases, the issue is the expression of those elements, rather than their existence. Tone-policing decries emotional rhetoric, slut-shaming decries the visual rhetoric with which a woman expresses herself, and mansplaining decries a rhetoric of disrespect. All three conclusively derail discussion, and they testify to language's power to wreak broadscale harm.

These rhetorical derailments aren't the sole property of the mainstream. The margin practices tone-policing too — often by deploying phrases like "precious fee-fees." Accusations of emotionality can be wielded by anyone, which serves as a reminder that it is nearly impossible to call out these behaviors without practicing them oneself. The only way to bring attention to someone's critique of rhetoric at the expense of content is… to critique their rhetoric. Perhaps because of the Enlightenment's downplaying of rhetoric, we do not have a valid and rigorous standard for criticizing not ideas but the expression of those ideas, but we desperately need one. By making rhetoric off-limits for discussion, we claim that it cannot affect us, but we are surrounded by evidence to the contrary. Without a viable means to bring rhetoric into the realm of critique, our conversations are permanently at risk of being shut down by uncommunicable emotion.

We also need to think about the extent to which emotion should have a voice in our scholarly discourse. As I mentioned

---

20 Baron Alfred Tennyson, *Tennyson's Guinevere and Other Poems* (Glasgow: Blackie & Sons, 1923), 8.

21 *Duhaime's Law Dictionary*, s.v. "Slut-Shaming," http://www.duhaime.org/LegalDictionary/S/Slut-Shaming.aspx.

above, the mainstream frequently targets marginal voices for expressing frustration and anger, while the margin can often be found denying mainstream voices the right to their emotions. Both of these urges seem predicated on the assumption that one's feelings are entirely controllable and subject to rational decision-making, but this cannot be the case. If emotions aren't fully under logical control, to what degree is it fair (or productive) to criticize someone for having them? Negative emotions, like classroom disruption, arise when needs are not being met, and the more diverse an environment, the more likely it is that needs will be ignored and strong feelings will be provoked. Sara Ahmed suggests in *The Cultural Politics of Emotion* and elsewhere that emotions — particularly negative ones — are analogous to and products of social inequality, oppression, and violence. As we work to restructure discourse in an inclusive world, we must give serious thought to the role emotion should play in the realm of academic discourse.

Part of the problem, or at least associated with it, is the fact that too often the mainstream thinks that persuasion is the only reason we express ourselves. A primary mode of expression by scholars speaking from the fringes of DH is affirmation and co-alition-building — which the mainstream critiques as "preaching to the choir" and therefore pointless. I learned this lesson abruptly and memorably, when I wielded this churchy phrase in a discussion of diversity and was rightly called on it. The affirmation mode might be pointless to those who see themselves as officiants, or at least standing in the chancel. But for those minority voices scattered across the nave, speaking to find each other and to ascertain their shared perspective can be far more important than speaking to persuade. The derogatory term "slacktivism" is often thrown at people who use Twitter and other social media to locate fellow thinkers, affirm community, and boost each other's signals, and this term crystallizes for me the communicative moat that separates mainstream and margin. Hegemony only needs one reason to speak, and rhetoric, historically rooted in and only in persuasion, is inherently hegemonic. Until we all acknowledge and respect multiple modes of expression, we are fated to talk past one another in at least some of our exchanges.

DISRUPTIVE COLORATION: noun. A pattern of coloration that breaks up the shape and destroys the outline of an object, hindering detection.[22]

So far, this essay seems like a series of laments about the rhetorical pickle we have gotten ourselves in and near-impossible challenges for getting out of said pickle — all structured around the word "disrupt." But that word offers hope too, or so it seems to me, in the form of disruptive coloration. This is a form of camouflage in which something (plant, animal, ship) does not try to conform itself to the environment but instead develops sharp color contrasts, in order to disguise its silhouette. There are many examples in nature; the herd of zebras is perhaps most iconic. The predator recognizes their presence, but in the absence of distinguishable silhouettes, it cannot tell how many zebras are present or which is the best target for attack — "to break up the outline and destroy identity," in the words of Peter Forbes.[23] (Here Forbes is using "identity" in the sense of individuality — a meaning closely related to the more familiar notion of identity as that aspect of ourselves that sets us apart or against the larger group.) Disruptive coloration was first identified by artist and zoologist Abbot Thayer, and beginning with the Spanish Civil War, Thayer and those influenced by him developed the artificial equivalent, razzle-dazzle camouflage. Although it is sometimes used for planes, razzle-dazzle camouflage is most often applied to ships, to achieve three purposes: to minimize detection at a distance; to obscure the identification of the ship

---

22 Rebecca Solnit, "Men Explain Things to Me," reprinted online in *Guernica: A Magazine of Art and Politics,* 20 August 2012.

23 Modified from Martin Stevens and Sami Merilaita, "Defining Disruptive Coloration and Distinguishing Its Functions," *Philosophical Transactions of the Royal Society B* 364, no. 1516 (2009): 481–88, at 481.

type and particularly its artillery; and to render masts invisible and thus make gunnery range-finding nearly impossible.[24]

This divagation into the history of camouflage is more than whimsy. In it, I find a useful model for how the fractured constituencies of the digital humanities can abide without abandoning their differences. Camouflage is a form of self-protection, and while I don't think DH is being stalked by lions or U-boats, it does seem to me very much in danger: in danger of disruption in its earliest sense, in danger of falling to pieces. Thus I propose that we camouflage ourselves with disruptive coloration. Let us play up our contrasts in order to simultaneously disguise and preserve the unity of the whole.

Do not mistake this for a can't-we-all-get-along plea or a call to flood onto the great DH dance floor and be excellent dancers to one another. "Can't we all get along?" is too often the voice of power pressuring the grumpy outsiders to hush up and assimilate. Enculturation and assimilation are, if anything, the enemy that disruptive coloration protects us from. Deep down, I suspect, most of us want everyone else in the discipline to agree with us and to share our intellectual priorities and commitments — but we know that ain't never gonna happen. This is perhaps the very definition of a postlapsarian world. Thoughtful, honest engagement with rhetoric — rhetorospection, if you will — will not solve all the problems in the postlapsarian digital humanities, but it is and must be the first step. By our rhetoric, you will know us, and by our rhetoric we will become ourselves.

---

24 Peter Forbes, *Dazzled and Deceived: Mimicry and Camouflage* (New Haven: Yale University Press, 2009), 97.

Bibliography

Academic Press Dictionary of Science and Technology. Edited
by Christopher G. Morris. New York: Harcourt Brace
Jovanovich Publishers, 1992.

Clayton Christensen. http:/www.claytonchristense.com/.

Crazy, Dr. "What Does It Mean to Be Collegial? What
Constitutes Civility? And How Do We Promote These
Things?" *Reassigned Time 2.0* (blog). 19 June 2012. https://
reassignedtime.wordpress.com/2012/06/19/what-does-it-
mean-to-be-collegial-what-constitutes-civility-and-how-do-
we-promote-these-things.

Diagnostic and Statistical Manual of Mental Disorders. 4th
edition. Washington, DC: American Psychiatric Association,
1994.

Duhaime's Law Dictionary. http://www.duhaime.org/
LegalDictionary/S/Slut-Shaming.aspx.

Elias, Norbert. *The Civilizing Process.* Vol. 2. Translated by
Edmund Jephcott. New York: Pantheon Books, 1982.

Faraday, Michael. "Experimental Researches in Electricity."
*Philosophical Transactions of the Royal Society of London
(1776–1886)* 122 (1832): 125–62. DOI: 10.1098/rstl.1832.0006.

Forbes, Peter. *Dazzled and Deceived: Mimicry and Camouflage.*
New Haven: Yale University Press, 2009.

Kirschenbaum, Matthew (@mkirschenbaum). Twitter post. 04
January 2013, 2:33 p.m. https://twitter.com/mkirschenbaum/
status/287280574815666177?ref_src=twsrc%5Etfw.

Koh, Adeline. "Niceness, Building, and Opening the Genealogy
of the Digital Humanities: Beyond the Social Contract of
Humanities Computing." *differences: A Journal of Feminist
Cultural Studies* 25 (2014): 93–106. DOI: 10.1215/10407391-
2420015.

Lepore, Jill. "What the Gospel of Innovation Gets Wrong."
The New Yorker. 23 June 2014. http://www.newyorker.com/
magazine/2014/06/23/the-disruption-machine.

Martinez, Alberto A. *Science Secrets: The Truth About Darwin's Finches, Einstein's Wife, and Other Myths.* Pittsburgh: University of Pittsburgh Press, 2011.

Oxford English Dictionary. http://www.oed.com.

Pocock, J.G.A. "Cambridge Paradigms and Scotch Philosophers: A Study of Relations Between the Civic Humanists and the Civil Jurisprudential Interpretation of Eighteenth Century Social Thought." In *Wealth and Virtue: The Shaping of Political Economy in the Scottish Enlightenment.* Edited by Istvan Hont and Michael Ignatieff, 235–52. Cambridge: Cambridge University Press, 1983.

Ramsay, Stephen. "DH Types One and Two." *Sitewide ATOM.* http://stephenramsay.us/2013/05/03/dh-one-and-two.
———. "On Building," *Sitewide ATOM.* http://stephenramsay.us/text/2011/01/11/on-building.

Scott, Joan W. "The New Thought Police." *The Nation.* 15 April 2015. http://www.thenation.com/article/new-thought-police.

Sen, Jai. "The Power of Civility: Some Critical Reflections on Global Civil Society." *Development Dialog* 49 (2007): 51–67.

Solnit, Rebecca. "Men Explain Things to Me." Reprinted online in *Guernica / A Magazine of Art and Politics.* 20 August 2012. https://www.guernicamag.com/daily/rebecca-solnit-men-explain-things-to-me.

Stevens, Martin, and Sami Merilaita. "Defining Disruptive Coloration and Distinguishing Its Functions." *Philosophical Transactions of the Royal Society B* 364, no. 1516 (2009): 481–88. DOI: 10.1098/rstb.2008.0216.

Tennyson, Baron Alfred. *Tennyson's Guinevere and Other Poems.* Glasgow: Blackie & Sons, 1923.

Ulrich, Karl. "The Fallacy of 'Disruptive Innovation.'" *The Wall Street Journal.* 6 November 2014. http://blogs.wsj.com/experts/2014/11/06/the-fallacy-of-disruptive-innovation.

Woodman, Thomas. *Politeness and Poetry in the Age of Pope.* Rutherford: Fairleigh Dickinson University Press, 1989.

4

The Public Digital Humanities

Jesse Stommel

I am going back and re-reading a handful of articles from the 1970s published in the journal "Computers and the Humanities."[1] I am struck by how innovative and imaginative so many of these articles are. I am also struck and saddened by how little the field has progressed. R.L. Widmann, one of my dearest mentors, wrote (five years before I was born), "the increasing numbers of works which do use computers with intelligence and perceptiveness demonstrate the versatility to which the scholar's imagination can be put. It was a very good year."[2] The year was 1971. The pieces I'm reading outline the 6 or 7 or 8 new (and revolutionary) approaches being used for thinking about the intersections between computers and the humanities. These pieces also map something very close to the current philosophical terrain of digital humanities pedagogy. They do so with an intense curiosity about what else might be possible. In R.L.'s work, there is no entrenchment. There is no unwillingness to account for edge cases. The limits of what counts and doesn't count are described repeatedly as "arbitrary." I'm certain the field wasn't entirely rosy in 1971, but I have an incredible amount of hope reading these

---

1   R.L. Widmann, "Computers and Literary Scholarship," *Computers and the Humanities* 6, no. 1 (1971): 3–14.

2   Ibid., 12.

now and thinking about what we might be able to discover in the next 45 years.

In 2011, I co-founded *Hybrid Pedagogy*: an open-access journal of learning, teaching, and technology.[3] The project was from its origin focused on faculty development but also on asking hard questions of technology and education. These two do not always sit well together. And when teachers and scholars converse at this edge, the discussion is often strained. I have worked at several institutions where their merely brushing against one another has led to scholars expressing a kind of derision that should have no place in education. Ultimately, the problem is less about specific individuals looking askance and more the result of a system that pits academics against one another in conflict over scarce resources. This invariably privileges those who are already privileged, those who (because of their race, class, gender, sexuality, ability) already have protected positions within institutions.

Hybrid Pedagogy has been since its formation staunchly extra-institutional, because we felt better poised to comment on higher education from that vantage. Our goals have been to: (1) interrogate academic publishing practices by making them transparent; (2) share models that could be duplicated, reconfigured, and reworked by other digital publishing projects; (3) offer scholars strategies for making their pedagogical, editorial, and outreach work legible as scholarship; (4) make publishing more overtly pedagogical; (5) and to make pedagogy more public, an open dialogue not a monologue. Christopher Long writes in "To Be Published or To Be Read," "Although publications with reputable university presses or journals continue to be the cornerstone of the tenure and promotion process, many remain inaccessible to a broad audience, bound up, as they often are, in paper volumes or locked behind paywalls required by the outmoded business practices of scholarly publishers."[4]

---

3   See *Hybrid Pedagogy*, http://hybridpedagogy.org/.
4   Christopher Long, "To Be Published or To Be Read," blog post, 23 March 2014.

Hybrid Pedagogy uses a collaborative peer-review process, in which editors engage directly with authors to revise and develop articles. Editorial work is done both asynchronously and synchronously in a Google Doc that evolves through an open dialogue between author and editors (a process very similar to the one used to edit this volume). We also encourage co-authored and multiple-author submissions; we invite the community into conversation around articles; and we link articles directly to their sources, creating a web of influence and dialogue. *Hybrid Pedagogy* is less focused on publishing articles as content repositories and more on reimagining scholarship as pedagogical, publishing as a way to create conversations and bridge academic and non-academic communities. We are a group of mostly humanists using digital tools to build a network of teachers and students helping to rethink the what, why, and who of public scholarship. Our primary aim has always been to make space for voices that might not otherwise be heard within academia.

The public digital humanities starts with humans, not technologies or tools, and its terrain must be continuously co-constructed. There is no place within the public digital humanities for exclusion or anti-intellectualism. No place for hierarchies: inside the academy–outside the academy; teacher–student; senior scholar–junior scholar; tenure-track–adjunct; all too distant past–inaccessible future. And we mustn't stare or snarl in derision at people mid-thought, expecting that only the final draft of an idea should be public. Or mistake mean-spirited criticism and closing down of conversations for critical engagement.

The public digital humanities is a Venn diagram at the point where public work, digital work, and humanities work intersect. And these points of intersection are always shifting, so I won't attempt to neatly map them here. Making scholarly work legible to the public and helping it find audiences is a form of outreach, community building, and advocacy. This is the work I've endeavored to do with *Hybrid Pedagogy* and all my other projects.

* * *

I long identified as a "digital humanist," but the digital humanities, as a discipline, has not endeared itself to me. I find considerably more solace in the feminist community, in the queer community, and in motley subsets of digital humanities outliers. I have never found DH particularly welcoming, for all its proclamations of being "nice." Adeline Koh argues that "niceness/civility [...] play important gatekeeping roles within the digital humanities public sphere."[5] What I've discovered is that "nice" translates too often into quiet, unassuming, staid, and scholarly (in the worst senses of the word). All the while daggers are brandished behind people's backs, as is far too common in the scarcity economy of academia. Nevertheless, there are countless people in the digital humanities community that have won my love. My allegiances are to people and communities, not to disciplines. And it is, I think, the desire to legitimize the digital humanities itself as a discipline that brings out the daggers — and, perhaps even more detrimental, the urge to keep one eye always over our shoulders.

This question of "discipline" has been something my career has incessantly circled around. I've been in many rooms where humanities scholars have recited in the round a litany of fields, subfields, periods, major authors, and disciplinary affiliations. For 15 years, I made up something almost completely new at every one of these scholarly show-and-tells. I have been a 19th-century Americanist, a queer feminist, a film scholar, a new media specialist, a Shakespearean, a digital humanist. I still am most of these things, but none of them describes me or circumscribes me. And I've had the boundaries of these disciplines drawn on the other side of my work, to my exclusion, more often than not. With regard to my work on *Hybrid Pedagogy,* I was told very publicly by a senior digital humanist, "You claim the mantle of scholarship while avoiding the actual work of scholarship." And, at conferences, in blind peer-reviews, on

---

5   Adeline Koh, "Niceness, Building, and Opening the Genealogy of the Digital Humanities: Beyond the Social Contract of Humanities Computing," blog post, 24 April 2014.

social media, this kind of gatekeeping is not at all unusual, both within the field and from outside of it. The Graduate Education Research Committee at University of Wisconsin-Madison wrote these words to me in a letter, "The Committee wants to send a clear message that what matters is tenure, what matters for tenure is peer review, and work posted on the web is not considered peer-reviewed." Sentences like these take a toll. But they also make me even more resolute.

Ultimately, what counts as digital humanities is work that doesn't try to police the boundaries of what counts as digital humanities.[6]

While I have played in and around many disciplines, my primary scholarly interest has always been pedagogy. I have devoted nearly my entire professional life to teaching, to the collaborative work I do with students — and with other teachers as students. Pedagogy is not synonymous with teaching or talking about teaching, nor is it entirely abstracted from the acts of teaching and learning. It is both my discipline and threaded through all of my disciplines. Pedagogy is praxis, the place where philosophy and practice meet. Most of my pedagogies, including my digital ones, are rooted in critical pedagogy — in thinkers like Emerson, Elbow, hooks, Dewey, and Freire. And like Freire, I am a hopeful critical pedagogue. In *Pedagogy of Hope,* he writes, "I am hopeful, not out of mere stubbornness, but out of an existential, concrete imperative."[7] But, also like Freire, I recognize that hope must be balanced with action and struggle.

My digital humanities has focused less on reading humanities texts with digital tools, and more on using humanities tools — humane tools — to read and make digital texts. Since I started teaching in 2001, I've become more and more concerned with thinking about ways to make public what I do in the classroom and what I do in the safe confines of a word-processing window. The impetus for my scholarly work and publishing is to

---

6   See Jesse Stommel (@Jessifer), Twitter post, 4 January 2015, 10:46 a.m.

7   Paulo Freire, *Pedagogy of Hope: Reliving Pedagogy of the Oppressed,* trans. Robert R. Barr (New York: Continuum, 1992), 2.

do my pedagogy in much larger and more open spaces — teaching, and teaching teachers, as a form of activism. What I call "public digital humanities" is built around networked learning communities, not repositories for content, and its scholarly product is a conversation, one that blurs distinctions between research, teaching, and service.

As Brett D. Hirsch writes in the introduction to *Digital Humanities Pedagogy: Practices, Principles, and Politics,* "To bracket pedagogy in critical discussions of the digital humanities or to completely exclude it from these discussions reinforces an antagonistic distinction between teaching and research, in which the time, effort, and funding spent on the one cannibalizes the opportunities of the other."[8] Our work must be collaborative. Even when our work is not produced by multiple authors/artists, it becomes collaborative when it is given generously to its readers. And, it's not just that we need to find and celebrate new modes of digital scholarship, but that we must allow our new digital environments to influence all forms of scholarship.

The public digital humanities must be rooted in a genuine desire to make the work legible to a broader audience inclusive of students, teaching-focused colleagues, community college colleagues, and the public. I believe pedagogical work should be honored as the best kind of research, and our scholarship should be pedagogical. This is the voice I speak in, the voice I write in, and it is a voice that chooses at strategic moments to generalize. It is a rigorous voice because it is a hybrid voice, attempting to balance the nuanced analysis of a scholarly approach with a desire to make the work accessible. Scholarly writing for broad publics must invite readers and students (once its mere satellites) into a more intimate, more provocative dance. This scholarship cannot be static, traditional, or staid. It must resist the deadening impulse of much so-called "academic rigor."[9]

---

8   Brett D. Hirsch, ed., *Digital Humanities Pedagogy: Practices, Principles and Politics* (Cambridge: Open Book Publishers, 2012), 5.

9   Sean Michael Morris, Pete Rorabaugh, and Jesse Stommel, "Beyond Rigor," *Hybrid Pedagogy,* 9 October 2013.

And the gathering together of work into collections like this one demands a move away from the metaphor of the static bibliographic record toward hyperlinked ones — our work is not just influenced by but connected metonymically to its sources, to the other works that gather around it, a cacophony of sometimes disparate voices.[10]

And my sources for the work here are myriad:

I hear echoes of my friend and colleague R.L. Widmann in the words I've written in this chapter. Well beyond her early articles in "Computers and the Humanities," it has been through conversation with her over two decades that I've found my way to these sentences.

I also hear echoes of Bonnie Stewart who writes, in "What counts as academic influence online?": "The work of research that is not legible to others always feels, rhetorically, like lifting stones uphill: constantly establishing premises rather than moving on to the deep exploration of that one particular thing."[11] Doing public work is different from making academic work public. Available is not always accessible.

Sean Michael Morris and I write, in "Hybrid Pedagogy, Digital Humanities, and the Future of Academic Publishing": "Postprint publishing keeps its focus on moving objects: digital artifacts and networked conversations that can be plumbed at the level of the code behind them, tracked in their progress through the web, or catalogued next to works beside which they would not normally sit."[12] For over 15 years, Sean's words have come to inhabit my sentences.

---

10 See Adeline Koh, "A Close Reading of The DHThis Cat: Policing/Disrupting the Boundaries of the Digital Humanities and Strategic Uses for Cat GIFs," *Disrupting the Digital Humanities: Digital Edition,* 9 January 2015; Kathi Inman Berens, "Want to 'Save the Humanities'? Pay Adjuncts to Learn Digital Tools," *Disrupting the Digital Humanities: Digital Edition,* 5 January 2015; and Élika Ortega, "Multilingualism in DH," *Disrupting the Digital Humanities: Digital Edition,* 31 December 2014.

11 Bonnie Stewart, "What Counts as Academic Influence Online," *the theoryblog* (blog), 27 April 2014.

12 Sean Michael Morris and Jesse Stommel, "Hybrid Pedagogy, Digital Humanities, and the Future of Academic Publishing," *Hybrid Pedagogy,* 29

Describing how words on a page can also be action, Maha Bali writes, "if our writing works on the world by striving to challenge it, to change it, to influence it, our writing can be praxis."[13]

One of my other early mentors, Martin Bickman, writes in "Returning to Community and Praxis: A Circuitous Journey through Pedagogy and Literary Studies": "We often ignore the best resource for informed change, one that is right in front of our noses every day — our students, for whom the most is at stake."[14] This has been (and will continue to be) the focus of my work, bringing students to tables where talk of education is underway.[15] In this, I have had my greatest successes and my greatest failures. The work is hard. We have built an almost ironclad academic system — and I acknowledge myself as one of its privileged builders — a system which excludes the voices of students, which calls students "customers" while monetizing their intellectual property, which denigrates the work of learning through assessment mechanisms and credentialing pyramid schemes.[16]

Finally, Steven Lubar writes, in "Seven Rules for Public Humanists": "The work of public engagement comes not after the scholarship, but as part of the scholarship."[17] A public digital humanities is constantly interrogating itself, but never at the expense of bringing non-academic, non-specialist voices into the conversation. Lubar's work on the public humanities has moved from a model that brings humanities work to the public to "a re-

---

April 2014.

13 Maha Bali, "Yearning for Praxis: Writing and Teaching Our Way Out of Oppression," *Hybrid Pedagogy,* 21 October 2015.

14 Martin Bickman, "Returning to Community and Praxis: A Circuitous Journey through Pedagogy and Literary Studies," *Pedagogy: Critical Approaches to Teaching Literature, Language, Composition, and Culture* 10, no. 1 (2009): 11–23, at 21.

15 See Sean Michael Morris and Jesse Stommel, "Co-Intentional Education: A #digped Discussion," *Hybrid Pedagogy,* 30 January 2013.

16 See Jesse Stommel, "Who Controls Your Dissertation?" *Vitae,* 07 January 2015, https://chroniclevitae.com/news/852-who-controls-your-dissertation.

17 Steven Lubar, "Seven Rules for Public Humanists," *On Public Humanities* (blog), 5 June 2014.

alization that our work was not about us, or for us."[18] It's not that we need to do this work in bigger and bigger tents but that we need to move outside tents altogether. This is what I have called *humongous tent digital humanities.*

* * *

When I wrote a draft of what became the first paragraph for the introduction to *Disrupting the Digital Humanities,* I wondered where it would fit inside this collection. I wondered how the rest of the prose would bear the weight of these assertions: "Academic turf wars have no place in a world of mass-shootings, fear-mongering, xenophobia, and white supremacy. Demanding fellow scholars do a literature review before speaking their mind has no place in a world of AR-15 assault rifles and weaponized algorithms. When something as basic as going to the bathroom lacks dignity for so many, we have no use for double-blind peer review." As I put the final touches on this chapter, I realize how much — and for how long — my own prose has been weighed down by the trappings and bureaucracies of academic work. I realize how much bigger this conversation must be. Words are indeed action, but paragraphs alone, no matter how strident, can't possibly make the necessary space for the work we must do.

At the center of the digital humanities should be an emphasis on individual and collective agency, which means advocating for marginalized teachers, scholars, and students. I'm arguing for the exact opposite of objectivity — for an intense subjectivity. If the digital humanities is going to innovate, it can not be through competition, clearcutting, and hype cycles, but by listening intently to more (and more diverse) voices. The digital humanities needs to be about generosity — about breaking brains not hearts.

---

18   Steven Lubar, "Applied? Translational? Open? Digital? Public? New Models for the Humanites," *On Public Humanities* (blog), 5 June 2014.

Bibliography

Bali, Maha. "Yearning for Praxis: Writing and Teaching Our Way Out of Oppression." *Hybrid Pedagogy.* 21 October 2015. http://www.digitalpedagogylab.com/hybridped/yearning-for-praxis/.

Berens, Kathi Inman. "Want to 'Save the Humanities'? Pay Adjuncts to Learn Digital Tools." *Disrupting the Digital Humanities: Digital Edition.* 5 January 2015. http://www.disruptingdh.com/want-to-save-the-humanities-pay-adjuncts-to-learn-digital-tools/.

Bickman, Martin. "Returning to Community and Praxis: A Circuitous Journey through Pedagogy and Literary Studies." *Pedagogy: Critical Approaches to Teaching Literature, Language, Composition, and Culture* 10, no. 1 (2009): 11–23.

Brito, Marisol, Alexander Fink, Chris Friend, Adam Heidebrink-Bruno, Rolin Moe, Kris Shaffer, Valerie Robin, and Robin Wharton, "Love in the Time of Peer Review." *Hybrid Pedagogy.* 22 November 2014. http://www.digitalpedagogylab.com/hybridped/love-time-peer-review/.

Freire, Paulo. *Pedagogy of Hope: Reliving Pedagogy of the Oppressed.* Translated by Robert R. Barr. New York: Continuum, 1992.

Hirsch, Brett D., ed. *Digital Humanities Pedagogy: Practices, Principles and Politics.* Cambridge: Open Book Publishers, 2012.

Hybrid Pedagogy. http://www.digitalpedagogylab.com/hybridped/.

Koh, Adeline. "Niceness, Building, and Opening the Genealogy of the Digital Humanities: Beyond the Social Contract of Humanities Computing." Blog post. 24 April 2014. http://www.adelinekoh.org/blog/2014/04/24/niceness-building-and-opening-the-genealogy-of-the-digital-humanities-beyond-the-social-contract-of-humanities-computing/.

———. "A Close Reading of the DHThis Cat: Policing/Disrupting the Boundaries of the Digital Humanities and Strategic Uses for Cat GIFs." *Disrupting the Digital*

Humanities: Digital Edition. 09 January 2015. http://
www.disruptingdh.com/a-close-reading-of-the-dhthis-
cat-policingdisrupting-the-boundaries-of-the-digital-
humanities-and-strategic-uses-for-cat-gifs/.

Long, Christopher "To Be Published or To Be Read." Blog
post. 23 March 2014. http://www.cplong.org/2014/03/to-be-
published-or-to-be-read/.

Lubar, Steven. "Seven Rules for Public Humanists." *On
Public Humanities* (blog). 5 June 2014. https://stevenlubar.
wordpress.com/2014/06/05/seven-rules-for-public-
humanists/.

———. "Applied? Translational? Open? Digital? Public? New
Models for the Humanities." *On Public Humanities* (blog). 5
June 2014. https://stevenlubar.wordpress.com/2014/06/05/
applied-translational-open-digital-public-new-models-for-
the-humanities/.

Morris, Sean Michael, and Jesse Stommel. "Co-Intentional
Education: A #digped Discussion," *Hybrid Pedagogy.*
30 January 2013. http://www.digitalpedagogylab.com/
hybridped/co-intentional-education-digped-discussion/.

———, Pete Rorabaugh, and Jesse Stommel. "Beyond
Rigor." *Hybrid Pedagogy.* 9 October 2013. http://www.
digitalpedagogylab.com/hybridped/beyond-rigor/.

——— and Jesse Stommel. "Hybrid Pedagogy, Digital
Humanities, and the Future of Academic Publishing."
Hybrid Pedagogy. 29 April 2014. http://www.
digitalpedagogylab.com/hybridped/hybrid-pedagogy-
digital-humanities-future-academic-publishing/.

Ortega, Élika. "Multilingualism in DH." *Disrupting the Digital
Humanities: Digital Edition,* 31 December 2014. http://www.
disruptingdh.com/multilingualism-in-dh/.

Stewart, Bonnie. "What Counts as Academic Influence
Online." *the theoryblog* (blog). 27 April 2014. http://theory.
cribchronicles.com/2014/04/27/what-counts-as-academic-
influence-online/.

Stommel, Jesse (@Jessifer). Twitter post. 4 January 2015, 10:46
a.m. https://twitter.com/Jessifer/status/551766513292939264.

———. "Who Controls Your Dissertation?" *Vitae*. 7 January 2015. https://chroniclevitae.com/news/852-who-controls-your-dissertation.

——— "The Public Digital Humanities." *Disrupting the Digital Humanities: Digital Edition.* 9 January 2015. http://www.disruptingdh.com/the-public-digital-humanities/.

Widmann, R.L. "Computers and Literary Scholarship." *Computers and the Humanities* 6, no. 1 (1971): 3–14. https://www.jstor.org/stable/30199443.

5

Universal Design and Its Discontents

Richard H. Godden and Jonathan Hsy

This collaborative essay offers two perspectives on disability and universalism in the fields of Digital Humanities (DH) and Universal Design (UD). One of the authors, Richard H. Godden, considers how a particular experience of disability shapes his use of media and also informs his reactions to proscriptive statements about the use of technology; the other author, Jonathan Hsy, writes as a nondisabled ally who considers some of the discursive and practical complications that arise in efforts to make the web more accessible to people with disabilities. While we each come from different perspectives, both of us seek to interrogate what it exactly means for a community to establish a set of "best practices" for the use of technology, and we both reveal how even the most well-intentioned universalist discourses can risk effacing crucial particularities of embodied experience.

Richard H. Godden: As an entry point to my reflections on Universal Design, I want to first think about some of the ways that Digital Humanities (DH), Disability Studies (DS), and Universal Design (UD) productively converge using recent discussions about the physical act of hand-written notes as an opening example. This is not unusual in a bid to consider the necessity of

UD; however, I also want to use this example in order to begin to disorient some of our understandings of UD. Although UD arose out of a real social and political response to the disabling aspects of everyday life for People with Disabilities, I want to suggest that the "Universal" in UD can carry with it some unintended and unexpected assumptions about normalcy and our physical orientation to the world.

Over the last few years, it has become a regular occurrence to see someone post on social media about a study concerning student note-taking. You know the one. Studies have confirmed, it would seem, that the pen has slain the keyboard. One such article from www.sciencenews.org begins "When it comes to taking notes, the old-fashioned way might be best."[1] I will come back to this old-fashioned-ness in a moment. The article then goes on to say "People taking notes on laptops have a shallower grasp of a subject than people writing with their hands, and not just because laptops distract users with other activities such as web surfing, the new study suggests."[2] I am not a scientist, so I am not going to fully challenge the ultimate findings of this study in this space.[3] What has me so irritated, though, is the often triumphant (explicit or implicit) attitude on display when people post such articles. Additionally, there is also often a sense of relief, or of "I told you so." I understand the nostalgia people feel for physical books and for pen and paper. There is enormous pleasure to be had in the tactile engagement with such storehouses of knowledge. The only problem, however, is that I am often excluded from such pleasures. A book sitting on my shelf in my office might as well be a continent anyway.

The short articles that I repeatedly see posted on the subject focus on the superiority of old-fashioned technologies versus

---

1   Laura Sanders, "Students Retain Information Better with Pens than Laptops," *ScienceNews: Magazine of the Society for Science and the Public,* 30 April 2014.

2   Ibid.

3   For a preliminary discussion of the potentially flawed nature of the study, see Kevin Gannon, "Let's Ban the Classroom Technology Ban," *The Tattooed Professor,* 15 May 2016.

newer digital tools. However, and unsurprisingly, looking at the actual study that spawned these articles tells a slightly different tale. In a recent issue of *Psychological Science,* Pam A. Mueller and Daniel M. Oppenheimer, in an article called "The Pen is Mightier than the Keyboard," conclude that students taking notes longhand do better in terms of knowledge retention than their laptop using peers, even when the distracting qualities of web surfing and other forms of multitasking are controlled for. The difference, perhaps counterintuitively, is that laptop users can record information faster. Because of this, they tend to transcribe almost verbatim what they hear, and this becomes a mindless task. Longhand note-takers, on the other hand, must be selective, and therefore end up processing information better. As Mueller and Oppenheimer state at the close of their article, "Although more notes are beneficial, at least to a point, if the notes are taken indiscriminately or by mindlessly transcribing content, as is more likely the case on a laptop than when notes are taken longhand, the benefit disappears."[4]

Now, I must admit to being somewhat unfair. Not everyone who recently posted this article, or variations of it, were doing so in the hopes of validating their own technological preferences. And, I should note that the initial article that I began discussing does acknowledge, albeit at the very end, that the issue is how information is processed and not the actual tool being used. What I take issue with, for the moment, is the title of the original article ("The Pen is Mightier Than the Keyboard") and the article's reference to "old-fashioned." The real heft of the original study focuses on information processing, but the advertising focuses on a binary between new and old, between the physical and the digital. When someone suggests that the "old-fashioned" is best, they are not only professing a preference for a physical book over a Kindle or iPad, but they are also revealing an anxiety about or suspicion toward the unavoidable ramifica-

4   Pam A. Mueller and Daniel M. Oppenheimer, "The Pen Is Mightier Than the Keyboard: Advantages of Longhand Over Laptop Note Taking," *Psychological Science* 25, no. 6 (2014): 1159–68, at 1166.

tions of the digitization of knowledge. But what they are also doing, whether intended or not, is participating in "compulsory able-bodiedness," where "normal," "best," and "able-bodied" ultimately occupy the same subject position.[5]

Another article that has made the social media rounds, sometimes with affirmation and at times with consternation, is Adam Kirsch's "Technology Is Taking Over English Departments: The False Promise of the Digital Humanities" in *The New Republic*. (Never mind that I read this piece because it is posted to the magazine's website.) After surveying and critiquing (sometimes justifiably) the triumphant tone that often accompanies Digital Humanities, Kirsch offers the following appraisal in his next-to-last paragraph: "The best thing that the humanities could do at this moment, then, is not to embrace the momentum of the digital, the tech tsunami, but to resist it and to critique it. This is not Luddism; it is intellectual responsibility. Is it actually true that reading online is an adequate substitute for reading on paper? If not, perhaps we should not be concentrating on digitizing our books but on preserving and circulating them more effectively. Are images able to do the work of a complex discourse? If not, and reasoning is irreducibly linguistic, then it would be a grave mistake to move writing away from the center of a humanities education."[6] There are many things going on here for Kirsch. One is certainly a nostalgic embrace of the old-fashioned, veiled in the trappings of "intellectual responsibility." More troubling to me, however, is the insistent refusal to engage with questions of accessibility. We can curate books and circulate them more, but does that always help the physically disabled? And, aside from the alarmist notion that writing is going to be removed from the humanities curriculum, what about the fact that multimodal objects may be a great help to some students who

5   See Robert McRuer, "Compulsory Able-Bodiedness and Queer/Disabled Existence," in *Disability Studies: Enabling the Humanities,* eds. Sharon L. Snyder, Brenda Jo Brueggemann, and Rosemarie Garland-Thomson, 88–99 (New York: Modern Language Association of America, 2002).

6   Adam Kirsch, "Technology Is Taking Over English Departments: The False Promise of the Digital Humanities," *New Republic,* 2 May 2014.

process information differently, and therefore feel excluded by linguistic-only expression? Within his nostalgic move he also expresses a normate position, thinking that we all learn, process, and engage the world in the same way.[7] What is good for Kirsch is good, apparently, for the rest of us.

Snark aside, "old-fashioned" often stands in for a wistful invocation of privilege, be it gendered, racial, or ableist. Kirsch articulates a sense of "best practices," and in doing so enshrines a particularly privileged orientation. If taking notes longhand is better for student retention than typing, then, the logic goes, professors and universities are correct to limit or ban laptops in the classroom. As a teacher, I too am concerned by the problematic qualities of laptops for student use, but as someone who is disabled, I know that if such a policy were in effect when I was a student, I would need to be an exception because handwritten notes are simply not something that I can do. My body works differently.[8]

As a corrective to such a retreat to the "old-fashioned" Humanities, I would look to George Williams, who, in his "Disability, Universal Design, and the Digital Humanities," observes that "Digital knowledge tools that assume everyone approaches information with the same abilities and using the same methods risk excluding a large percentage of people. In fact, such tools actually do the work of disabling people by preventing them from using digital resources altogether."[9] To address this exclusion, Williams advocates that the field of Digital Humanities adopts the principles of Universal Design. As is widely known, UD began as movement in architecture. Ron Mace developed

7   For my use of the term, "normate," see Rosemarie Garland-Thomson, *Extraordinary Bodies: Figuring Physical Disability in American Culture and Literature* (New York: Columbia University Press, 1997), 8–10.

8   For a fuller discussion on laptop bans and accessibility in the classroom, see Anne-Marie Womack and Richard H. Godden, "Making Disability Part of the Conversation," *Hybrid Pedagogy,* 12 May 2016.

9   George Williams, "Disability, Universal Design, and the Digital Humanities," in *Debates in the Digital Humanities,* ed. Matthew K. Gold, 202–13 (Minneapolis: University of Minnesota Press, 2012), 202.

"the concept of designing all products and the built environment to be aesthetic and usable to the greatest extent possible by everyone, regardless of their age, ability, or status in life."[10] Williams points to an oft-cited example, the sidewalk curb cut: "initially created to allow people in wheelchairs to cross the street more easily, curb cuts became recognized as useful also to other people such as someone making a delivery with a dolly, a traveler pulling luggage on wheels, a parent pushing a child in a stroller, or a person walking beside their bicycle."[11] While not an architect, as the user of a power-wheelchair I've experienced firsthand how significant UD can be for the built environment that I must navigate on a daily basis. For example, an out-of-the-way ramp leading from a university quad up to the rest of the campus can be frustrating and problematic and laborious, compared to the entire walkway being turned into a gently sloping ramp that is better for everyone.

I very much agree with Williams, and I think that he makes several important and necessary interventions into Digital Humanities. However, while the nostalgic (and ultimately hierarchical) expression of normativity we see in Kirsch's call to arms seems to stand in direct contrast to the more open principles of UD, I want to suggest that both positions engender a sense of "best practice" that could obscure the specific sociopolitical and embodied orientation of an individual user. For the remainder of this essay, I want to consider further the ramifications of the call toward a design principle that speaks to and accommodates the maximum amount of people.

In his critique of UD, Rob Imrie interrogates what he describes as "the philosophical basis of UD, that is, the universalistic rationalism of enlightenment philosophy."[12] In this analysis,

---

10 This is Ron Mace's definition as provided by the Center for Universal Design at North Carolina State University. See "About the Center: Ronald L. Mace," *Center for Universal Design,* http://www.ncsu.edu/ncsu/design/cud/about_us/usronmace.html.

11 Williams, "Disability, Universal Design, and the Digital Humanities," 205.

12 Rob Imrie, "Universalism, Universal Design and Equitable Access to the Built Environment," *Disability and Rehabilitation* 34, no. 10 (2012): 873–82,

UD would share some philosophical perspectives with the Enlightenment views of the universal subject. At first glance, then, this seems like a surprising avenue of analysis for Imrie. Much of the important work that Disability Studies scholars have undertaken is to dismantle the Enlightenment subject, revealing how its status as whole and independent is illusory. Lennard Davis, for example, introduces the idea of Dismodernism as a challenge to just such a subject position. In charting the terrain of a Dismodern orientation, Davis argues "[i]mpairment is the rule, and normalcy is the fantasy. Dependence is the reality, and independence grandiose thinking. Barrier-free access is the goal, and the right to pursue happiness the false consciousness that obscures it."[13] He then argues that "Universal design becomes the template for social and political designs."[14] Although Williams does not specifically cite the work of Davis, I would argue that Dismodernism and UD are philosophical cousins. Both approaches seek to universalize disability as opposed to treating it like a particular. As Williams describes of UD, "Devoting efforts to accessibility might improve the built environment for disabled people, but devoting efforts to universal design improves the built environment for all people."[15] Something built specifically for the disabled might be prohibitively costly and aesthetically displeasing, whereas something built for everyone, both able-bodied and disabled, will be accessible and preferable to the maximum amount of people. Similarly, Davis has famously argued that normal parking ought to be viewed as a subset of handicap parking, and not the other way around. Therefore, accessibility becomes the norm, the universal, not the exception or specific instance.

Universal Design, like any principle or system, has both positive (often intended) and negative (often unintended) outcomes. In terms of positive outcomes, UD, according to Imrie, should

---

at 879.

13 Lennard J. Davis, *Bending over Backwards: Disability, Dismodernism, and Other Difficult Positions* (New York: New York University Press, 2002), 31.

14 Ibid.

15 Williams, "Disability, Universal Design, and the Digital Humanities," 204–5.

be viewed "as distinctive to conventional development and design philosophies and processes, which are otherwise seen as hierarchical and insensitive to the variations in human capabilities to interface with, and use, different features of the designed environment."[16] Further, as Imrie continues to observe, "UD rejects design that fails to respond to, and interact with, everyone irrespective of their socio-cultural status and bodily capabilities and capacities."[17] While maximum accessibility is a laudable goal, in practice UD often fails to attend to the particular as it espouses the universal. As an example, he describes an instance of a wheelchair-user unable to use a hydraulic lift on a bus. This particular user wished to board forward because she was not able to do so backward, whereas the bus driver insisted the user could only board backward. While policies existed to allow citizens to board in either direction, the driver insisted on one particular direction, and this slowed down the overall progress of the bus, creating a tense and frustrating social experience. The design was, in theory, a good one — a bus is made accessible to all by the addition of a ramp, but the highly individualized experience of a particular user and her own social and physical situatedness unexpectedly made this design untenable. In other words, the theory appears to be sound in principle, but in practice the drive for universalism obscures the embodied particularity of individuals.

Another relatively recent example of unexpected outcomes would be the Reachability feature introduced on the iPhone 6 and iPhone 6 Plus. Because of the screen size of the Plus, Apple developed this feature where two light taps on the home button will bring the top half of the screen down to the bottom half. The problem that this feature addresses is the fact that, even for able-bodied consumers, this screen on the Plus was too big for a user to navigate one-handed. This seems to me like an excellent example of UD in action — this feature is not only useful

---

16 Imrie, "Universalism, Universal Design and Equitable Access to the Built Environment," 879.

17 Ibid.

to someone using the bigger phone, but it can also be useful to a disabled user even on the smaller phone, as it can often be difficult for someone with a physical impairment to reach the top of the screen if they are holding it near the bottom. But, in my own very specific situation, I'm not always able to bend my head downward comfortably, and so sometimes my line of sight for the lower half of the screen might be obstructed. This is a highly specific and I am sure unanticipated problem with this particular functionality. I raise this only to suggest that while UD is far, far preferable to the head-in-the-sand quality of Kirsch and others, both orientations toward technology evince surprisingly similar limitations when it comes to the highly localized experience of embodied difference. Kirsch expresses a normative, privileged position, whereas UD proponents express a universalism (objects used by all, able-bodied and disabled), yet, despite these differences both perspectives have the capacity to overlook the ways that the distinctiveness of sociocultural embodiment can affect usability.

In closing this essay, I want to briefly interrogate the utopian promise of technology, especially as a fundamental quality to UD (and Digital Humanities). In Imrie's critique of UD, he notes that the "focus on technical innovation may underestimate how far design outcomes are dependent on use and fail to recognize that far from technology being a prop of/for social action, it is influenced, and mediated, by its emplacement in specific social and cultural contexts."[18] Imrie's example of the wheelchair-user boarding a bus speaks to the ways that use can fail in practice. Closer to the world of Digital Humanities, Dominika Bednarska offers an example concerning blind students using assistive technology that raises some important questions that all advocates for People with Disabilities need to consider. Although Bednarska is writing about the limitations of technology and not specifically about UD, I do think that her cautions are salient. She argues that "[a] greater emphasis on technology can often overlook the drawbacks of technological reliance [...]. A focus

18 Ibid., 877.

on these technologies as primary or exclusive means for solving accessibility issues also makes prior accommodations and accessibility modifications less available."[19] To illustrate this, she examines how voice recognition software for the visually impaired could be seen to eliminate the need for assistants and note-takers. This is, in fact, one of the great benefits of assistive technology and UD — by building environments, physical and digital, that provide barrier-free access, then People with Disabilities can function more independently, and with less reliance on other people. As someone with a disability, I feel deeply and urgently the need to be less reliant on other people, but sometimes existing technology can be inadequate — it can break down, be unreliable, or may just be a poor substitution for human help (even if I don't want that help). Bednarska relates how, at her own institution, the University of California at Berkeley, funding for disabled students to have assistants became more restricted and limited because of the promise of available technologies. So, a student who did in fact work best with someone providing note-taking services would need to first demonstrate that available technologies were inadequate. This can provide an unnecessarily difficult bar to clear for some.

While my above discussion does articulate some ways that the effects of Universal Design may run counter to its hoped for aims, I am not suggesting a firm rejection of UD as it is applied to DH. However, I do think we need to move forward by balancing the Universalist and utopian aims of UD with a more local, attentive approach to individual use. As Imrie would describe it, advocates for UD need to specify how we conceive of the universal and the particular in terms of design.[20] As a medievalist also working in the field of Disability Studies, I have been trained to look for the particular and the local, the anomalous and the perplexing. In contrast with Davis's sweeping notion of

19 Dominika Bednarska, "Rethinking Access: Why Technology Isn't the Only Answer," in *The Culture of Efficiency: Technology in Everyday Life,* ed. Sharon Kleinman, 158–69 (New York: Peter Lang, 2009), 160.

20 Imrie, "Universalism, Universal Design and Equitable Access to the Built Environment," 879.

Dismodernism where disability stands in for the postmodern subject, Rosemarie Garland-Thomson describes the "extraordinary bodies" of the disabled,[21] and in my own field of medieval literature, Christopher Baswell has referred to nonstandard bodies as "eccentric."[22] Eccentric and extraordinary bodies have the potential to puncture the illusion of the universal that UD champions, disorienting and, more importantly, reorienting how we conceive of access and equality. Williams himself cites the work of Garland-Thomson in his work on UD, and I do think that his analysis attends to the particular in better ways than the more architecture-based UD that Imrie critiques. For example, Williams encourages a reciprocity between user and designer, arguing that "by working to meet the needs of disabled people — and by working with disabled people through usability testing — the digital humanities community will also benefit significantly as it rethinks its assumptions about how digital devices could and should work with and for people."[23] In response, I would suggest that the goals that animate UD should be and will continue to be a powerful principle in DH, but such a design principle needs to accompany, not supplant, the attention to the particular. Reciprocity could mean mutual care, of and for each other, but it should not need to flatten us out into a universal subject in the process.

Jonathan Hsy: In my reflections, I'd like to interrogate the role of overtly utopian discourses in Universal Design (UD) endeavors and the Digital Humanities (DH). Like any other collective movements, both UD and DH offer dreams of world-transformation that can, at times, enact proselytizing (if not activist) impulses. Both UD and DH advocates often invoke an unrealized and idealized conception of collective space (physical or

---

21 Garland-Thomson, *Extraordinary Bodies,* 5–9.

22 Baswell uses this term in a series of talks. See Christopher Baswell, "The Felicity Riddy Lecture: Kings and Cripples: Royal and Eccentric Bodies in Thirteenth-Century England" (lecture, University of York, 25 November 2010).

23 Williams, "Disability, Universal Design, and the Digital Humanities," 210.

online) in order to challenge dominant beliefs and practices and to encourage people to join in a newly reconfigured sense of common purpose.

In her analysis of UD discourses in the US, media theorist Jane Bringolf explains that the "vision for [UD] is to cultivate the creative minds of designers to consider the whole of the population bell curve in their designs."[24] Designating not so much a discrete goal but a "Utopian ideal," UD "is posed as an intellectual challenge for designers" or people developing other projects and products.[25] While the term "Universal Design" was coined in the US by architect and designer Ron Mace and originally applied to the configuration of physical space, UD has since broadened to include online media and digital environments.[26] In Europe, UD is more commonly called "Design for All," while in the UK the term "Inclusive Design" is preferred.[27] While all these terms differ slightly, Bringolf observes that "the same underpinning concept" underlies each one: the drive to "[design] for the whole of the population bell curve" and to "[create] maximum utility for the maximum number of people regardless of age, culture, and education or ability level."[28]

These ideals are wonderful in theory, but there are some unanticipated drawbacks to UD discourses as they inform actual practice. Do UD endeavors in their efforts to embrace the totality of all humanity actually seek to accommodate difference or rather to eradicate it? Mainstream UD discourses, especially as appropriated by designers in technology companies in the US, have a tendency to render UD synonymous with the creation of "accessibility features" and other kinds of products to be used by people with disabilities. For instance, a mobile phone's capac-

---

24 Jane Bringolf, "Universal Design: Is It Accessible?" *Multi: The RIT Journal of Plurality and Diversity in Design* 1, no. 2 (2008): 45–52, at 47.

25 Ibid.

26 Carlos Nunes Siva, "Universal Design," in *Green Cities: An A-to-Z Guide*, eds. Nevin Cohen and Paul Robbins (Thousand Oaks: SAGE Publications, 2011), 433–6.

27 Bringolf, "Universal Design," 48.

28 Ibid.

ity for voice dictation and or tactile magnification of text may be touted as "accessibility features" for people who are blind or visually impaired, but *nondisabled* people readily use such features too. Bringolf notes that a pervasive discursive practice of casting "accessibility" as a *subset* of UD (or even conflating "accessibility" and UD entirely) limits the scope of UD's original intent. In its broadest sense, UD promotes much more than creating a "special subset" of accommodations for disabled people but rather embraces a capacious orientation toward design that might work for as many people as possible, disabled and nondisabled alike.

To rework Bringolf's arguments a bit, I wonder if a general discursive tendency to conflate UD with narrower discourses of "accessibility" risks enacting the reverse of what UD initially envisions. Rather than attending to embodied variance as a way to multiply and sustain diverse modes of interaction with physical or digital environments, a narrowly conceived notion of UD as a set of separate (or supplemental) "accessibility features" conceives the challenge of UD as one of integrating disabled people into an existing set of nondisabled norms.

The complex operations of universalist discourses in promoting DH projects offers another example of how utopian thinking has the potential for unanticipated drawbacks insofar as they can reinforce a set of "best practices" that in itself asserts a new normative force. In arguing that information be made available to everyone through digitization efforts and other online media, DH endeavors can invoke a dream of a shared repository of knowledge that anyone can use, or (to adapt various UD discourses) such discourses suggest an idea of fully "inclusive content" or "scholarship for all" in a grand vision of public "outreach" and collective participation.[29] George Williams justifiably observes that "people with disabilities will benefit significantly

---

29 See, for instance, the online discussion hosted by HASTAC Scholars Bridget Draxler, Jentery Sayers, Edmond Y. Chang, and Peter Likarish, titled "Democratizing Knowledge in the Digital Humanities: Making Scholarship Public, Producing Public Scholarship," *HASTAC*, 21 September 2009.

if the digital humanities community pursues projects [that] take seriously the need to adopt universal design principles."[30] While such discourses are earnestly striving to respect human diversity and embodied variation, a future-oriented utopianism articulates an under-examined desire for some conformity (or alignment with, or participation along) a shared baseline: a set of collective values, "best practices," or shared cultural expectations.

As a medieval literature scholar, one way I try to think about this tension between an imagined universalism and the messiness of embodied diversity is through literary fiction. Fictional works often express cultural hopes or desires while also promoting a political ideology, whether or not that ideology is overtly disclosed. *The Book of John Mandeville* (most likely first composed in French the mid-fourteenth century) was a medieval "bestseller" with wide appeal: it was translated into Latin and many European vernacular languages and enjoyed a long life in many manuscripts and print media. Part pilgrimage manual, travel narrative, and proto-ethnography, the work narrates an English knight's journey from home to the Holy Land and back again, and along the way the narrator moves through diverse social environments. One modern translator describes the work as a "mash-up" or dynamic "recombination of sources [...] characterized by a shifting mix of genres,"[31] with its narrator breathlessly announcing his travel across "many countries and many different provinces and many different regions and different islands" and "many different peoples with diverse laws and diverse customs,"[32] including social groups of varied religions, languages, races, genders (including hermaphrodites), and other extraordinary modes of embodiment.

Despite its clear discursive interest in (if not desire for) embodied diversity, Mandeville's *Book* transmits its own fantasies of universalist polity. A prologue makes a call to "reclaim the

---

30 Ibid.

31 Iain Macleod Higgins, ed. and trans., *The Book of John Mandeville: With Related Texts* (Indianapolis: Hackett, 2011), xi.

32 Ibid., 5.

[Holy Land] and wrest it from the hands of the foreigners" (i.e., Muslims) and a chapter on "Saracen" beliefs, emphasizing what beliefs they share with Christians, transmits the fantasy that they can be easily converted and assimilated into a Christian worldview.[33] Such modes of thinking were not without precedent in the medieval West, tracing themselves back to a Biblical passage asserting that "there is no longer Jew or Greek, there is no longer slave or free, there is no longer male and female; for all of you are one in Christ Jesus" (Galatians 3:28). This formative passage of Scripture discursively embraces divergent modes of difference (cultural, linguistic, gendered) but only to assert an ardent wish for oneness of shared social belonging (in this case, a Christian universalism).

There's a vast historical chasm between the medieval West and modern digital contexts, but I would suggest that UD, like Mandevillean discourse, has a clear eschatology (an ultimate destination for networked humanity) — and its arrival is always-already deferred. If we just take the example of a website as a project that could enact UD principles, it's hard to imagine that one user interface could be equally accessible to everyone across every human language (spoken, written, or signed), every form of media, and every form of embodied variance (sensory, motor, cognitive). Joe Clark, a journalist and author specializing in media technologies intended to make information accessible to people with disabilities (such as captioning and audio description), contends in a provocative blog posting that UD is a myth.[34] I might reshape Clark's observation to say that UD is a motivating fiction or tantalizing impossibility: a unicorn, Holy Grail, earthly Paradise, pick your metaphor. In its association with temporal deferral, UD suggests a close association with the very concept of disability as unrealized futurity. As cultural critic and theorist Robert McRuer has astutely noted, disability is not a "special" category or subset of humanity but a "spectral" prospect that haunts us all: "If we live long enough, disability is

---

33 Ibid., 4.
34 Joe Clark, "Universal Design Is a Myth," *fawny.blog* (blog), 15 October 2009.

the one identity that we all inhabit."[35] In its deferred arrival, UD can be considered, like disability itself, an intellectual and theoretical concept that evinces an elusive futurity: a prospect that is always receding on the horizon.

This notion of deferred futurity informs how mainstream social justice discourses of access and inclusion can reassert notions of a shared norm or space even as they acknowledge the attractive vitality of the very idea of sustaining social and embodied variety. The engaging *Accessible Futures* workshop series held at five different universities from 2013 through 2015 embraced a utopian discourse with the laudable mission to educate DH practitioners in how to make their projects accessible to internet users who have disabilities.[36] This series of workshops sponsored by the Office of Digital Humanities (ODH) at the National Endowment for the Humanities (NEH) has brought together scholars, archivists, and design practitioners to address disability and access issues relating to DH projects. Having attended one iteration of the series on February 28–March 1, 2014, at the University of Texas-Austin, I can say these workshops (and its associated website) are informative, lucid, and productive. Perhaps in line with the expertise of the organizers Jennifer Guiliano, George Williams, and Tina Herzberg, most of our time in the Austin workshop addressed improving the accessibility of websites for people with visual impairments, and we considered strategies for incorporating captions and alt-tags for images as well as ensuring that website architecture can be read and navigated by people using screen readers that voice online text

---

35 Robert McRuer, *Crip Theory: Cultural Signs of Queerness and Disability* (New York: New York University Press, 2006), 200.

36 According to the Accessible Future website: "Building an Accessible Future for the Humanities Project is organizing four 2-day workshops during which participants will learn about technologies, design standards, and accessibility issues associated with the use of digital environments." See *Accessible Future,* http://www.accessiblefuture.org.

aloud.[37] One session included an audit of various DH projects to discuss how well they integrated such accessibility elements.

One of the websites we discussed was the *Deaf Studies Digital Journal* or *DSDJ*, founded in 2009 and published by the ASL [American Sign Language] & Deaf Studies Department at Gallaudet University in Washington, DC.[38] What makes this online publication so intriguing as a "case study" is the how the use of non-textual digital media shapes its linguistic and cultural content. ASL is a fully expressive language with as much potential as any other spoken language for artistic and intellectual expression, and *DSDJ* is the first peer-reviewed academic and creative arts journal to use ASL for all of its content (it also publishes some material in English, as I will discuss below). Since ASL is inherently a kinetic language that uses embodied actions including manual gestures and facial expressions for its grammar, recorded video clips in Adobe Flash Player are crucial for the presentation of ASL content. The embodied physicality of sign language perhaps lends an unintended meaning to the word "digital" in the journal's English title — suggesting first the electronic or online medium of the publication and secondarily a "spectral" reference to fingers and the embodied labor visually showcased in the videos themselves. An online video produced by Deaf scholars Jill Bradbury and Tyrone Gioradano (debuted at the #TransformDH conference at the University of Maryland in October 2015) explores facets of Shakespeare performance in ASL and addresses the historical exclusion of Deaf people from sound-centered forms of theater and scholarship, and video — presented online with English captions and a full online transcript and description of its visual contents — deftly exploits the manifold valence of the "digit" in its pluralized title *Digit(al) Shakespeares.*[39] *DSDJ* and other digital media such as the *Digit(al)*

---

37 For an excellent overview of the Austin workshop, see Susan Floyd, "Thinking About Accessibility: Accessible Future 2014 at UT-Austin," *Texarchivist* (blog), 14 March 2014. See also Floyd's writing on Twitter (@Texarchivist).

38 See *Deaf Studies Digital Journal* (*DSDJ*), http://dsdj.gallaudet.edu.

39 See *Digit(al) Shakespeares,* http://transformdh.org/2015-video-showcase/digital-shakespeares-tyrone-giordano-and-jill-marie-bradbury.

Shakespeares project increasingly provide expanded opportunities for Deaf communities to connect with each other within the US and across the globe.

In a technological gesture towards universality, *DSDJ* displays a number of important strategies for reaching different kinds of people including Deaf communities beyond the US. It provides abstracts (summaries) of each contribution, most often presented in sign language by the author. Some, but not all, of the content features a downloadable PDF presenting the equivalent content in English (other times the site features a previously published English-language article now translated into ASL). *DSDJ* also includes academic contributions in sign languages around the world such as International Sign (IS), a conventionalized transcultural Deaf contact language used in contexts where people use mutually unintelligible sign languages. By incorporating sign languages beyond ASL, the journal's content is made at least partially accessible to Deaf users around the world who might not use ASL or written English.[40]

An intriguing aspect of the group discussion of *DSDJ* in the *Accessible Future* workshop in Austin in 2014 was the sense that the lack of audio or captions in these videos make the content "inaccessible" by one set of embodied norms (that is, a set of UD principles that would call for embedded features for internet users who have visual impairments). As I reflect on this conversation afterwards, I have come to realize that the uneven media functionality of the journal suggested a discomforting social reality for those of us who were present at that particular workshop: much of the content of this Deaf-oriented journal was at the time rendered *inaccessible to a hearing majority* (or, to put things more precisely, the online journal's content was only partially accessible to non-ASL users).

The question of whether an ASL journal *should* provide equivalent English-language content for all its material is com-

---

40 Peter C. Hauser's article, "Deaf Eyes: Visual Learning and Deaf Gain," *DSDJ* 2 (Fall 2010), is presented by the author in ASL as well as IS. As of 2 February 2018, the PDF of an English language translation is forthcoming.

plex not only for its sociopolitical ramifications but also in terms of the labor and logistics involved: captioning content for any video requires more than mere *transcription* of language (as would be the case in videos using spoken languages); these particular videos require a process of translation from ASL into written English that necessitates a close engagement with Deaf culture. In some cases, a link to a PDF with equivalent English text or at least an informative summary in English is provided as a link beside the video, but the question of how (or if) the online journal can provide non-ASL users with access to all of its ASL content (especially ASL poetry) is a more challenging prospect.[41] As a hearing person with only some basic knowledge of ASL, I find it intriguing that an extensive commentary on an academic article about audism or "audiocentric privilege" does not provide a link to a PDF of the commentary that I can read in written English (perhaps one in the future might be provided).[42] In this case, the current user interface appropriately forces me to confront my own audiocentric (and Anglophone) privilege and I find myself navigating an online linguistic environment that is only unevenly or partially configured for my use.[43]

In my reflections on the utopian prospects of UD and its un-intended limits or exclusions, I hope to encourage a more nu-anced orientation to disability and embodied diversity as we continue to create, rework, engage, and critique DH projects. We need more flexibility in how we conceive of UD and not assume a unidirectional delivery or translation of content, informa-tion, or experience. It's attractive to maintain a utopian dream of some "universal design concept" that could bring all kinds of

41 For instance, Justin Jackerson's ASL poem "uses handshapes of the letters within the name 'Gallaudet University' twice [to tell] the fast paced expe-rience of being a student at Gallaudet." See "Gallaudet University," *DSDJ* 4 (Spring 2014).

42 Amy June Rowley and Richard Eckert, "Audism: A Theory and Practice of Audiocentric Privilege," *DSDJ* 4 (Spring 2014).

43 For an excellent disability-centered analysis of the uneven accessiblity of digital media, see Elizabeth Ellcessor, *Restricted Access: Media, Disability, and the Politics of Participation* (New York: New York University Press, 2016).

embodied variance into one shared physical space or digital environment, but we should be more careful about the presumed set of cultural and embodied norms and "best practices" that such initiatives might unthinkingly promote.[44] A multidirectional approach to how we all engage with digital media and content can open up both the U and the D in new ways — and humanist engagement with the arts, rhetoric, and critical theory must continue to play an active role in shaping these endeavors.

Concluding thoughts

The two essays assembled here, one by a disabled user of various types of assistive technology and the other by a nondisabled ally who engages with aspects of Deaf culture, bring together particular sets of embodied experience in order to probe and interrogate the assumptions and inhibiting freight that the "Universal" in "Universal Design" draws in its wake. In our critical evaluations of UD, we share several conclusions and concerns with the contributors to the webtext *Multimodality in Motion: Disability and Kairotic Spaces,* and we wish to close this essay with a brief discussion of the important insights they articulate.[45] In their opening "Access Statement," Yergeau et al. immediately acknowledge that "Universal design is a process, a means rather than an end. There's no such thing as a universally designed text. There's no such thing as a text that meets everyone's needs. That our webtext falls short is inevitable."[46] They go on to caution that the inevitable failure of UD "is not a justification for failing to consider what audiences are invited into and imagined as part of a text."[47] Rather, the recognition of failure at the heart

---

44 On the conceptual limitations to "technology-led" approaches to UD and disability in the context of physical space, see Imrie, "Universalism, Universal Design and Equitable Access to the Built Environment."

45 Melanie Yergeau et al., "Multimodality in Motion: Disability and Kairotic Spaces," *Kairos: A Journal of Rhetoric, Technology, and Pedagogy* 18, no. 1 (2013).

46 Ibid.

47 Ibid.

of Universalist paradigms can enable us to attend more closely to the particular embodied orientation of users and stakeholders. We would embrace this emphasis on process over product, on becoming and emergent technologies over closed systems of top-down provisions for accommodation. While we agree that Universal Design is an unachievable goal, we would go further and argue that the goal itself is problematic and ultimately inadequate to the continuously evolving situation of not only the inclusion of more and more disabled/extraordinary/eccentric bodies into "normal" society but also the ever-shifting able-ness of any body as it moves toward inevitable failure.

In his section "Over Here" in *Multimodality in Motion,* Michael J. Salvo discusses a possible successor to UD, the concept of Resonant Design as developed by Graham Pullin. As Salvo describes it, Resonant Design "offers designers and culture-at-large a phrase for the kind of responsive, use-centered, stakeholder-involving, context-sensitive artifact creation methods [Pullin] advocates."[48] Yet, while being more responsive to difference than UD, Resonant Design itself is an illusory goal because it "does not explore the potential contribution to culture that would come from further interrogating the relationships that make society a powerfully disabling force, limiting to physical, social, and lifeworld potentials for millions. In other words, it calls for change without fully recognizing how disruptive the needed changes may be."[49] For Salvo, the inadequacy of Pullin's model lies in its failure to reconfigure the terms by which society defines normality, simply putting embodied difference at the center as opposed to the margins. However, we would contend that substituting Universalism (despite its potential for inclusiveness) for normativity would achieve less than what we expect or desire, and such a principle of design would similarly fail to cause any significant or re-orienting disruption. We would advocate the continued emphasis of multimodality and multi-

---

48 Michael J. Salvo, "Resonant Design," in "Multimodality in Motion."
49 Ibid.

directionality in DH endeavors, and to do so we may need to abandon the aims of Universalism.

Bibliography

Accessible Future. http://www.accessiblefuture.org.

Center for Universal Design. http://www.ncsu.edu/ncsu/design/ cud/about_us/usronmace.html.

Baswell, Christopher. "The Felicity Riddy Lecture: Kings and Cripples: Royal and Eccentric Bodies in Thirteenth-Century England." Lecture, University of York, 25 November 2010.

Bednarska, Dominika. "Rethinking Access: Why Technology Isn't the Only Answer." In *The Culture of Efficiency: Technology in Everyday Life,* edited by Sharon Kleinman, 158–69. New York: Peter Lang, 2009.

Bringolf, Jane. "Universal Design: Is It Accessible?" *Multi: The RIT Journal Of Plurality and Diversity In Design* 1, no. 2 (2008): 45–52.

Clark, Joe. "Universal Design Is a Myth." *fawny.blog* (blog). 15 October 2009. http://blog.fawny.org/2009/10/15/universal-design-myth.

Davis, Lennard J. *Bending Over Backwards: Disability, Dismodernism, and Other Difficult Positions.* New York: New York University Press, 2002.

Deaf Studies Digital Journal (DSDJ). http://dsdj.gallaudet.edu.

Digit(al) Shakespeares. http://transformdh.org/2015-video-showcase/digital-shakespeares-tyrone-giordano-and-jill-marie-bradbury.

Draxler, Bridget, Jentery Sayers, Edmond Y. Chang, and Peter Likarish. "Democratizing Knowledge in the Digital Humanities: Making Scholarship Public, Producing Public Scholarship." *HASTAC.* 21 September 2009. http:// www.hastac.org/forums/hastac-scholars-discussions/ democratizing-knowledge-digital-humanities.

Ellcessor, Elizabeth. *Restricted Access: Media, Disability, and the Politics of Participation.* New York: New York University Press, 2016.

Floyd, Susan. "Thinking About Accessibility: Accessible Future 2014 at UT-Austin." *Texarchivist* (blog). 14 March

2014. https://texarchivist.com/2014/03/14/thinking-about-accessibility-accessiblefu-2014-utaustin-2/.

——— (@Texarchivist). Twitter feed. https://twitter.com/Texarchivist.

Gannon, Kevin. "Let's Ban The Classroom Technology Ban." *The Tattooed Professor.* 15 May 2016. http://www.thetattooedprof.com/2016/05/15/lets-ban-the-classroom-technology-ban/.

Garland-Thomson, Rosemarie. *Extraordinary Bodies: Figuring Physical Disability in American Culture and Literature.* New York: Columbia University Press, 1997.

Hauser, Peter C. "Deaf Eyes: Visual Learning and Deaf Gain," *Deaf Studies Digital Journal* 2 (Fall 2010). http://dsdj.gallaudet.edu/index.php?view=entry&issue=3&entry_id=81.

Higgins, Iain Macleod, ed. and trans. *The Book of John Mandeville: With Related Texts.* Indianapolis: Hackett, 2011.

Imrie, Rob. "Universalism, Universal Design and Equitable Access to the Built Environment." *Disability and Rehabilitation* 34, no. 10 (2012): 873–82. DOI: 10.3109/09638288.2011.624250.

Jackerson, Justin. "Gallaudet University." *Deaf Studies Digital Journal* 4 (Spring 2014). http://dsdj.gallaudet.edu/index.php?view=entry&issue=5&entry_id=190.

Kirsch, Adam. "Technology Is Taking Over English Departments: The False Promise of the Digital Humanities." *New Republic.* 2 May 2014. https://newrepublic.com/article/117428/limits-digital-humanities-adam-kirsch.

McRuer, Robert. *Crip Theory: Cultural Signs of Queerness and Disability.* New York: New York University Press, 2006.

———. "Compulsory Able-Bodiedness and Queer/Disabled Existence." *Disability Studies: Enabling the Humanities,* edited by Sharon L. Snyder, Brenda Jo Brueggemann, and Rosemarie Garland-Thomson, 88–99 (New York: Modern Language Association of America, 2002).

Mueller, Pam A. and Daniel M. Oppenheimer. "The Pen Is Mightier Than the Keyboard: Advantages of Longhand Over

Laptop Note Taking." *Psychological Science* 25, no. 6 (2014): 1159–68. DOI: 10.1177/0956797614524581.

Nunes Siva, Carlos. "Universal Design." In *Green Cities: An A-to-Z Guide,* edited by Nevin Cohen and Paul Robbins, 433–36. Thousand Oaks: SAGE Publications, 2011.

Rowley, Amy June, and Richard Eckert. "Commentary: 'Audism: A Theory and Practice of Audiocentric Privilege.'" *Deaf Studies Digital Journal* 4 (Spring 2014). http://dsdj. gallaudet.edu/index.php?issue=5&section_id=3&entry_ id=250.

Salvo, Michael J. "Resonant Design." In "Multimodality in Motion: Disability and Kairotic Spaces." *Kairos: A Journal of Rhetoric, Technology, and Pedagogy* 18, no. 1 (2013). http:// kairos.technorhetoric.net/18.1/coverweb/yergeau-et-al/ pages/here/res.html.

Sanders, Laura. "Students Retain Information Better with Pens than Laptops." *ScienceNews: Magazine of the Society for Science and the Public.* 30 April 2014. https://www. sciencenews.org/article/students-retain-information-better-pens-laptops.

Williams, George. "Disability, Universal Design, and the Digital Humanities." In *Debates in the Digital Humanities,* edited by Matthew K. Gold, 202–13. Minneapolis: University of Minnesota Press, 2012. http://dhdebates.gc.cuny.edu/ debates/text/44.

Womack, Anne-Marie, and Richard H. Godden. "Making Disability Part of the Conversation." *Hybrid Pedagogy.* 12 May 2016. http://www.digitalpedagogylab.com/hybridped/ making-disability-part-of-the-conversation.

Yergeau, Melanie, et al. "Multimodality in Motion: Disability and Kairotic Spaces." *Kairos: A Journal of Rhetoric, Technology, and Pedagogy* 18, no. 1 (2013). http://kairos. technorhetoric.net/18.1/coverweb/yergeau-et-al/pages/ access.html.

6

DH as "Disruptive Innovation" for Restorative Social Justice: Virtual Heritage and 3D Reconstructions of South Africa's Township Histories

Angel David Nieves

Much of this essay constitutes an initial effort to frame an open access digital-first publication tentatively entitled, *Apartheid Heritage(s): A Spatial History of South Africa's Black Townships* with Stanford University Press as its designated publisher. The project involves the development of a multi-modal 3D qualitative geospatial archive and platform for research into the apartheid-era "South Western Townships," better known as Soweto, outside Johannesburg, South Africa.

Some background is needed to better situate this research. In 2011, the United Nations issued a report that declared internet access a human right.[1] (Interestingly, on that same day two-

---

1   United Nations General Assembly, *Report of the Special Rapporteur on the promotion and protection of the right to freedom of opinion and expression, Frank La Rue,* A.HRC.17.27 (Geneva, 2011).

thirds of Syria's internet access had gone dark — something that was likely the work of the Assad regime in response to unrest in that country.) In 2013, a group of scholars from across the United States approached the Alliance of Digital Humanities Organizations (ADHO) with a proposal for a new special interest group (or SIG) with a focus on social justice and human rights. As one step forward, the group developed an *"advisory document* for building collaborative projects, conducting events, gathering sensitive data, and composing scholarly communications with social justice issues and human rights in mind."[2] As part of the initial group of scholars who worked to develop this special interest group, I was much more invested in the potential praxis-based strategies we might develop — perhaps a list of ethical guidelines or even a kind of *social justice toolkit* — for engaging in community-centered digital humanities projects. Some of what I will be discussing in this essay is in itself filled with some controversy and is worthy of further debate with regard to issues of ethics, "reality," and truth as applied to historical reconstructions — specifically computer-generated visualizations of historic landscapes and buildings within contested areas in certain fields — as issues of power and representation cannot be overlooked.

In a series of email exchanges in 2014 between Willard McCarty, Professor of Digital Humanities at King's College London and Andrew Taylor, Associate Curator in the Department of Art History at Rice University, the debates over historical simulations/recreations and questions concerning the scholars' research goals in creating accurate representations were made all too apparent to subscribers of the Humanist Discussion Group's listserv (and also help to highlight what presently engages me across several disciplines). McCarty writes:

---

2   "Digital Humanities as Restorative Social Justice: Virtual Heritage, 3D Reconstructions and South Africa's Township Histories," *Digital Humanities Initiative.* See also "Update on Proposed ADHO SIGs," *Alliance of Digital Humanities Organizations.*

I hear colleagues who work in visualization [*sic*] talk about the problem of how photo-realism, say in a VR representation of an ancient building which survives only in fragments, can be dangerously misleading. What then is an accurate representation? The most obvious response, I suppose, is one that informs the viewer somehow of the difference between that which survives and that which is inferred, ideally representing degrees of certainty. [...] I'd suppose that you do not want an *inaccurate* representation, but in the circumstance I am imagining, accuracy is just a stepping-stone. [...] I wonder further if this isn't quite close to the historian's tricky question of getting to "what actually happened" (von Ranke's famous phrase). Even if counterfactual history is your thing, I'd think you'd be doing it in order better to illumine what did (in some sense actually) happen. Historians are quite sensitive about counterfactual studies and about the degree to which history-writing is creative. At the same time an accurate, let us say complete, chronological account is not a history, only the beginning of one.[3]

McCarty is essentially questioning the kinds of principles and methodologies of practice that help guide scholars through the many complex issues involved with creating historical reconstruction. As some archaeologists have argued, "One of the most significant consequences of the introduction of digital 3D modeling in the *Cultural Heritage field* is the possibility to use 3D models as highly effective and intuitive means of communication as well as interface to share and visualize information collected in databases."[4] I would also argue that 3D reconstructions

---

3   Willard McCarty, email to "Online seminar for digital humanities," 12 February 2014. See http://lists.digitalhumanities.org/pipermail/humanist/2014-December/012484.html.

4   Anna Maria Manferdini and Fabio Remondino, "Reality Based 3D Modeling, Segmentation and Web-Based Visualization," in *Digital Heritage: Third International Conference, EuroMed 2010 Lemessos, Cyprus, November 2010 Proceedings,* eds. Marinos Ioannides et al., 110–24 (Berlin: Springer-Verlag, 2010), 110. Emphasis added.

have the potential for the building-up of more robust — and potentially on-line — textual and visual historic archives. The use of new media offers an enormous creative potential for marginalized communities to disrupt official history and reclaim aspects of their lost or difficult heritage if digital technologies (through digital humanities) are harnessed for their use. Verne Harris, longtime archivist of the Nelson Mandela Foundation makes clear, archives are "far from being a simple reflection of reality [...] [they] are constructed windows into personal and collective processes. They at once express and are instruments of prevailing relations of power."[5] Shifting that power into the hands of township residents requires a social justice framework whereby archival practices become an inherent part of a human rights agenda across the African Diaspora.

I was reminded when preparing this essay, of Hayden White's work for a 2005 issue of *Rethinking History,* "Historical Fiction, Fictional History, and Historical Reality," in which he quotes Ralph Ellison from a 1958 essay, "Some Questions and Some Answers."[6] There Ellison writes, "Men cannot unmake history, thus it is not a question of reincarnating those cultural traditions which were destroyed, but a matter of using industrialization, modern medicine, modern science in general to work in the interest of these peoples rather than against them."[7] In some small ways I see so much of our work, perhaps in an Afro-futurist sense, as taking full advantage of "modern science" and technology to question our narrative practices in the digital realm. My work, I would argue, also raises questions about the persistence of a digital divide that now exists between the Global North and South. Radical change is therefore necessary

---

5   Harris Verne, "The Archival Sliver: Power, Memory, and Archives in South Africa," *Archival Science* 2 (2002): 63–86, at 63.

6   Hayden White, "Introduction: Historical Fiction, Fictional History, and Historical Reality," *Rethinking History: The Journal of Theory and Practice* 9, nos. 2–3 (2005): 147–57, at 157. See also Ralph Ellison, "Some Questions and Some Answers," in *The Collected Essays of Ralph Ellison,* ed. John F. Callahan (New York: Random House, 2003), 291–302.

7   Ibid., 157.

along the many social, economic, political, regulatory, and in-frastructural barriers that continue to disadvantage many of the world's "informational peripheries" — to aid those people who remain invisible or unheard in the African Diaspora. In the field of Africana or Black Studies, Abdul Alkalimat of the University of Illinois at Urbana-Champaign has made clear that "the impact of the information revolution can lead to a renaissance of community development, cultural creativity, and liberation politics."[8]

Much of my work-to-date explores the building of a multi-modal information environment to discuss Soweto's past, pre-sent, and future redevelopment — one part of a new series of cultural practices of remembrance, reconciliation, and empow-erment with a view towards an integrative approach to social justice and the practice of digital humanities scholarship. My digital scholarship comprises several works *in process,* already several years in the making and touching upon several discrete, but ultimately inter-related, areas of inquiry in apartheid-era South Africa. As it stands today, "virtual heritage" projects re-quire multi-disciplinary teams of historians, writers, design-ers, software developers, cultural heritage managers, and local community informants who would collaborate in the design, development, and management of an immersive 3D virtual heritage landscape. This emerging digital research paradigm is quite unlike that of the archetypal solitary scholar toiling alone in a dusty archive. In particular, my projects would not be possible without a team of scholars and practitioners from Hamilton College's Digital Humanities Initiative (DHi), where I was Co-Director with Janet T. Simons.[9] Digital humanities, as a field, as a discipline, and a new knowledge community, is by its very nature a collaborative and iterative process that cannot be undertaken without a cohort of "experts" from all sides of the "learning through making and doing" spectrum that includes librarians, undergraduate student interns, designers, and soft-

---

8   Abdul Alkalimat, "eBlack: a 21st century challenge," *Mots Pluriels* 19 (2001).
9   *Digital Humanities Initiative,* http://dhinitiative.org/.

ware engineers. In other words, as a professor and researcher I am, in many ways, a project manager of a team of "expert practitioner scholars" upon whom I rely to help me tell this particular spatial narrative. I would be remiss if I did not acknowledge, or at least situate myself, in an intellectual place (or space). Thus, DHi is a *collaboratory* — a humanities lab — where new media and computing technologies are used to promote humanities-based teaching, research, and scholarship across the liberal arts. There at Hamilton, a small Northeast us college, the liberal arts environment places a strong emphasis on the teaching curriculum and the integration of humanities-based research questions into undergraduate scholarship.

In South Africa the legacy of apartheid has meant a constant engagement with cultural trauma and its impact on all aspects of social life, particularly for township residents beginning in the early twentieth century. I have been working on various preservation efforts in Soweto outside of Johannesburg with local residents and former student activists for over a decade beginning in 2004. Located some 30km from downtown Johannesburg, the township of Soweto has been a site of both historical contestation and numerous state-sponsored heritage projects. Soweto was also where my first area of digital inquiry and recovery began at the Hector Pieterson Memorial and Museum in Johannesburg, South Africa, a heritage site that preserves the history and memory of all those who were involved in the Soweto Uprising of 16 June 1976. The museum is named in honor of 13 year-old Hector Pieterson, among the first student victims to die in the Uprisings. On that fateful day, Soweto students gathered to protest against the use of Afrikaans language as a medium of teaching and learning in black schools. Shortly thereafter, police began shooting at the assembled marchers, violently disrupting what was to be a peaceful protest. Hector Pieterson's death, and the subsequent killing of 575 other protestors in the Uprisings that would help bring about the first democratic elections of 1994, are memorialized at this National Heritage Site.[10]

That first South African research project, eventually entitled *Soweto '76* had a scope that provided for the digitization and preservation of the archival collections of the Hector Pieterson Museum with the intention of providing on-line access to its holdings for broad public use.[11] The holdings were considered endangered due to a lack of available resources for their care and preservation. The project was initially proposed to convert to digital format some twenty audio-cassette tapes of interviews conducted with students involved in the Uprisings of 1976. Even after years of working on community-based projects, I some-what foolishly thought I could digitize the audio-cassette tapes, at the Hector Pieterson Museum over the course of six months while also developing a front-end interface for accessing the interviews on-line. As my first digital humanities project, I was very naïve about the many challenges facing the archive-making process for community-based township museums. Nonetheless, between 2006–2007, while at the Maryland Institute for Technology in the Humanities (or MITH), the project team began the process of digitizing a broader selection of the Hector Pieterson Museum's multi-media collections and holdings.

As cultural studies scholar Chela Sandoval argued in her book *Methodology of the Oppressed* the world inhabited by wired, technologized, privileged subjects requires a shift in educational preparation and cultural expertise so that "the technologies developed by [and with] subjugated populations to negotiate this realm of shifting meanings can prove indispensable."[12] Placing various technologies in the hands of "subjugated populations" allows for new kinds of engagements to occur. The rise of network technologies (Web 2.0) has now allowed a diverse group of users to actively express and interrogate their racial,

---

10 "Hector Pieterson Memorial and Museum," *Gauteng*, http://www.gauteng.net/attractions/hector_pieterson_memorial_and_museum/.

11 *Soweto '76*, http://www.soweto76archive.org.

12 Chela Sandoval, *Methodology of the Oppressed* (Minneapolis: University of Minnesota Press, 2000), 87.

gendered, national, and class identities. We have seen the power of the internet to transform the political, social, and economic future of a nation (for better or for worse) — for example, here in the US with Barack Obama's and Donald Trump's presidential elections, with the many "Arab Springs" that have occurred across the Middle East over the past handful of years, and within countries across the African continent including Liberia, Rwanda, and South Africa. However, in my own work with township residents in Soweto, Johannesburg, I have witnessed the emancipatory potential of the internet and new digital technologies for disclosing as yet untold stories about the anti-apartheid movement which not only impacts South Africans, but which is a worldwide movement itself. In South Africa those post-apartheid identities have largely been mediated through what Deborah Posel, professor of sociology at the University of Cape Town, sees as the "avowedly normative, officializing project of the truth commission [or Truth & Reconciliation Commission]."[13] The various projects I have been involved with in South Africa were developed, in part, to address the failures of the Truth & Reconciliation Commission of the mid-1990s, thus disrupting the "officializing" narrative.

A common thread throughout all of my research has been a focus on the experiences of women across the African Diaspora, who have not only struggled against the forces of the state and nation, but who have also sought innovative ways to tell their stories and provide the testimony needed to begin the processes of historical recovery, rebuilding, and reconciliation. I want to relate the story of one woman, Pauline Mohale — a woman whose story was referred to me by the then director of the Hector Pieterson Museum — and who was detained because of The General Law Amendment Act of 1963 which "authorized [sic] any commissioned officer [of the South African government] to detain, without a warrant any person suspected of political

---

13   Deborah Posel, "History as Confession: The Case of the South African Truth and Reconciliation Commission," *Public Culture* 20, no. 1 (2008): 119–41, at 120.

activities and to hold them without access to a lawyer for 90 days."[14] On April 30, 1996, before the Truth and Reconciliation Commission (TRC), a representative of the commission read the following, "June 16 1976 saw the outbreak of violence on a larger scale than has ever been experienced in South Africa. During this time police were engaged in countrywide arrests, both adults and children were arrested. Quite a number of children went missing and most of them were not being held by police but had gone into hiding following the house-to-house raids. It was during this time that Pauline [Mohale] got arrested and suffered all kinds of human rights violations."[15] Mohale was held for almost two-years and tortured by SA-Police to reveal information about student protestors/activists in the mid-1970s. In her own words, Pauline recounts the events that led to her arrest:

In 1976 I wasn't working. [...] I was a member of SCM, Student Christian Movement. I was working with the students, that's during the time we were fighting the Afrikaans issue and the equal rights as far as education was concerned. We marched in 1976. We used to march to John Vorster [Square — Police Headquarters in downtown Johannesburg]. When we arrived at New Canada [nearby Police Headquarters] they started throwing teargas at us. Some of our friends died there and others were arrested. But that day I managed to escape. I wasn't arrested. Some of them were being looking after by friends but I was travelling to Swaziland. I was helping the other children to escape the country. They used to sleep under the table and throughout and then we used to take a combi [truck] so that they could go to Swaziland, get further training in Swaziland. They wanted to cross the border of Swaziland. But on that it happened that when we left, it was on the 16th [day of the "Uprising"], but I was also booked to

---

14 South Africa Department of Justice, Truth and Reconciliation Commission, *Human Rights Violations Submissions — Questions and Answers: Gotla Paulina Mohale,* Johannesburg, Day 2, 30 April 1996, http://www.justice.gov.za/trc/hrvtrans%5Cmethodis/mohale.htm.

15 Ibid.

go, because I realized it was beginning to hot-up. When we arrived at the border gate — before we arrived at the border gate there was a road-block. We just saw a huge light and they stopped the driver. They told us we know that you are going to cross the border. You are going to get military training so that you can come back and start killing White people. We said no, we were lost. They arrested us and they put us in a cell in a prison near the border gate. [...] I was the only girl among them, the rest were the boys, so they closed me separately from that group. [...] The following day the police came in a truck. They came from Krugersdorp to fetch us. They were from the Special Branch.[16]

Black feminist scholarship during the past thirty years or so, I would argue, has made so many of us more conscious of the importance of letting women speak about their experiences as a legitimate way of questioning dominant paradigms of knowing and even *unknowing*.[17] The popularity of oral histories, on the web, in recent years, reflects an attempt to capture the voices of immediate experiences, but as has been pointed out, many of these "so-called voices [...] are mediated (edited, translated, corrected) by 'intellectuals' working in the academy" or even in libraries and repositories already strapped for resources. As seen across South Africa, the resources needed to preserve intangible heritage and even extant cultural heritage (artifacts, buildings, material objects, etc.), remain out of reach for a whole host of reasons.

Few studies have considered the historical significance of these townships — townships that the disenfranchised such as Paulina Mohale called home — as extant physical artifacts of a difficult past; however, they now face complex heritage issues and the concurrent pressures of the international tourist market.

---

16 Ibid.

17 Bonnie Thornton Dill, Amy McLaughlin, and Angel David Nieves, "Future Directions of Feminist Research: Intersectionality," in *Handbook of Feminist Research: Theory and Praxis,* ed. Sharlene Nagy Hesse-Biber (New York: Sage Publications, 2007), 629–37.

Do the meaning and significance (as sites of trauma, resistance, and empowerment for residents) of these planned communities defer to the competing interests of urban redevelopment, large-scale heritage planning, and globalization? A blog post sent to me by a Hamilton colleague, reminded me that much of the work I have been engaged in is what the Lesbian Herstory Archives calls "radical archiving."[18] I am sure we can all deconstruct and take issue with the term "radical," but in many ways those of us working at the intersections of archive-making, virtual environment development, and historical reconstructions have grappled with a series of complex social justice issues while working within communities that have been adversely impacted by the work of architects and planners in service of the apartheid state. Apartheid planning and architecture were the direct results of serious human rights violations perpetuated by a state that was based entirely on racial violence against anyone other than those labeled "white" and European.

Unfortunately, accounts by women such as the aforementioned Pauline Mohale remain at the margins. Even Hector Pieterson's narrative of martyrdom, although featured prominently at the Museum named in his honor, does not, however, tell a full account of the events of that day. What is often overlooked is that the site marked by the City of Johannesburg, as the location of where Hector was actually shot by police, is not where that violation of basic human rights occurred. In fact, there is still much debate as to where Hector actually fell as a result of being shot by police, because the "official" accounts suggest that he was one of many students hurling rocks at the police as they stood by and attempted to contain the "Uprising." I would argue, that a 3D reconstruction, of the events of that day — something that could conceivably emerge from our ongoing efforts at Hamilton's DHi — could provide unique insights into what occurred and perhaps even act as a form of restorative social justice if all of these conflicting accounts were told through a spatial history

---

18 "A Brief History," *Lesbian Herstory Archives,* http://www.lesbianherstoryarchives.org/history.html.

process that allowed for digital testimony and "digital witnessing" to occur.

In South Africa, growing concern over the preservation of documents related to the liberation struggle of the 1970's against apartheid has spurred new theoretical, methodological, and pedagogical questions over the making of web-based archives for local community-based township museums. I see the potential of an African digital history to interrogate the conditions of life histories of human rights violations, by examining those conditions for their "emancipatory potential and their capacity for instituting dialogical forms of historical consciousness between the testimony donors and possible communities of witness" on the internet.[19] In other words, can those stories about life under apartheid actually effect change among their viewership on the internet? Can "radical archive" making act as a form of advocacy, alongside efforts to promote a form of restorative social justice? Can historical reconstructions — immersive virtual environments — provide opportunities for reconciliation, new forms of "truth-telling" and archive-making in countries ravaged by colonial empire building? I would argue that this is possible, although filled with its share of much-needed interrogation.

I would also argue that the link between "human rights" and the preservation of cultural heritage resources — particularly those in the built environment — is often misunderstood. If we are truly seeking social justice, we must remember these historical injustices and recognize how they continue to shape identities even today. It is therefore essential to understand cultural heritage resources as a part of peoples' efforts to maintain and construct their own identity within a reconciliation process. Historic sites are critical elements in the struggle for equality and democracy, and new technologies can be used to increase access to the information kept in these important spaces. For example, recent work of Edward Gonzalez-Tennant, assistant professor at Monmouth University in New Jersey, provides an

---

19   Allen Feldman, "Memory Theaters, Virtual Witnessing, and the Trauma-Aesthetic," *Project Muse* 21, no. 1 (2004): 163–202, at 164.

example of efforts to "utilize new media to open (digital) spaces thus encouraging candid reflection on the connections between historical, face-to-face violence and present social inequality."[20] In recreating the historical development and destruction of Rosewood, Florida, culminating in the 1923 massacre and leveling of the town, Gonzalez-Tennant has used geographic information systems (GIS) to provide a deeper contextualization of its history. In much the same way, my team and I employed geospatial tools and technologies to assist in succeeding phases of my South African research.

I would argue that new digital technologies can help to challenge and disrupt how one has interpreted and used various forms of historical evidence and testimony concerning the legacy of apartheid. In particular, digital technologies can work to fill in those many absences in the historical record, particularly as they relate to everyday citizens and their roles in social movements across the Global South. The development cycle of *Soweto '76* (and particularly its follow up, *Soweto '76 3D*: the 3D recreation and simulation component of the overall project) was an ambitious technological goal when we first took it up in 2007. Although we had sufficient tools to create a sample 3D model of Soweto, the goal at the time was ultimately to let our users explore it in real time — without feeling limited by their operating system or web browser, and without having to download and install a desktop program simply to access what was ultimately a web-based archive. Technological obstacles in 2007 (a bit too complex to detail here), led to an intermediate solution: to create a proof-of-concept demo. Using Google SketchUp's own animation support to create "flythroughs," we rendered both models and transition videos to create an "on-the-rails" 3D interactive environment that we could deliver through (the widely available and, at the time, well-supported) Flash, a type of animation software.

20 Edward González-Tennant, "Intersectional Violence, New Media, and the 1923 Rosewood Pogrom," *Fire!!!* 1, no. 2 (2012): 65.

We were immediately satisfied with the result, but more so with the ability it gave us to demonstrate and describe our intended feature set to our audience and potential project partners. Of course, this approach was ultimately limited in its extensibility, and required a great deal of hands-on work to add new content. Given the extensive work the project required in other areas of its technical infrastructure — particularly the monumental task of assembling and implementing the database of locations, assets, and relationships between and amongst them — the decision was made to focus on these technology developments while waiting for the development efforts of various WebGL projects (for showing interactive 3D graphics, or whatever else might come along as alternatives) to catch up to the standard web browser such as Firefox, or Chrome.

A decade later we find ourselves in a much more viable technological position, with numerous solutions in place for realistic development in a browser-based 3D space. Forward-thinking browsers such as Google Chrome and Mozilla Firefox today include support for WebGL environments as part of their standard public releases. WebGL is an API, or interface, that specifies how software of this kind — rendering interactive 3D graphics and 2D graphics — operates within any compatible web browser without the use of plug-ins. And now, alternative 3D game engines are emerging as commercial alternatives for use in higher education research. With so much of the academic aspect of the project now better developed and more fully realized, the possibilities for utilizing these technologies seem all the more exciting, and have opened the doors to other valuable technical inquiries, such as the affordances of GIS data in *Soweto '76*'s sibling project, the *Soweto Historical GIS Project*.[21]

The *Soweto Historical GIS Project* (SHGIS) began as a collaborative research project with three students in the Department of Geography at Middlebury College (along with Professor Anne Knowles, then on the faculty at Middlebury, as Project Consult-

---

21 "Soweto Historical GIS Project," *Digital Humanities Initiative,* http://www.dhinitiative.org/projects/shgis.

ant) to build a historical GIS database drawn from a collection of thirty-eight largely unseen maps, architectural plans, and drawings that were recovered by our research team from the National Archives Repository in Pretoria, South Africa. These thirty-eight maps, architectural plans and drawings were drawn from the holdings of the apartheid-era Public Works Department. The documents, developed by architects, engineers, and city planners, and dating from the period of the 1890s to the 1950s and '60s, provide unique insights into the design and construction of model township communities for the City of Johannesburg during the apartheid era. That these existing idealized township designs were never realized in total for a variety of political, social, and economic factors is a topic that no researcher has yet to fully investigate in the fields of historical GIS and historical geography.

The research question developed by the team considers the following: How were apartheid policies constructed in the Soweto landscape? Our early findings demonstrate and chronicle how a research question can inspire a methodology for historical GIS through collaboration across disciplines and knowledge communities by working with undergraduate research assistants as collaborators. However, this study is different from most previous scholarship on the history of South Africa's all-Black townships because few, if any, of these sources have been available to a wider public until now. To demonstrate this point, I am now suggesting that a more familiar American landscape, that of *Virtual Williamsburg,* running on the game engine Unity 3D and providing a model of the town as it appeared in 1776, can be usefully examined by way of contrast.[22]

SHGIS seeks to build a multi-layered historical geographic information system database that explores the social, economic and political dimensions of urban development under South African apartheid. SHGIS is a unique project because it combines both an intense methodological engagement with a wide variety

---

22 "Begin Exploring Williamsburg," *Virtual Williamsburg 1776,* http://re-search.history.org/vw1776/start/.

of analog and digital archival materials and, at the same time, is a pedagogical effort to write the *spatial history* of a *traumascape* shaped by the legacy of apartheid, racial terror, and political violence in Southern Africa. More simply, our research has allowed us to see how population density or overpopulation — just one social factor of the many we are exploring — over time became a determining factor in the kinds of resistance employed by township residents over subsequent decades against government sanctions.

Building a virtual heritage suite of tools and accompanying research methodologies for the purpose of designing, developing, and displaying authentic virtual heritage knowledge in a geospatially accurate environment is not by any means a simple task, especially when working so closely with community stakeholders who want to see this work used and implemented further locally. Questions about the efficacy of developing platforms and digital archives in communities-of-color in the Global South cannot be ignored and need to be further explored through a collaborative process with community stakeholders. Staff from the Hector Pieterson Museum, members of the Soweto '76 Foundation (former students who took part in the liberation struggle) and local community-based scholars have all participated in various ways over the past decade. This is, admittedly, a somewhat haphazard process given the many intra- and inter-community struggles that often mirror the changes in local and national government leadership.

The use of virtual heritage applications in museums and cultural institutions is becoming more and more commonplace and is used as a vehicle for attracting younger visitors or "digital natives." Sites of "difficult heritage" across the US and Europe have been able to take advantage of the recent proliferation and affordability of digital scanning devices to provide virtual and physical replication of objects or entire landscapes. The State Museum of Auschwitz-Birkenau offers a virtual tour in part of Auschwitz/Birkenau through Quick-Time Virtual Reality clips or Flash panoramas because so much of the site now fails to look anything as it did during the Allied bombing campaigns.[23] Since 2012, German architects and Israeli historians have been working to "produce 3-D computer visualizations based on detailed blueprints and architectural plans of each of the hundreds of structures located in the three central parts of the camp."[24] However, with re-creations such as these has also come controversy. Historical geographer Tim Cole has labeled much of the visitor experience at Auschwitz/Birkenau as "Auschwitz-land [...] a 'Holocaust theme-park' rather than a 'Holocaust concentration camp.'" Cole writes, "Walking through 'Auschwitz-land' we do not see an authentic past preserved carefully for the present. We don't experience the past as it really was, but experience a mediated past which has been carefully created for our viewing."[25] For some, acts of reconstructing sites of tragedy and establishing memorial landscapes only contribute to "historical inauthenticity, trivialization and a commercial exploitation of death and violence."[26] Much of these same criticisms could be lobbied at the memorial practices of a white-majority heritage industry across South Africa that has carefully scripted the fight to end apartheid as a narrative of *good vs. evil*. Many of those narratives depict the African National Congress (ANC) as a multi-racial social movement that did not perpetuate systems of oppression, as did the white minority-led Nationalist Party. Indeed, oppression was carried out on the part of the ANC, particularly along gender-lines. Instead, the ANC continues to foster a national narrative, largely resulting from the proceedings of the televised Truth and Reconciliation Commission of a miraculous morality tale.

23 Alan Jacobs and Krysia Jacobs, "A Virtual Tour of Auschwitz/Birkenau," *The State Museum of Auschwitz-Birkenau & Remember.Org,* http://www.remember.org/auschwitz/.

24 Ofer Aderet, "A German-Israeli Team Undertakes the Reconstruction of Auschwitz," *Haaretz,* 17 August 2012.

25 Tim Cole, "Auschwitz," in *Selling the Holocaust: From Auschwitz to Schindler: How History if Bought, Packaged and Sold,* 97–120 (New York: Routledge, 2000), 111.

26 Joy Sather-Wagstaff, "Unpacking 'Dark' Tourism," in *Heritage that Hurts: Tourists in the Memoryscapes of September 11,* 67–88 (New York: Routledge, 2011), 79.

Institutions such as the Apartheid Museum, the District Six Museum, the Robben Island Museum, the Red Location Museum, and Freedom Park, all outgrowths of the ANC's 1996 Legacy Project, were built to challenge and disrupt colonial museum narratives and provide new forums for formerly marginalized voices to emerge. These newer institutions — and others built across South Africa's many townships, including those in Soweto — played a major role in advancing forms of reconciliation and helped to formulate a shared national identity for collectively dismantling apartheid. However, for some township residents these museum and national memorial sites only further compounded the inaccuracies, and distorted the everyday realities of how apartheid was lived day-to-day.

Despite the abundance of 3D virtual environments for historic sites that have emerged over the past decade, the impact of historical character modeling in the digital humanities has received little scholarly attention. Instead, when characters (or avatars) are used in virtual environments, the emphasis often tends to be on the constructed space with less attention paid to the modeling of the characters themselves and how these virtual embodiments impact the viewer/player. While this lapse may be due, in part, to a lag in technology, avatars now have the potential to become increasingly realistic. This presents us with many conceptually significant choices vis-à-vis avatar creation, each with important cultural and historical implications.

Maurizio Forte, the William and Sue Gross Professor of Classical Studies at Duke University, has called attention to "the use of 3D representations [that] has been completely random and thus has not had a great impact on the development of research methodologies and protocols."[27] However, the Unity 3D game engine has literally been a game changer in this regard, bringing in its wake increasing benefits and pitfalls. While on the one

---

27 Maurizio Forte, "Virtual Archaeology: Communication in 3D and Ecological Thinking," in *Beyond Illustration: 2D and 3D Digital Technologies as Tools for Discovery in Archaeology,* eds. Bernard Frischer and Anastasia Dakouri-Hild, 20–34 (Oxford: British Archaeological Reports, 2008), 22.

hand these realistic 3D environments seem to engage directly with what Mark Gillings of the University of Leicester has called "dynamic interactive visualization," little research has been done on the impact of character representations within these environments or how those choices might shape the narrative for the viewer.[28] The development of Web 2.0 and the ability to deliver interactive content, and the creation of virtual world environments most assuredly now permits visitors to interactively explore past landscapes.

In the autumn of 2013, persons drawn from five different institutions were awarded a National Endowment for the Humanities Office of Digital Humanities Start-Up Grant. The Hamilton team — which it was my honor to help represent — was one of the awardees. This project, "Dangerous Embodiments: Theories, Methods, and Best Practices for Historical Character Modeling in Humanities 3D Environments" (with Co-Principal Investigator, Alyson Gill) has worked towards the development of a comprehensive typology for avatar creation — an essential, new, and potentially valuable contribution to the field — and the deployment of different possible representative avatars in two virtual "difficult heritage" environments (Soweto, Johannesburg and the Lakeport Plantation, Arkansas). We are now studying viewer responses to different representative avatars within these environments using tools drawn from experimental philosophy, and are working to publish the results with interpretation by scholars across a diverse array of fields.

As new technologies emerge, scholars are testing all available possibilities and alternatives for user interaction, remaining cognizant of the continuing chasm of the digital divide. Recently, Hamilton's DHi began working as developers for Oculus Rift, a mobile virtual reality wearable device/headset that allows users to step inside immersive environments. Our lead developer, Greg Lord, has integrated Oculus support into our Unity development pipeline via the Oculus SDK (Software Develop-

---

28 Mark Gillings and Alicia Wise, eds., *Guides to Good Practice in the Creation and Use of Digital Resources* (Oxford: Oxbow Books, 1999).

ment Kit). The Oculus SDK allows us to render our Unity virtual environment project through a special image rendering technique that automatically creates a stereoscopic view, splitting the scene's rendered output into two left- and right-eye camera positions. These images are processed through a barrel distortion optical effect, warping the image to into a kind of fisheye lens effect that will be corrected to appear spherical by the Oculus lenses. This creates a 90-degree field of view, with accurate depth effects, that allows the scene to appear fully three-dimensional within the headset. The result is a striking realism, lending a true-to-life immediacy to the virtual environment, as if the viewer were really standing inside the scene. The Oculus, and our Unity application, also makes use of a head-tracking camera that allows the viewer to freely tilt and rotate their head within the scene, updating their view in perfect sync with their motions. Although this "true 3D" effect requires the special head-mounted display to work, the Oculus is expected to become the leading virtual reality technology in the years ahead, with a significant industry buy-in and an active, growing community of users and developers. Our current application will also have the capability to fall back on traditional 3D rendering, for use with standard computer monitors for all other users.

The dangers of iconic representations and the powerful influence that images exert over us have long been recognized by scholars working in the humanities, with Brian Molyneaux of the University of South Dakota eloquently noting that, "The reinforcement of ideas in some images is very powerful. [...] Pictures and other visual representations — have a tremendous inertia, or staying power, that may persist long after the ideas behind the images have gone out of fashion."[29] The preamble to the *London Charter for the Computer-based Visualization of Cultural Heritage* notes that "a set of principles is needed that will ensure that digital heritage visualization is, and is seen to be, at least as intellectually and technically rigorous as longer established cul-

---

29 Brian Leigh Molyneaux, "Introduction," in *The Cultural Life of Images: Visual Representation in Archaeology,* 1–8 (New York: Routledge, 1997), 6.

tural heritage research and communication methods."[30] This an important issue for the humanist as the delivery of interactive, high quality 3D content via the web is becoming feasible in a way that was virtually impossible just a handful of years ago.

In conclusion, I would like to quote John Fleckner, then senior archivist at the Smithsonian's National Museum of American History, from a 1990 address before the Society of American Archivists. Fleckner noted, "without the documentary record there could have been no calling to account, no investigation, no prosecution. And that record, the tapes, the documents, and all the rest-stands as witness in the future to those who would forget or rewrite that past."[31] Today, South Africa's former all-Black townships could easily be erased/destroyed as a result of ill-conceived tourism initiatives, irresponsible local politicians or neglect. As I have argued elsewhere, I ultimately believe that these sites of tragedy and dissonant heritage, such as those related to the Soweto Uprisings of 1976, can be used — in part through the use of 3D immersive virtual environments coupled with documentary evidence of institutional racial violence as experienced in the built environment through historical recreations — as models for community-based education and renewed political and social inclusion.[32] Today, scholars of *difficult heritage* are often confronted with the challenge of producing meaningful engagements with diverse audiences through the use of new digital technologies. With this engagement we often face risks as we represent serious, often painful and controversial, historical content through a medium once closely aligned with popular entertainment. In closing and despite these caveats, it is

---

30 *London Charter for the Computer-based Visualization of Cultural Heritage,* http://www.londoncharter.org/introduction.html.

31 John A. Fleckner, "'Dear Mary Jane': Some Reflections on Being an Archivist," (speech, Seattle, 30 August 1990), http://www.archivists.org/governance/presidential/fleckner.asp.

32 Angel Nieves, "Engagements with Race, Memory, and the Built Environment in South Africa: A Case Study in Digital Humanities," in *The Routledge Companion to Media Studies and Digital Humanities,* ed. Jentery Sayers, ch. 40 (New York: Routledge, 2018).

inevitable that the transformative and disruptive promise of historical 3D models and virtual environments will be more fully realized.

Bibliography

"A Brief History." *Lesbian Herstory Archives.* http://www. lesbianherstoryarchives.org/history.html.

Aderet, Ofer. "A German-Israeli Team Undertakes the Reconstruction of Auschwitz." *Haaretz.* 17 August 2012. http://www.haaretz.com/israel-news/a-german-israeli-team-undertakes-the-reconstruction-of-auschwitz-1.458948.

Alkalimat, Abdul. "eBlack: a 21st century challenge." *Mots Pluriels* 19 (2001). http://motspluriels.arts.uwa.edu.au/MP1901aa.html.

"Begin Exploring Williamsburg." *Virtual Williamsburg 1776.* http://research.history.org/vw1776/start/.

Cole, Tim. "Auschwitz." In *Selling the Holocaust: From Auschwitz to Schindler: How History if Bought, Packaged and Sold,* 97–120. New York: Routledge, 2000.

Digital Humanities Initiative. http://dhinitiative.org/.

"Digital Humanities as Restorative Social Justice: Virtual Heritage, 3D Reconstructions and South Africa's Township Histories." *Digital Humanities Initiative.* http://www. dhinitiative.org/node/184.

Dill, Bonnie Thornton Amy McLaughlin, and Angel David Nieves, "Future Directions of Feminist Research: Intersectionality." In *Handbook of Feminist Research: Theory and Praxis,* edited by Sharlene Nagy Hesse-Biber, 629–37. New York: Sage Publications, 2007.

Ellison, Ralph. "Some Questions and Some Answers." In *The Collected Essays of Ralph Ellison,* edited by John F. Callahan, 291–302. New York: Random House, 2003.

Feldman, Allen. "Memory Theaters, Virtual Witnessing, and the Trauma-Aesthetic." *Project Muse* 21, no. 1 (2004): 163–202. DOI: 10.1353/bio.2004.0030.

Fleckner, John A. "'Dear Mary Jane': Some Reflections on Being an Archivist." Speech. Seattle, 30 August 1990.

Forte, Maurizio. "Virtual Archaeology: Communication in 3D and Ecological Thinking." In *Beyond Illustration: 2D and 3D Digital Technologies as Tools for Discovery in Archaeology,* edited by Bernard Frischer and Anastasia Dakouri-Hild, 20–34. Oxford: British Archaeological Reports, 2008.

Freedberg, David. *The Power of Images: Studies in the History and Theory of Response.* Chicago: University of Chicago Press, 1991.

Gillings, Mark, and Alicia Wise, eds. *Guides to Good Practice in the Creation and Use of Digital Resources.* Oxford: Oxbow Books, 1999.

González-Tennant, Edward. "Intersectional Violence, New Media, and the 1923 Rosewood Pogrom." *Fire!!!* 1, no. 2 (2012): 64–110. DOI: 10.5323/fire.1.2.0064.

"Hector Pieterson Memorial and Museum." *Gauteng.* http://www.gauteng.net/attractions/hector_pieterson_memorial_and_museum/.

Jacobs, Alan, and Krysia Jacobs. "A Virtual Tour of Auschwitz/Birkenau." *The State Museum of Auschwitz-Birkenau & Remember.Org.* http://www.remember.org/auschwitz/.

London Charter for the Computer-based Visualization of Cultural Heritage, http://www.londoncharter.org/introduction.html.

Manferdini, Anna Maria, and Fabio Remondino. "Reality Based 3D Modeling, Segmentation and Web-Based Visualization." In *Digital Heritage: Third International Conference, EuroMed 2010 Lemessos, Cyprus, November 2010 Proceedings,* edited by Marinos Ioannides et al., 110–24. Berlin: Springer-Verlag, 2010.

McCarty, Willard. Email to "Online seminar for digital humanities." 12 February 2014. http://lists.digitalhumanities.org/pipermail/humanist/2014-December/012484.html.

Mohale, Gotla Paulina. Interview by Angel David Nieves. Johannesburg, 2008.

Molyneaux, Brian Leigh. "Introduction." In *The Cultural Life of Images: Visual Representation in Archaeology,* 1–8. New York: Routledge, 1997.

Nieves, Angel. "Engagements with Race, Memory, and the Built Environment in South Africa: A Case Study in Digital Humanities." In *The Routledge Companion to Media Studies and Digital Humanities,* edited by Jentery Sayers, ch. 40. New York: Routledge, 2018.

Posel, Deborah. "History as Confession: The Case of the South African Truth and Reconciliation Commission." *Public Culture* 20, no. 1 (2008): 119–41. DOI: 10.1215/08992363-2007-019.

"Preamble." *The London Charter for the Computer-Based Visualisation of Cultural Heritage.* 7 February 2009. http://www.londoncharter.org/preamble.html.

Sandoval, Chela. *Methodology of the Oppressed.* Minneapolis: University of Minnesota Press, 2000.

Sather-Wagstaff, Joy. "Unpacking 'Dark' Tourism." In *Heritage that Hurts: Tourists in the Memoryscapes of September 11,* 67–88. New York: Routledge, 2011.

South Africa Department of Justice, Truth and Reconciliation Commission. *Human Rights Violations Submissions — Questions and Answers: Gotla Paulina Mohale.* Johannesburg, Day 2, 30 April 1996. http://www.justice.gov.za/trc/hrvtrans%5Cmethodis/mohale.htm.

Soweto '76. http://www.soweto76archive.org.

"Soweto Historical GIS Project." *Digital Humanities Initiative.* http://www.dhinitiative.org/projects/shgis.

United Nations General Assembly, *Report of the Special Rapporteur on the promotion and protection of the right to freedom of opinion and expression, Frank La Rue,* A.HRC.17.27 (Geneva, 2011). http://www2.ohchr.org/english/bodies/hrcouncil/docs/17session/A.HRC.17.27_en.pdf.

"Update on Proposed ADHO SIGs." *Alliance of Digital Humanities Organizations.* http://adho.org/announcements/2013/update-proposed-adho-sigs.

Verne, Harris. "The Archival Sliver: Power, Memory, and Archives in South Africa." *Archival Science* 2 (2002): 63–86. DOI: 10.1007/BF02435631.

White, Hayden. "Introduction: Historical Fiction, Fictional History, and Historical Reality." *Rethinking History: The*

Journal of Theory and Practice 9, nos. 2–3 (2005): 147–57.
DOI: 10.1080/13642520500149061.

7

Lowriding through the
Digital Humanities

Annemarie Perez

Note: The title of this piece is shamelessly borrowed from Barbara Noda's "Lowriding Through the Women's Movement," a piece which creatively addresses the power a group made up of women of color could have on individuals during the women's movement. Noda's essay was published in This Bridge Called My Back: Writings by Radical Women of Color.[1]

> *My "low rider" laptop is decorated with a 3-D decal of the Virgin of Guadalupe, the spiritual queen of Spanish-speaking America. It's like a traveling altar, an office and a literary bank, all in one.*
> — Guillermo Gómez-Peña

Lowriders are customized automobiles with a specific aesthetic that first appeared in the United States southwest in the mid-to-late 1940s, a product of a southwestern Mexican American cul-

---

1   Barbara Noda, "Lowriding Through the Women's Movement," in *This Bridge Called My Back: Writings by Radical Women of Color,* 4th edition, ed. Cherríe Moraga and Gloria Anzaldúa, 136–37 (Albany: State University of New York Press, 2015).

ture.[2] Lowriders are also the individuals who make these modifications. In contrast to the jacked up "hot rods" raced by Anglos of the period, lowriders are cars modified to be low to the road. The point isn't to speed, but to cruise as low and slow as possible, disrupting traffic. In this way, the lowriders resisted the period's white youth culture and made their creators visible to each other's gaze. Lowriding also functions as social practice, claiming and constructing specific spaces in streets and parks where lowriders can, as Ben Chappell writes, "cruise, display their rides, and socialize."[3] In digital humanities, a discipline that publishes quickly, this essay has been written slowly and with much modification since its frame began in 2011 in the comments of a series of #TransformDH blog posts.[4]

There has been wonderful work recently on #TransformDH by the TransformDH collective discussing how racial/gender/sexual/disabled bodies in the academy are and always have been doing digital humanities work, calling on the digital humanities to be "center[ed] on the intersection of digital production and social transformation."[5] Yet, because hegemony constantly replicates the dominant discourse, there needs to be a consistent and constant engagement with issues of race, ethnicity, gender, sexuality, class, and able-bodiedness as its counter. To do discuss this hegemony, I'm going to fall back on Chicana feminist praxis, the construction of something new out of found objects, a hybrid writing of Gloria Anzaldúa's theory of Nepantla, locating myself between Chicanx studies and the digital humanities. I speak from that in-between space with the hope that from this self-situated ethnography some insight into my concerns may

2   Ben Chappell, *Lowrider Space: Aesthetics and Politics of Mexican American Custom Cars* (Austin: University of Texas Press, 2012).

3   Ibid., 3.

4   See *#TransformDH* (blog), http://transformdh.org/.

5   "2015 Conference and THATCamp," *#TransformDH* (blog), http://transformdh.org/2015-conference-thatcamp/. See also Moya Bailey, Anne Cong-Huyen, Alexis Lothian, and Amanda Phillips, "Reflections on a Movement: #transformDH, Growing Up (forthcoming)," in *Debates in the Digital Humanities 2016*, ed. Matthew K. Gold and Lauren F. Klein, 71–80 (Minneapolis: University of Minnesota Press, 2016).

come. I use autoethnography to specifically discuss issues and effects of racial absence in the digital humanities community and what the costs of that may be. This piece begins to discuss how the discourse surrounding racial and racialized bodies and their absence in digital humanities spaces replicates the discourse surrounding the invisibility/absence of women of color from second-wave feminism. Much as American literature and feminism are haunted by race, issues of race and hegemony are the ghost in the machine of the digital humanities.[6]

Those who think Twitter is a waste of time, as opposed to it being a time-waster, are failing to see its potential. Or perhaps they're not following the right people. Twitter is the main way I keep up on what's going on in two areas: digital humanities and ethnic studies, especially Chicanx and Latinx studies. My point of entry into both of these online communities is through Twitter, though in the case of ethnic studies, these areas also reflect my disciplinary background and my areas of research: Chicana feminist textual communities and editorship. Both the DH and Chicanx Twitter streams are very active, some days more than others, but even so, more than I can read most days. Both use hashtags to discuss important issues, though Chicanx studies somewhat less often than DH. Twitter is also the intersection of my Chicanx studies classes work with the digital world and Latinx artists and authors as my students annotate their reading and viewing through their use of tweets and course hashtags.

> *I say*
> *My typewriter sticks in the wet.*
> *I have been using the same ribbon*
> *Over and over and over again.*[7]

---

6   See Tara McPherson, "Why Are the Digital Humanities So White? Or Thinking the Histories of Race and Computation," in *Debates in the Digital Humanities,* ed. Matthew K. Gold, 139–60 (Minneapolis: University of Minnesota Press, 2012) and David Columbia, *The Cultural Logic of Computation* (Cambridge: Harvard University Press, 2009).

7   Cherríe Moraga, "It's the Poverty," in *Loving in the War Years: lo que nunca pasó por sus labios,* 62–64 (Boston: South End Press, 1983), 62.

In 2011, as I struggled with the final revisions of my dissertation between adjuncting gigs, feeling the absence of intellectual community, I found academic Twitter. Although I had been active in online communities going back to Usenet alt.* groups in the mid-late 1990, except for a little work on a department listserv, my online communities never intersected with my academic ones. Questions of why should probably be reserved for the digital equivalent of psychoanalysis (would that be blogging?). However, by the end of 2011, I had been online as a Chicana doctoral student/recent PhD for more than a year. I had engaged with an online community of Chicanos, of other academics, of *Doctor Who* fans (communities that frequently overlapped). I had also participated in creating a hashtag — #AztlanReads — as a response to the general lack of knowledge about Chicanx authors and books. It took on a life of its own and become a small but vocal movement as a website and then anthology. Being part of this made me imagine how and where the fields of digital humanities and digital pedagogy were intersecting with Chicanx studies specifically, and ethnic studies/critical race theory more generally. As such, I began to research the genealogies of digital humanities, seeking to find where it intersected with critical race theory and cultural studies.

> *Yes, we both agree I could use*
> *A new ribbon. But it's the poverty*
> *The poverty of my imagination, we agree.*
> *I lack imagination, you say.*[8]

Because of this Twitter experience engaging with and in technology with my digital community, I attended the 2012 Modern Language Association (MLA) convention in Seattle with plans of branching out from attending mainly Chicanx panels and into this DH community I'd grown to (virtually) know through social media. I'm taking a long time telling this. It is because the memory is painful. The panels and workshops I attended were

---

8   Ibid., 62.

a shock. Not only because the work was so exciting, especially, for me, the pedagogy, the mapping and timelining and other amazing projects with the potential of involving our students in concrete and useful research. But because even at MLA, even at a literature conference, I had never experienced a stronger sense of being racially/ethnically other. The rooms, crowded to bursting, were visibly, notably white spaces. This was a bit jarring, but what was even more so was that no one was talking about this. No one was asking where the brown people were — yet I knew, as Moya Bailey wrote in 2011, that people of color were engaged in digital work.[9] Where were they? The absence of racialized bodies was un-noted. Let me underline that this was a kind space, welcoming to questions, eager to teach new skills. However the niceness, the civility, which Koh has noted and seems to underpin much of DH as a community, was one deployed in what seemed a racially homogeneous community, whose very homogeneity made this civility possible.[10] The very issue of "civility" is a fraught one in Chicanx studies –a discipline born out of the decidedly "uncivil" protests of the Chicano Movement. For a recent example connected to the use of the term "civility" one can look at the 2014 protest directed at the leadership of the National Association of Chicana and Chicano Studies when it proposed as its conference theme "Exploring Civility within the Chicana & Chicano Studies Discipline" (Soto).[11] The degree to which Chicana/o studies has been civilized (or, to use a more coercive term, disciplined) and made part of the academy is seen by some as taking it away from its radical roots in the communities that agitated for its creation. How then to place Chicanx studies beside or bring it into a new discipline which sees itself as defined by civility?

---

9   Moya Bailey, "All the Digital Humanists Are White, All the Nerds Are Men, But Some of Us Are Brave." *Journal of Digital Humanities* 1, no. 1 (2011).

10 Adeline Koh, "Niceness, Building, and Opening Genealogy," *differences: A Journal of Feminist Cultural Studies* 25, no. 1 (2014): 93–106, at 93.

11 Sandy Soto, "When Civility Is Brown," *Bully Bloggers* (blog), 13 February 2014.

No. I lack language.
The language to clarify
My resistance to the literate.
Words are a war to me.
They threaten my family.[12]

The degree to which I was unnerved by the whiteness of the MLA workshops is hard to overstate. On the one hand, here were all these wonderful ideas and practices, ways of thinking about literature and community and its intersection — merger even — with the digital that I had never considered. On the other, there was seemingly a lack of awareness of race, of the hegemonic replication of whiteness the bodies in the room represented, as though critical race studies did not exist or speak to this group. I left with nothing to say, unable to say anything, something that's unusual as I'm generally a loud-mouthed sort of woman. I was unnerved to realize I had felt a fear of participating, of speaking. Most people understand that it's hard being the only woman in a room of 50 to 100 men. For people of color, most of us know it's just as hard to be the lonely only. That's how I felt. Alone and painfully self-conscious, aware of my difference. When I'm one of the only, however kind and welcoming the environment, I experience physical and emotional stress. There's a fear of asking questions lest I be seen as speaking for my race/culture and somehow reinforcing biases of ignorance. I left those DH sessions with the thought of attending the Chicanax/Latinx/Asian American/African American literature sessions as a form of decompression.

To gain the word to describe the loss,
I risk losing everything.[13]

But on leaving the DH sessions, I went and found coffee, sat in the hotel lobby and tried to sort out my thoughts and emotions.

---

12 Moraga, "It's the Poverty," 62–63.
13 Ibid., 63.

On a personal level the moment was hard. After enduring the alienating damage of being one of a very few graduate students of color in my PhD program, I had been enfolded, and to an extent, healed, by Mujeres Activas en Letras y Cambio Social (MALCS), a Chicana/Latina/Indigenous association of women scholars and community activists who had welcomed me and my scholarship, including my budding DH work. Did I want to leap back into the world of unthinking micro and macro racial aggressions, especially with my limited conference time, and as an adjunct, paying for it with my own money? As a scholar of color there are few things as rare and wonderful as getting to be in a room with a multitude of scholars of color. Such spaces are precious. For me, there's a feeling of intellectual safety, of being able to take risks without risking being found intellectually naive, or worse still, reflecting badly on all Chicanxs, of representing only myself. I feel I can be wrong, that we can build theoretical castles in the air, find their flaws, send them crashing down. Alexis Lothian, someone I knew from graduate school, came into view and kindly came over. I poured out what I experiences in my somewhat limited exposure to the digital humanities at MLA. She affirmed what I'd seen and felt and we began discussing issues of racial, gender and sexuality hegemony, the ablest rhetoric, within the DH community, something which is clear from her own work on marked bodies.[14] It seems this feeling was something of the zeitgeist of the moment and soon tweets began to appear with the #TransformDH hashtag, beginning in 2011 at the American Studies Association (ASA) conference.

I may create a monster,
The word's length and body
Swelling up colorful and thrilling
Looming over my mother, characterized.[15]

---

14 Alexis Lothian, "Marked Bodies, Transformative Scholarship, and the Question of Theory in Digital Humanities," *Journal of Digital Humanities* 1, no. 1 (2011).

15 Moraga, "It's the Poverty," 63.

There was and continues to be pushback, an insistence that DH is welcoming to all and has no need to transform itself or to be transformed, or, perhaps that this may have been case once, but now everything is better, is fine. In the five years since the Seattle MLA, I have witnessed the hostility and impatience that greets discussions of how DH could/should imbue itself with critical race theory and feminist praxis and that these theories be a starting point for critical DH work, not something added or stirred into projects already conceived. Enumerating DH projects by or about communities of color or women often substitutes for engaging with the white male hegemony (as well as neoliberal capitalism) being reproduced by and from our academic institutions into DH structures and communities. For me, these draw eerie parallels to the experience of women of color with second-wave feminism, constantly being told they would be welcome, if only they would come. And when feminists of color came, they experienced discomfort and hostility. Like white feminism, the digital humanities is haunted by racial discourse, which erupts like an uncivil poltergeist, highlighting awkwardness and discord. Yet there seems little willingness to change the shape of the discussion and be shaped by a different issues and voices.

> *Her voice in the distance*
> *Unintelligible   illiterate.*
> *These are the monster's words*[16]

At the same time, the potential for change, and my excitement for the field continues. A common metaphor between Chicana feminism and digital humanities is the notion of "making" the discipline, of building with one's own hands. As a symbol in Chicana studies, this construction or building connects to Lorna Dee Cervantes poem, "Beneath the Shadow of the Freeway," where readers, like the granddaughter are told to "trust only

---

16 Ibid., 63.

what [they] have built with my own hands."[17] This is such a powerful symbol that when Chicana studies wrote its own anthology, it drew its title, *Building with Our Hands,* from Cervantes'
poem.[18] Likewise, lowriders take the old, late-model cars, mod-
ifying and remaking them bright as jewels so they can cruise
through their neighborhoods, glitteringly visible. As Chicanas
lowride through predominantly Anglo disciplines, there are
constantly modifications — what Chela Sandoval calls the ever-
changing "differential mode" feminism and consciousness of
women of color.[19] If DH can allow itself to be disrupted, to "go
low and slow," being modified by and learning from critical race
theory to recognize and unsettle its privilege, starting from the
position of the differential rather than bringing diversity in as an
afterthought, demanding diversity of itself and its communities,
this would go a long way in bridging these discourses. It means
not waiting for scholars of color to find DH and ask about it,
but going to them, understanding and listening to their theories
and practices and discussing with them how the digital works
in connection with the work they're already doing. The price
of admitting to not knowing or "getting" something, whether
it is Python or critical race theory, is high and even dangerous
and should not be taken lightly. If race could be the ghost in the
machine of code, of data as some suggest, then maybe modifications to this machine can turn it into something wonderful and
unimagined by its original makers.[20]

Meanwhile, I find myself wanting. As a Chicana, I know I
speak English only. As a digital humanist my computer is almost ten years old. My code is as stumbling and ungraceful as

---

17 Lorna Dee Cervantes, "Beneath the Shadow of the Freeway," in *Emplumada,*
    11–14 (Pittsburgh: University of Pittsburgh Press, 1981).

18 See Adela de la Torre and Beatríz Pasquera, *Building with Our Hands: New
    Directions in Chicana Studies* (Berkeley: University of California Press,
    1993).

19 Chela Sandoval, *Methodology of the Oppressed* (Minneapolis: University of
    Minnesota Press, 2000), 54.

20 See McPherson, "Why Are the Digital Humanities So White?" and Colum-
    bia, *The Cultural Logic of Computation.*

my Spanish, sticking like an old typewriter ribbon. I feel like Cherríe Moraga, trying to be a bridge and not being adequate to the task.

Bibliography

Bailey, Moya Z. "All the Digital Humanists Are White, All the Nerds Are Men, But Some of Us Are Brave." *Journal of Digital Humanities* 1, no. 1 (2011). http://journalofdigitalhumanities.org/1–1/all-the-digital-humanists-are-white-all-the-nerds-are-men-but-some-of-us-are-brave-by-moya-z-bailey/.

———, Anne Cong-Huyen, Alexis Lothian, and Amanda Phillips. "Reflections on a Movement: #transformDH, Growing Up." In *Debates in the Digital Humanities 2016,* edited by Matthew K. Gold and Lauren F. Klein, 71–80. Minneapolis: University of Minnesota Press, 2016.

Cervantes, Lorna Dee. "Beneath the Shadow of the Freeway." In *Emplumada,* 11–14. Pittsburgh: University of Pittsburgh Press, 1981.

Chappell, Ben. *Lowrider Space: Aesthetics and Politics of Mexican American Custom Cars.* Austin: University of Texas Press, 2012.

Columbia, David. *The Cultural Logic of Computation.* Cambridge: Harvard University Press, 2009.

De la Torre, Adela and Beatríz M. Pesquera. *Building with Our Hands: New Directions in Chicana Studies.* Berkeley: University of California Press, 1993.

Gómez-Peña, Guillermo. "The Virtual Barrio @ the Other Frontier." *Zone Zero.* 10 May 1999. http://v2.zonezero.com/index.php?option=com_content&view=article&id=502%3Athe-virtual-barrio-the-other-frontier&catid=5%3Aarticles&lang=en.

Koh, Adeline. "Niceness, Building, and Opening Genealogy." *differences: A Journal of Feminist Cultural Studies* 25, no. 1 (2014): 93–106. DOI: 10.1215/10407391-2420015.

Lothian, Alexis. "Marked Bodies, Transformative Scholarship, and the Question of Theory in Digital Humanities." *Journal of Digital Humanities* 1 (2011). http://journalofdigitalhumanities.org/1–1/marked-bodies-transformative-scholarship-and-the-question-of-theory-in-digital-humanities-by-alexis-lothian/.

McPherson, Tara. "Why Are the Digital Humanities So White? Or Thinking the Histories of Race and Computation." In *Debates in the Digital Humanities,* edited by Matthew K. Gold, 139–60. Minneapolis: University of Minnesota Press, 2012.

Moraga, Cherríe. *Loving in the War Years: lo que nunca pasó por sus labios.* Boston: South End Press, 1983.

Barbara Noda, "Lowriding Through the Women's Movement." In *This Bridge Called My Back: Writings by Radical Women of Color.* 4th edition, edited by Cherríe Moraga and Gloria Anzaldúa, 136–37. Albany: State University of New York Press, 2015.

Sandoval, Chela. *Methodology of the Oppressed.* Minneapolis: University of Minnesota Press, 2000.

Soto, Sandy. "When Civility Is Brown." *Bully Bloggers* (blog). 13 February 2014. https://bullybloggers.wordpress.com/2015/02/13/when-civility-is-brown/.

8

Goldstar for You

Mongrel Coalition Against Gringpo[1]

GOLD STAR FOR NOTICING GENTRIFICATION — YA KNOW THIS VIOLENT RACIST TRADITION OF SEGREGATION THAT'S BEEN THE BUILDING BLOCK OF AMERICA — GOLD STAR FOR FINALLY NOTICING THIS AS A POTENTIAL REALITY

GOLD STAR FOR THE MOST PUBLIC WHITE TEARS AND

GOLD STAR FOR DEFENDING KG AND VP DUE TO INDIFFERENCE TO POC BODIES AND SELFISH CAREERIST CONCERNS

GOLD STAR FOR BEING WHITE AND REMEMBERING YOUR MIGRATION NARRATIVE. MINUS GOLD STAR FOR EQUATING THIS MEMORY AS EXCLUDING THE PRIVILEGES OF POSSESSIVE WHITENESS

GOLD STAR FOR READING ONE BOOK BY A BLACK POET AND POSTING ABOUT IT ONE TIME ON FACEBOOK GOLD STAR

---

1   Originally published in Jennifer Tamayo, "The Gold Star Awards… A message from The Mongrel Coalition Against Gringpo," *Poetry Foundation,* 2 April 2015, https://www.poetryfoundation.org/harriet/2015/04/the-gold-star-awards-a-message-from-the-mongrel-coalition-against-gringpo.

GOLD STAR FOR FEELING "MOVED" BY CLAUDIA RANKINE'S CITIZEN BUT BEING MOVED TO DO NOTHING IN AN ACTUALITY THAT MIGHT IMPACT YOU. GOLD STAR!

GOLD STAR FOR LOVING "HYBRID" BOOKS. BY HYBRID EVERYTHING THAT IS HODGEPODGE WHITEWASHED DISEMBODIED OH SO CLEVER HIGH FIVES FROM THE NYC POETRY FASHIONISTA CLIQUES

GOLD STAR FOR LOVING "HYBRID" BOOKS BUT MAINTAINING A NOT-SO-HYBRID CONSCIOUSNESS (ERASURE OF POETS OUTSIDE THE BLACK/WHITE BINARY, DELETION OF INDIGENEITY FOR THE UMPTEENTH TIME)

GOLD STAR FOR YOUR INDIGENT AND PROUDEST WHITE FEMINISM. GOLD STAR FOR YOUR TWEEN EMPOWERMENT FANTASIES SHATTERING THE CEILINGS WHILE SOMEONE ELSE SWEEPS THE GLASS FANTASIES: YOUR HELLO KITTY STAND IN DELUSIONS: HOW MANY STICKERS WILL IT TAKE TO ELEVATE YOUR BRAND. GOLD FUCKEN STAR FOR NO AWARENESS BUT GOOD INTENTIONS GOLD GOLD STAR!

GOLD STAR FOR YOUR DESIRE TO REACH COMPROMISE, TO SPLIT THE DIFFERENCE, REMAIN IN CHARGE: GOLD STAR FOR COMPROMISE!

GOLD STAR FOR IDENTIFYING YOUR WHITE PRIVILEGE BUT REFUSING TO GIVE IT UP — INABILITY TO CONCEIVE OF LIFE WITHOUT WHITE PRIVILEGE POLICE SMILES. IDENTIFICATION WITHOUT ABOLITION EQUALS MINUS TWO STARS. MOVED TO GOLD STAR RESERVE!

GOLD STAR FOR GOOD POLITICS WELL-SAID. YEY! GOLD STAR! JUST KIDDING NO GOLD STAR FOR LITERACY AND RHETORICAL SKILLS. HAR HAR!

GOLD STARZ FOR WHITE GUILT PERFORMED WITH PAIN(T)ED FACE

GOLD STAR FOR PROTECTING YOUR NETWORK AND REFUSING TO CALL OUT YOUR RACIST BUDS OR PUBLICLY SUPPORT THOSE WHO DO. PROPS TO WHITENE$$!

GOLD STAR FOR INVITING THAT SAME POC TO YOUR READING OR PANEL. DIVERSITY! YEY!

GOLD STAR FOR DESIRING THE OTHER (THEREBY BECOMING THE OTHER, OH MY! VENTRILOQUISM!)

SILVER STAR FOR HAVING 1/8TH SOMETHING OR OTHER OTHER (WE ARE ALL SO DIVERSE) AND USING THIS PORTION AS A SHIELD AGAINST CRITIQUE. BLOOD QUANTUM GAMES SILVER STAR

GOLD STAR WITH GLITTER FOR STATISTICS ABOUT BLACK ON BLACK CRIME AND USING THESE STATISTICS TO TO PEDDLE RAC-ISM EVEN TO POC WRITERS. SLOW GLITTER CLAP FOR THE AT-TORNEY!

GOLD STAR FOR USING THE WORD "PROBLEMATIC" TO OBFUS-CATE OR AVOID GETTING YR PAWS DIRTY

GOLD STAR FOR HAVING GARDENERS BUT NEVER HAVING TO BE GARDNER. GOLD STAR FOR ASKING RUBÉN HIS OPINION, DOÑA MARJORIE!

GOLDEN STAR ATOP A TREE FOR TAKING AN INTEREST IN BIG-NAME *THIRD WORLD* WRITERS (MULTICULTURAL FIBER OPTIC LIGHTS FOR BOLAÑO)

THREE GOLD STARS FOR SETTING DOVES FREE, BEING WELL-MEANING, AND "BEING AN IRRESPONSIBLE WHITE PERSON"

GOLDEN GLOBE FOR BEST PERFORMANCE OF LIBERAL DISAVOW-
AL OF PRIVILEGE WHILE DISCUSSING FRANZ FANON

GOLD OSCAR FOR SANCTIONING EUROCENTRIC EDUCATIONS
PROVIDED BY THE STATE AND PRIVATE INSTITUTIONS AS THE
ONLY WAY TO BE EDUCATED!

GOLD STAR FOR IGNORING POC WRITERS WHO COULDN'T GO TO
COLLEGE DUE TO SLAVERY RACISM AND POVERTY LIKE HARRIET
E. WILSON AND WANDA COLEMAN. GOOD FOR YOU!

GOLD STAR FOR VOTING, FEELING PROUD, AND FINGER WAGGING
AT UNENLIGHTENED ABSTAINERS — WE KNOW YOU ARE THE
REASON THE SYSTEM IS BROKEN

GOLD STAR FOR NOT HAVING READ ANYTHING BUT FEELING THE
NEED TO INSERT SELF AND EMOTIONS INTO EVERY CONVERSA-
TION AND THEN SHOUTING IDENTITY POLITICS WHEN A POC AP-
PEARS DISINTERESTED IN HONORARY WHITENESS. GOLD STAR!

GOLD STAR FOR HIGH FIVING PERLOFF IN DISCUSSIONS OF RACE:
WE ALL KNOW SHE INVENTED CRITICAL RACE STUDIES AND IN-
TERSECTED IT INTO POETRY!

GOLD STAR FOR IDENTIFYING YOUR WHITE CIS HETERO PRIVI-
LEGE. MINUS ALL THE STARS FOR NOT SELF-ABOLISHING. GO
AWAY! YEY!

GOLD STAR FOR THE UNIVERSAL LOVE OF THE UNIVERSAL AKA
THE SOUL OF WHITE FOLK

GOLD STAR SNOWFLAKE PRIZE FOR NEVER EVER EVER SHUTTING
UP ABOUT GERTRUDE STEIN AND/OR ANDY WARHOL

GOLD STAR FOR ACCUSING POC OF "SHAME CULTURE" OMFG.
JUST. STOP.

GOLDEN GLOBE FOR LOVING LINCOLN AND BEST SUPPORTING
ROLE AS THE SLAVE-FREEING COSMOPOLITAN SUBJECT

GOLDEN GLOBE FOR THE PERFORMANCE OF REASONABLE DIS-
TANCE AND DIALOGUE WHILE DISCUSSING HEATED TOPICS SUCH
AS "RACE"

GOLD STAR FOR THINKING OUTSIDE THE "GHETTO"

GOLD COMET FOR HOSTING POETRY DIVERSITY NITE AT THE IVY
LEAGUE THEN SITTING ON THE OTHER END OF THE TABLE DUR-
ING DINNER.

GOLD COMET FOR SAYING THE "POLICE ARE YOUR FRIENDS" ON
A CONTEMPORARY POETRY BLOG AND TELLING US WE SHOULD
BE WORRYING ABOUT THE ANTHROPOCENE INSTEAD

BLEACHED STAR FOR BELIEVING BOB MARLEY NINA SIMONE
ARETHA JAY-Z BIGGIE KANYE BEYONCE DRAKE & FRANK OCEAN
WILL SAVE YOUR SOUL

GOLD STAR FOR READING THAT ONE BLACK WRITER IN HIGH
SCHOOL MAYA ANGELOU TONI MORRISON OR RALPH ELLISON
WHICH ONE WAS IT????

GOLD STAR FOR BELIEVING TO KILL A MOCKINGBIRD IS THE
NOVEL ON CIVIL RIGHTS AND MASTURBATING TO THE FANTASY
OF WHITE SAVIOR ATTICUS FINCH

GOLD STAR FOR NEVER CHALLENGING YOUR RACIST PARENTS
AND RELATIVES DUE TO FEARS OF GETTING FINANCIALLY CUT
OFF SO YOU CAN'T GO TO ART OR MFA SKOOL AND/OR GOLD STAR
FOR USING YOUR RACIST GRANDPARENTS AS EXAMPLES OF WHY
POCS SHOULD SELF POLICE TONE

GOLD GLITTER FOR ADOPTING A NON-WHITE POET. GOLD GLIT-
TER FOR ORDERING FROM "THIRD WORLD" CANONS FOR NON-

WHITE POET SO THEY MAY HAVE A "BETTER LIFE" WITH YOU IN THE SUBURBS AND/OR GENTRIFIED/GATED NEIGHBORHOOD OF THE AVANT-GARDE

GOLD STAR FOR SHOWING UP AT THE PROTEST TO CORRECT EVERYONE'S GRAMMAR & YELL "I LIKE SENTENCES"

GOLD STAR FOR CLOSING YOUR BUSINESS OR OFFERING DISCOUNTS ON MARTIN LUTHER KING DAY

GOLD STAR FOR "JE SUIS GRINGPO"! SO RISQUE SO CLEVER SO SUBVERSIVE SO SMART! YEY!

GOLD FUCKEN STAR FOR KNOWING THE NAME OF THE LATEST PERSON OF COLOR THAT GOT MURDERED BY THE COPS

GOLD ESTRELLA FOR KNOWING ENOUGH SPANISH TO GET BY ON YOUR VACATION FROM THE CANON.

GOLD STAR FOR COMPLETELY IGNORING THE NEEDS OF UNDOCUMENTED POETS IN POETRY CONTESTS, LET ALONE ACKNOWLEDGING THEY EXIST! ESTRELLA DE ORO FOR ALL OF YOU AND

GOLD SUPERNOVA FOR THE END OF GRINGPO! YEY MONGREL STELLAR BLAST!

9

Mongrel Dream Library

Mongrel Coalition Against Gringpo[1]

mongrel cliff notes: literally shoving you off a cliff: the notes

MONGREL CLIFF NOTES LITERALLY SHOVING YOU OFF A CLIFF:
THE NOTES:

Some of these books have been published under different titles, but we understand it's hard to read anything conjured by the misshapen brains of POC. This is why we've decided to celebrate a longstanding Western literary tradition by starting a series of books with accessible titles that are reformulations of great "unknown" critical works, thereby making them more digestible and white-friendly.

BUT I ASSIGNED CITIZEN!: FINANCING AN ANTI-RACIST REPUTA-
TION WITH MINIMAL EFFORT ON A SHOESTRING BUDGET

THE FORCE OF A SNOWFLAKE: WHITE & WINDSWEPT 21ST CEN-
TURY POETICS

---

1   Originally published in Jennifer Tamayo, "TMCAG Presents... the MONGREL DREAM LIBRARY," *Poetry Foundation,* 1 May 2015, https://www.poetryfoundation.org/harriet/2015/05/mcag-presents-the-mongrel-dream-library.

MARTIN, MY MENTOR: HOW DR. KING DIED SO WHITE PEOPLE COULD QUOTE HIM

THE FAINTING COUCH: NOW IF YOU'LL EXCUSE ME, I'M GOING TO RE-OPEN THE VEINS OF LATIN AMERICA TO PRAISE THE WRITING OF WHITE MEN IN A POOL OF MY VICTIMHOOD

FOUNDATIONS OF POSTMODERN BROETRY 101: RAPE CULTURE EDITION

BRAZILIAN LITERATURE FOR DUMMIES: CLARICE LISPECTOR TELLS YOU TO SIT THE FUCK DOWN (purchased WITH NANCY DREW PRESENTS: THE GHOST OF CLARICE LISPECTOR DEMANDS A MINUTE OF WHITE SILENCE)

WHITE SUPREMACY & THE EPISTEMOLOGY OF VOGUING: A RE-SOURCE GUIDE ON THE AMERICAN EDUCATIONAL SYSTEM FOR POC WHO ALREADY KNOW HOW TO "READ"

WHY IS THIS WHITE WOMAN CRYING?: WHITE ~~FEMINISM~~ FRAGIL-ITY AND THE MACHINATIONS OF PRIVILEGE

CONCEPTUALISM AND SCIENTOLOGY: A COMPARATIVE STUDY OF NORTH AMERICAN DELUSION FOR THOSE WHO WISH TO TWIN-KLE

DON'T SHIT ON MY HEAD & TELL ME IT'S A HAT: THE CHAMBER PISS POT ANTHOLOGY OF TRADITIONAL AMERICAN POETRY

RUTABAGAS & THAT ONE LANA TURNER POEM: FRANK O'HARA & THE CULTS OF BROETRY & TWINKVERSE

DE-REDACTED: I INVENTED THE INTERNET: CONVER-SATIONS WITH CHRISTIAN BOK AND KENNY G ON AR-CHIVAL COMPENSATION FOR PHALLIC DEFLATION

LET MY PEOPLE GO: APPROPRIATING THE BLACK BODY AND RES-
URRECTING IDENTITY POLITICS FOR JEWISH ACADEMICS

VISION QUEST: A WHITE WOMAN'S TRAVEL GUIDE THROUGH THE
REDUCED COSMOS OF HER HURTWORLD

WHY IS MY PROFESSOR SUCH A COWARD: 100,000+ CASE STUD-
IES

THE CHALLENGES OF SHAMING: POISONING THE MOAT OF THE
WHITE EGO'S MEDIEVAL FORTRESS

TRIPPIN WITH NO MAP: STRATEGIES FOR POC INTERVENTIONS,
RESCUE MISSIONS & EXORCISMS OF HONORARY WHITENE$$

CHE GUEVARA HATES YOU: THE MAN BEHIND YOUR STUPID T-
SHIRT
[Editors' note: This is part of a series that includes other classics
such as: FRIDA KAHLO HATES YOU: THE WOMAN BEHIND YOUR
HALLOWEEN COSTUME AND SUBCOMANDANTE MARCOS HATES
YOU: NEOLIBERALISM AT WORK IN THE AMERICAN MIND]

IF IGLOO AZALEA WROTE POETRY: GRINGA ATTORNEY & THE
NEW WORLD OF MANAGERIAL POETICS

MY HEART WILL GO ON: THE QUEEN OF CONPO AND A LOYAL
CHEERLEADER IN CONCEPTUALISM'S FINAL HOURS

THE ETERNAL SIDE EYE: MONGREL POETICS IN A WHITE SUPREM-
ACIST WORLD

A KWH AND GARCÍA MÁRQUEZ COOKBOOK: HOW TO ROAST A
PIG IN THE AMAZONS AND STILL KEEP A WHITE FROCK

THERE'S ALWAYS BESTIALITY: A REHABILITATION GUIDE FOR
LIFE AFTER CONPO

WHO NEEDS PEOPLE?: THE DIGITAL HUMANITIES QUEST FOR TO-
TAL MIND MELD WITH INTERFACE

CONCEPTUAL BROMANCE & THE LONESOME DIGITAL PENIS

DE-REDACTED: MOMMIE DEAREST: PERLOFFIAN MA-
TERNALISM AND THE CONSTRUCTION OF AN ACA-
DEMIC EMPIRE (comes with complimentary map of PACIFIC
PALISADES)

DE-REDACTED: O JARDINERO, MI JARDINERO!: PER-
LOFF'S GUIDE TO FINDING GOOD HELP

WHY WON'T THEY LET ME WEAR MAKE UP??!!!: BABYCORE'S LAST
CRY (Also bought with THE SPICE GIRLS HAVE MORE POWER
THAN YOU: BABYCORE 101)

MY SOUL IS SO DARK I SHIT BATS: EMO(JI) ACTIVISM FOR THE
NEW MILLENIUM (Also bought with DARKS MCDEEPTEXT: HOW
TO COMPLETELY STALL FEMINIST CONVERSATIONS WITH IMES-
SAGE)

TOKEN DYNAMICS IN GROUP SETTINGS: HOW TO CHOOSE A
BROWN FRIEND TO DIVERSIFY YOUR ~~PORTFOLIO~~ PRESS, FACULTY,
READING, OR GROUP

GERTRUDE STEIN AND GATEKEEPING: THE MAKING OF WHITE
AMERICAN$

CANON: BRAIN HEART & WHITE SOUL SUPREMACY BOMBS: THE
WAR CONTINUES

DEPRESSION MONEY FOR A GRINGPO WORLD: PROPROPRO PRO-
ZAC

SURVIVING YOUR MFA/PHD: A YOUNG POC'S TALE OF.... OH,
FUCK IT, JUST QUIT.

TWEETING IS NOT HEREDITARY: HOW TO BE "HIP" WITH NEW POETS & STILL PRAY TO OLD GAWDS

YOU DON'T GET ME, I'M TOO DEEP: WHITE PEOPLE EXPLAIN THINGS TO US
[Editor's note: This title is part of our DEEP WHITE EXPERIMENTALISTS SERIES, which includes classics such as I'M NOT CONCEPTUAL, I'M INCONCEIVABLE; PASTY POET EXPLAINS THE REVOLUTION TO MCAG; and FORM IS INTERNATIONAL CODE FOR COLORBLIND.]

ONLY WHITE PEOPLE EXPERIMENT: A PUBLISHER'S GUIDE TO IGNORING ENTIRE LINEAGES OF POC INNOVATORS

WE ONLY PUBLISH EACH OTHER: BITCHIN' COLLECTIVES AND YOU

REDUCING THE STRUCTURAL TO THE PERSONAL: A GUIDE TO WHITE DEFLECT

THE MASTER SIGNIFIER IS A PERSON? ONLY THE SIGNIFIED HAS JOUISSANCE: WHITE ZOMBIE ENVY

EVERY WORD IS A BREAKING PETAL: THE PRECIOUSNESS OF WHITE AFFECT

A SINGLE WHITE TEAR: THE LONG WALK BACK TO MY SUBURB AND WHY I SEE PEOPLE NOT RACE

GRINGPO NIGHTMARES LAST SO LONG EVEN AFTER JEWEL IS GONE: HORROR STORIES FROM THE FIRES BY THE MONGREL COALITION

NONVIOLENCE MEANS HIGH FIVE THE POLICE CUZ WE ARE ALL HUMAN AFTER ALL (COMES SHRINKWRAPPED WITH KUMBAYA LIVE CD, AS PERFORMED BY KENNY G AND PAUL SIMON AT THE FREE AFRICA 2015 CONCERT)

WHITE PEOPLE EVERYWHERE AND NOT A DROP TO DRINK: PO-
LITICAL & SOCIAL DEHYDRATION IN 20TH & 21ST CENTURY ART

BLANQUITXS UNIDXS: WHITE LATIN AMERICANS COME TO THE
DEFENSE OF THEIR GRINGPO FRIENDS (BUT COULD CARE LESS
ABOUT THEM DEAD BLACKS)

DE-REDACTED: RIGID HOLES: A COLLECTION OF WHITE
GAY MALE POETRY EDITED BY THE WILDE BOYS

DON'T CRY FOR ME JOHN ASHBERY: THE TRUTH IS YOU'RE NOT
MY DADDY

VOULEZ-VOUS COUCHEZ AVEC YOURSELF: NEOCOLONIAL FRAN-
COPHILIA DOES NOT MEAN HAITI IS YOUR BLACK BUCK

THE RED OF NOIR: THE RED CARPET OF ENLIGHTENMENT IS
LINED WITH BLACK BUTLERS

DE-REDACTED: COEUR DE KA-CHING: CELEBRATING
AMERICAN APPAREL WHITE FEMINISM

CELEBRITY TRANSLATIONS: 101 WAYS TO MUDDLE YOUR WAY
THROUGH A LANGUAGE JUNGLE, RAID A VILLAGE, AND RETURN
TO "COLLOQUIAL ENGLISH" WITH WELL-EARNED SPOILS

USING A LITERARY JOURNAL AS TAX SHELTER AND OTHER PITY
PARTY STRATEGIES: A ONE PERCENTER MEMOIR FOR OUR TIMES

WHITE GIRL POETS IN "TRIBAL" TIGHTS: A COLLECTION OF
READINGS AT ANTHROPOLOGIE

FAST FOOD POLITICS FOR POETICS MAJORS, OR, WHAT CAN I DO
NOW THAT I HAVEN'T BEEN DOING MY WHOLE LIFE (Purchased
with FIVE EASY STEPS TO ENDING OPPRESSION)

WHY IS THIS MONGREL ANGRY?: 500 QUESTIONS TO AVOID IF YOU DON'T WANT POC TO THROW AN AUDRE LORDE BOOK IN YOUR FACE

I COME AS A REPRESENTATIVE OF YOUR PEOPLE: LEARNING ABOUT YOURSELF FROM NURTURING WHITE SPONSORS

BLACK LIFE, A H&G SPECIAL: HOW A RAPPER, BLACK ICON, OR ANY OL' BROWN BODY CAN SPRUCE UP YOUR POEM LIKE AN AC-CENT PILLOW

TOM-TOMS AND MACHETES: THE SOUND AND FURY OF THE MCAG REVOLUTION
(purchased with PROPRIETY AND PURITANISM: AN AMERICAN TRADITION)

WHITE PEOPLE PROBLEMS: A POC SAID I WAS RACIST ON FB
(comes with complimentary AUGMENTED WPP RULER!)

Don't forget to order your copy of the THE MONGREL ENCYCLO-PEDIA OF POETRY AND POETICS FOR A GRINGPO WORLD ;) and remember EVERYDAYISANEXCERCISEINFUTURITY

NAROPA STUDENTS OF COLOR, WE STAND WITH YOU.

DESEGREGATE ACADEMIA
DECOLONIZE THE CANON
DECOLONIZE YOUR MIND

Dressing the Wounds (a mongrel bookmark. literally slicing your book: our mark)

NEVER AGAIN applaud poems about progress while your people die at the hands of the state

NEVER AGAIN allow people to assert a lack of intersections be-tween experimentation and people of color.

experimentation is fundamentally rooted in blackness and wounded life: they don't get to tell you otherwise

your body is an experiment.

your memories: an experiment

your life: an experiment

every minute is an experiment in survival

everyday is an experiment toward liberation

10

Exceptionalism in Digital Humanities: Community, Collaboration, and Consensus

Michelle Moravec

In 2013 Diane Jakacki wondered "Is There Such a Thing as Digital Exceptionalism" and concluded that digital humanists have "spent so long being on the outside, marginalized, trying to convince anyone who would listen that what we do matter and that it is meaningful scholarship; now people are paying attention. [...] We can't quite shake that need to justify, to foreground, to compare, to privilege."[1] Digital exceptionalism is suffused with techno-optimism around what the digital can do and with the belief that the digital represents a marked, and presumably better, break with all that came before. Paul Fyfe has described the "hack versus yack" debate as digital humanities' particular variant on this theme.[2]

My inquiry focuses on the ways the rhetoric of exceptionality appears in the discourse produced by the community of digital

---

1 Diane Jakacki, "Is There Such a Thing as Digital Exceptionalism ..?" Blog post, 6 October 2013.
2 Paul Fyfe, "Digital Pedagogy Unplugged," *Digital Humanities Qarterly* 5, no. 3 (2011).

Table 1.

American Exceptionalism	US Digital Humanities[1]
City on a hill	Save the humanities
Legacy of conquests	Narratives of discovery
Sanctimonious tone	Love of Snark
White man's burden	Dark side of DH
Founding Fathers	Origin tales
Newness	Innovation
Building	Making

humanities. I first began hearing what I recognized as echoes of the rhetoric of what I know as "American exceptionalism" in digital humanities discourse in 2013. "American exceptionalism is the distinct belief that the United States is unique, if not superior, when compared to other countries."[3] Replace "countries" with disciplines and this seemed a pretty apt description of how digital humanities often presented itself. As Michael J. Kramer mused in 2013, "the digital is supposed to "transform" [...] through the pastoral dream of technological solutions to social and political problems (hello Leo Marx?)."[4] Like Kramer, I was reminded of those men who taught at or went to college in Boston and of their wilderness and hills, discovery and conquest, salvation and civilization (table 1).

While in the wake of 9/11, the term American exceptionalism gained political currency as a sort of short hand for jingoistic foreign policy, its older association within academia is with a particular strand of American studies. That scholarly narrative has been criticized for "produc[ing] an image of U.S. national unity in which the significance of gender, class, race, and ethnic differences was massively downgraded," a fate I fear digital humanities may replicate.[5] As an editor of a 2012 digital humani-

---

3 Jason A. Edwards and David Weiss, eds., *The Rhetoric of American Exceptionalism: Critical Essays* (Jefferson: McFarland, 2011), 1.

4 Michael J. Kramer, "Reviewing Lauren Frederica Klein's Review, 'American Studies after the Internet,'" *Digital American Studies*, 17 January 2013.

5 Donald E. Pease, "Exceptionalism," in *Keywords for American Cultural Studies*, eds. Bruce Burgett and Glenn Hendler, 108–11 (New York: NYU Press, 2007).

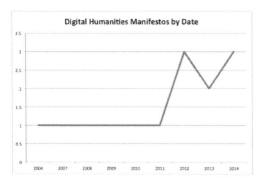

Fig. 1

ties anthology noted, criticisms of digital humanities include, "a lack of attention to issues of race, class, gender, and sexuality; a preference for research-driven projects over pedagogical ones; an absence of political commitment; an inadequate level of diversity among its practitioners."[6] These problems — identity, research v. pedagogy, disengagement, and diversity — offer ways to operationalize the topics I want to explore computationally. I build here on many analyses that come before me of the community known as digital humanities, careful counts of citations, of content, of grant recipients, of conference participants, listserv subscribers. I also place myself in the tradition of #transform-DH, #DHPoco, Hybrid Pedagogy, FemTechNet, and the many individual writers who have called attention to similar concerns.

Methodology

The nexus of identity and power in digital humanities is a place fraught with difficulty and has already resulted in more than a few kerfuffles. Because of that, my approach here is to turn digital humanities methods on the field itself, and rather than exploring scholarly literature about digital humanities to focus on various expressions of digital humanities by many different people.

---

6   Matthew K. Gold, "The Digital Humanities Moment," in *Debates in the Digital Humanities,* ed. Matthew K. Gold (Minneapolis: University of Minnesota Press, 2012).

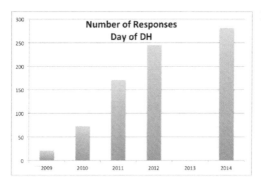

Fig. 2

Three bodies of texts are used to represent various levels of community in digital humanities. *The Digital Manifesto Archive,* compiled by Matt Applegate along with Graham Higgins and Yu Yin (Izzy), represents the most visionary aspect of digital humanities (fig. 1).[7] Using the "collections" created by the site's authors, I created several corpora.[8] The first contains fourteen digital humanities manifestos. For comparison, I used twenty-two manifestos categorized as digital composition and then twelve manifestos under the heading "digital feminisms." Definitions offered by participants in the annual Day of Digital Humanities, for the years 2009–2014, excluding 2013, offered both a broader base of authors, as well as more prosaic efforts to delineate digital humanities (fig. 2).[9] Finally, from its inception in March of 2009 through the end of 2012, the Twitter hashtag #digitalhu-

---

7   I'm grateful to Sharon Leon for alerting me to this excellent site. See *The Digital Manifesto Archive* homepage at http://digitalmanifesto.omeka.net/.

8   My corpora are based on the content of the site as of 16 April 2015. A list of these manifestos appears as Appendix 1. The site is currently in a process of revitalization and new content is once again being uploaded. Circa 2011–2012 four successive individually authored manifestos appear. In 2013 collaborative authorship resumes.

9   I am grateful to Jason Heppler for compiling these and making them available. He created a website that rotates through the definitions. It contains a link to his github repository. The 2013 definitions are on the web but in a way resistant to easy scraping. In keeping with my use of others' datasets as much as possible I did not attempt to pull them myself. See *What Is Digital Humanities?*, http://whatisdigitalhumanities.com/.

Fig. 3

manities represents many voices in conversation as its community was developed on that platform (fig. 3).[10]

Data analysis was performed using three software packages. Wmatrix, created by Paul Ryerson, is a "tool for corpus analysis and comparison," that produces a semantic analysis at the word level.[11] Antconc, written by Laurence Anthony, is a concordancing software that provides detailed information about patterns of word occurrences in a body of texts, such as n-grams as well as collocations.[12] Finally, Sci2 (Science of Science 2), offered ways to both process data and visualize it.[13]

My guiding principles for the data analysis presented in this paper are as follows:

10 Because Twitter has entered into third party agreements to sell access to the historical archive of tweets, scraping it is quite difficult. After 2012 the volume of tweets increased so much that I gave up. I also felt that since Twitter represents the most individual and perhaps least carefully crafted expressions of digital humanities community that a temporal buffer was necessary. I am reasonably confident I've grabbed the vast majority of tweets for each of the years.

11 "Wmatrix Corpus Analysis and Comparison Tool," *Lancaster University: University Center for Computer Corpus Research on Language,* http://ucrel.lancs.ac.uk/wmatrix/.

12 "Software," *Laurence Anthony,* http://www.laurenceanthony.net/software.html.

13 Information can be found at the *Sci2* website: https://sci2.cns.iu.edu/user/index.php.

1. I have relied on machine reading whenever possible before shifting to user-driven inquiries.
2. I have utilized datasets from other scholars before creating my own.
3. I have, following a methodology I first encountered in Clare Hemmings's excellent *Why Stories Matter,* left all quotations from my corpora unattributed.[14]
4. I have also gone to various lengths to anonymize quotations from tweets and the Day of Digital Humanities definitions.[15]

Digital Humanities as community

At its most utopian, digital humanities appears as a community akin to those groups in American history that have set themselves apart because of their unique values, much as the Puritans did. Note, while this resonates through my own nationalist rhetoric of American exceptionalism, this is not a geographic divide. Utopian leanings appear in a 2010 document issued at THATCamp Paris: "We, professionals of the digital humanities, are building a community of practice that is solidary, open, welcoming and freely accessible."[16] The spirit of this manifesto is

---

14 Clare Hemmings, *Why Stories Matter: The Political Grammar of Feminist Theory* (Durham: Duke University Press, 2011). Since I have chosen to write about digital humanities community without attribution to the authors in my corpora, I decided not to provide extensive footnotes to secondary literatures in DH as that would defeat my purpose of moving away from a focus on individuals to the community as a whole.

15 Manifestos are by their very nature aimed at a public, and those collected in *The Digital Manifesto Archive* are ones that were explicitly circulated online. While Day of Digital Humanities participants were certainly aware that the definitions they submitted in the process of completing an online sign-up form would be made public, I considered that they might be less carefully crafted than the texts of a manifesto and therefore by used very brief quotes and employed judicious use of ellipses to confound online searches. Tweets, also completely searchable due to the recent agreement between Twitter and Google to feed them directly into the top of the search engine's results, do not appear as direct quotes at all beyond the pervasive patterns identified using the previously described software.

16 "Manifesto for the Digital Humanities," *THATCamp Paris.*

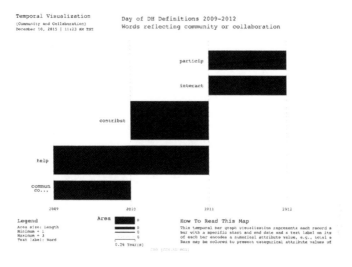

Fig. 4. Burst analysis of Day of DH Definitions 2009–2012.

just one of many that expresses a commitment to creating an intentional community predicated on shared values that are at least implicitly distinctive and new.

Digital humanities manifestos reflect the importance of community in higher semantic content, as compared to digital composition and digital feminisms manifestos, under tags such as "belonging to a group" (through words like *community, public, institutions, society,* and *collective*).[17] The central values of this community are participating (expressed as variations on the word *collaboration*) and helping (which includes *help, encourage, support,* and *promote*). One manifesto succinctly combines all three concepts in the exhortation to "[e]ncourage personal expression, collaboration, and community."

Declarations of digital humanities as a community are found not only in the visionary manifestos, but also in the definitions

---

17 Italics denote words that are results from computational analysis. Words in quotations are quotes from sources under analysis.

offered by Day of Digital Humanities (Day of DH) participants (fig. 4). Burst analysis, which identifies sudden increases in the frequency of words, shows *commun* bursting in 2009 definitions, the first year of the annual event, while words that indicate collaboration burst in subsequent years, such as *help* (2009–2011) and *particip* (2011–2012). *Commun* is a tokenized version of both "community" as well as "communication," which fits nicely with the idea of digital humanities as a proselytizing community. Digital humanities is both "a burgeoning community" but also a "social utopia." It is also described as a community of communicators, as in "open communication collaboration and expression." Few definitions point to a specific audience for this communication or give such a detailed a description of participants as this one: Digital humanities is "the transformation of [...] communication" not only by "academics [and] students" but also by "other experts" via "communication technologies" that amplify the "individual's power" to both "communicate" and to "create new [public] spaces."

While character count restrictions tend to limit the eloquence and expansiveness of tweets, Twitter offers an important arena for analyzing the formation of digital humanities as a community. Twitter is often pointed to as the platform where the digital humanities community coalesced: "Twitter has played an important and occasionally transformative role at every academic gathering I have attended since early 2008" by among other ways, allowing "key, already well-networked community members to participate" virtually.[18] Although the word community appears relatively infrequently in tweets, comparing unique bigrams and trigrams between the years of (March 2009 to December 2012), #digitalhumanities tweets provides evidence of how community formed and functioned.

---

18 Bethany Nowviskie, "Uninvited Guests: Twitter at Invitation-Only Events," in *Hacking the Academy: New Approaches to Scholarship and Teaching from Digital Humanities,* eds. Daniel J. Cohen and Tom Scheinfeldt, 124–31 (Michigan: University of Michigan Press, 2013).

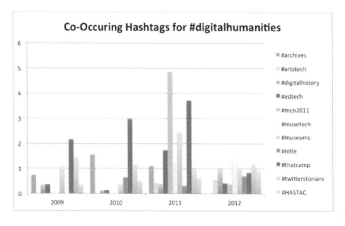

Fig. 5. Co-occuring hashtags #digitalhumanities.

In 2009, Twitter functioned primarily as a networking platform, as evidenced in ways of connecting people, such as "just
added," via "twitter directory" or "wefollow," and the first mention of HASTAC, as well as a handful of references to face-to-face
connections made at "the MLA" or "digitalhumanities thatcamp."
By 2010, Twitter began to evolve into a platform for scholarly
communication, such as blogs ("blog post," "post on" "digital
humanities blog") perhaps as a way of disseminating "digital
scholarship" or highlighting a "digital humanities project" or
"digital humanities research." Community institution building is also evident both through centers in the US such as the
Humanities Lab ("at humlab" "digitalhumanities humlab") and
internationally "thatcamp Paris" and "thatcamp Switzerland,"
but also fields like education and museums "edtech digitalhumanities" and "digitalhumanities musetech." Finally, Twitter
also becomes a conversational platform between individual participants in the community using the hashtag. Terms of direct
address become more frequent, such as "do you" and "for you,"
as do acknowledgements of others "thank you" and "thanks to."
By 2012, references to external social media networking platforms have all but disappeared (who even remembers wefollow
or Klout?), as Twitter has taken on that function, even as digital

humanities professionalizes, as evidenced by trigrams, such as "journal of digital humanities," and "digital humanities book."

Looking at hashtags that co-occur with #digitalhumanities provides insights into which groups or organizations were important in the early days of community formation (fig. 5). Three that pop beginning in 2011 relate to #museums, which appears in 2009, then moving into 2010 tweets about the 2011 MCN conference, which predates digital humanities but offered an early important venue, and the related hashtag #musetech. All three of these hashtags continue in 2012, but are smaller than a fourth hashtag (#artstech) that largely got traction due to its use by Neil Stimler, a museum professional at the Metropolitan Museum of Arts, who has been one of the major voices in the intersection of digital humanities and museology.

These co-occuring hashtags highlight aspects of digital humanities community formation earlier that are perhaps less known now. As one scholar recently lamented "No one imagined digital humanities — as a constructed field of practice–centered" outside academic institutions, despite the precedent of fields like museum technology.[19] The same might be said of digital humanities projects. A darker band in the dispersion of *communit** tweets, indicating points of greater density, turns out to align with references and retweets of an October 2011 blog post "is creating community a primary function of the digital humanities?" from Editing Modernism in Canada, a research group aimed to foster collaboration, offers training in "experiential-learning pedagogies," and develops relationships beyond institutions of higher education with "public libraries, and nonprofit cultural organizations (book clubs, reading groups, reading series, literary festivals)."[20] EMiC played multiple roles, not only producing digital scholarship, but also disseminating it to the public, while at the same time shaping the conversation

19 Sheila Brennan, "DH Centered in Museums?" *Lot 49*, 16 March 2015.

20 Reilly Yeo, "Is Creating Community a Primary Function of the Digital Humanities?" *Editing Modernism in Canada*, 18 October 2011, and "About Us," *Editing Modernism in Canada*, 3 May 2010.

around how that scholarship was produced and what it meant to be involved in digital humanities.

Digital Humanities as collaborative

Increasingly, what digital humanities meant was collaboration. What we do, we do together. "Collaboration is widely considered to be both synonymous with and essential to digital humanities" write one pair of researchers.[21] Collocates of *collabor** in tweets reveal that the ideal is invoked in all parts of the scholarship cycle, including how we develop a "project," conduct "research" and in reference to "publishing" the results. Despite analyses that question how truly collaborative digital humanities is, such as Julianne Nyhan and Oliver Duke-Williams' analysis of co-authorship as one marker of collaboration, the most recent day of digital humanities definitions (2014) show a statistically significant increase in *collab**: "DH is collaborative" and relies on a "spirit of collaboration."

However, it is in the manifesto corpus that perhaps, most predictably, the greatest enthusiasm is expressed (and in the most exuberant prose) for collaboration. Although one manifesto cautions, "the Internet didn't invent collaboration or solve all the problems," concordancing software reveals an overall positive attitude towards what another manifesto describes as the "collaborative turn." Of thirty-four lines containing *collab**, about 25% are preceded by positive verbs. The optimistic embrace of collaboration is captured in one manifesto, which cites it as a marker of the revolutionary nature of digital humanities. "The digital humanities revolution is about integration: the building of bigger pictures out of the tesserae of expert knowledge. It is not about the emergence of a new general culture, Renaissance humanism/humanities, or universal literacy, but on the contrary, promotes collaboration across domains of

---

21 Julianne Nyhan and Oliver Duke-Williams, "Is Digital Humanities a Collaborative Discipline? Joint-Authorship Publication Patterns Clash with Defining Narrative," *Impact of Social Sciences,* 10 September 2014.

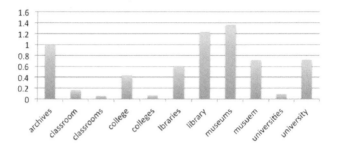

Fig. 6. Locations in #digitalhumanities tweets 2009–2012.

expertise."[22] References to "bigger pictures" and "digital human-ities revolution" all suggest that something new or distinctive is afoot in doing things digitally.

While there has been some pushback on digital humanities as "essentially different" or "different than" other traditional ways of doing scholarship, there are still plenty of people defin-ing digital humanities by its differences from other academic endeavors, especially in terms of its collaborative aspects. In digital humanities definitions, *collabor** co-occurs about 15% of the time with this sense of "new" or "different" or as part of a "shift" or a "trend," indicating some sort of transition. Further-more, "new" shows a burst in the definitions from 2014 indicat-ing that the idea of digital humanities as some sort of innovative thing is not fading.

Within academia, collaborative-produced knowledge is nothing new, as scientists and social scientists have long relied on this approach, and even in the humanities we have these his-tories. However, for most tenure track folks in the humanities disciplines at the dawn of the twenty-first century, collaboration appears innovative and that novelty is often emphasized. This enthusiasm for digital humanities as a novel seems to have also led to a sort of amnesiac disregard for the long history of col-laboration that has occurred in museums, libraries and archives,

22 "A Digital Humanities Manifesto," *ex.pecul.ando* (blog), 6 January 2009.

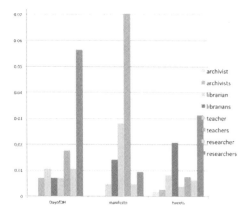

Fig. 7. Relative frequencies of roles in Day of DH definitions, DH Manifestos, and #digitalhumanities tweets.

as well as other scholarly endeavors that predate a self-identified digital humanities community, which lead me to try to determine who precisely is involved in collaboration? Looking at roles, locations, and actions provides ways to relate with whom, where, and why we say we collaborate.[23]

Not surprisingly, given the persistently recurring hashtags of #museums and #archives in #digitalhumanities tweets, these locations are most likely to be mentioned on Twitter as compared to in the Day of Digital Humanities definitions (which are more concerned with what we do digitally) or manifestos (largely about why we should do what we do), and therefore rarely point to specific locations. In tweets, teaching locations comes out lowest, even below institutions for professors (fig. 6). The disparity in classroom(s) is mirrored in roles and actions. A researcher is the most frequent role mentioned in both tweets and Day of Digital Humanities definitions, with manifestos mentioning teachers the most (fig. 7). (How are libraries and archives central, but librarian and archivists not?) Looking at actions, and this disparity between teaching and research is replicated again, with only manifestos given almost even weight to

---

23 Although Wmatrix calculates "roles" it assigns locations by their function so that I had to hand calculate these variables using my own inquiry terms. I derived locations and roles from word frequencies in the corpora.

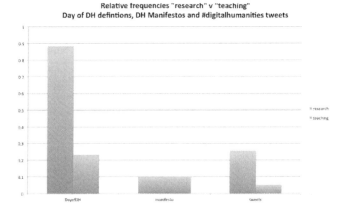

Fig. 8. Relative frequencies "research" v "teaching" in Day of DH definitions, DH Manifestos, and #digitalhumanities tweets.

the two activities, and this is largely due to one manifesto about
art education (fig. 8).

This emphasis on digital humanities as a community of researchers as opposed to teachers educating obscures the inter-
connectedness of pedagogy and scholarship in digital humanities. As illustrated by an early co-occurring hashtag #edtech, for
example, instructional technology has long been an important
site of digital humanities work within higher education. Even at
its most technical, the tools of digital humanities are often in-
debted to pedagogy. Much of this is related to rhetoric and composition, early bastions of "learning technologies" in many institutions. My own software of choice, antconc, is a repurposed use
software initially developed for aid in teaching second-language
acquisition.

Teaching and research are not mutually exclusive practices or
activities. However, in speaking volumes about research and relatively little about teaching, the community is very loud about
one thing and very quiet about another, which has implications
for people who are located institutionally in different places. Not
only are some of the members at teaching-intensive institutions
while others are at research-driven institutions; some people are

direct classroom instructors while others are in related fields, but not in classrooms; and still others are in institutions that may be affiliated with institutions of higher education, but are not always (i.e., libraries, archives, museums).

The relative sidelining of pedagogy as opposed to research highlights an uncomfortable aspect of community, mainly that the word implies some sort of boundaries between those within and those outside. That tension carries through to the ideal of collaboration as a second hallmark often noted of digital humanities' distinctiveness in the discipline. Roy Rosenzweig, as early as 2007, lamented that "academic historians have — unfortunately — grown distant from librarians, archivists, and museum professionals since World War II, a reversal of the close association that existed in the first half century of professional history in the United States."[24] The corpora I'm analyzing here provide evidence of very little reference to these practitioners of digital humanities work.

Digital Humanities as consensus

The relative sidelining of some people over others is indicative of power differentials, even as we pride ourselves on being a community that always gets along. The much discussed niceness problem is a description of consensus culture. This is most evident in "standards" like TEI, which as Julia Flanders notes, rests on consensus that "arise(s) from power structures" and represents the "functional homogenization" of "community."[25]

At its most promising digital humanities may transform, the word that signals most strongly its utopian promise, but

---

24 Rose Rosenzweig, "Collaboration and the Cyberinfrastructure: Academic Collaboration with Museums and Libraries in the Digital Era," *First Monday* 12, no. 7 (2007).

25 Julia Flanders, "Collaboration and Dissent: Challenges of Collaborative Standards for Digital Humanities," in *Collaborative Research in the Digital Humanities: A Volume in Honour of Harold Short, on the Occasion of His 65th Birthday and His Retirement, September 2010,* eds. Marilyn Deegan and Willard McCarty, 67–80 (Burlington: Ashgate Publishing, 2012), 71.

there are many tensions that impeded this promise. In 2011, a tweet questioned whether "scholars" should "collaborate" with social movement activists. Binary implied here is that scholars are not activists, and yet the very first #digitalhumanities tweet, by George H. Williams expressing support for Alan Turing, a founding father in digital humanities, is accompanied by a second hashtag #LGBTQ.

Origin tales going back to people like Rosenzweig, a committed "digital democratizer," link digital humanities to activist agendas: "By democratizing history, I mean democratizing the audience — reaching wide and diverse audiences; democratizing the content-incorporating diverse voices, especially the voices of ordinary people; and democratizing the practice — making history open and collaborative."[26] This raises the conundrum of how digital humanities as a community practices or reflects this ideal. If digital humanities understands itself as a community that always gets along — helping and collaborating — how can it resist homogenizing participants under the identity "digital humanist" and turn itself instead to addressing more explicitly and persistently issues of power and identity?

What goes by the "niceness" problem in digital humanities was originally coined to describe a perceived lack of dissent over the digital humanities theory. However, the niceness problem has been unpacked as more of a distinction of "saying and doing" than a description of affect, or as code for "civility" or the (unwritten) rules of engagement. This brings us to the intersection of a well-known discourse in traditional academia that of "collegiality." Collegiality is too often a coded discourse of power in academia and here I am drawing on my work with Tricia Matthew, who I first encountered on Twitter and presented with at a THATCamp. In 2012, Matthew tagged me in a tweet linked to a blog post that hits on the themes of both collaboration and community as they intersect with various identities. As she wrote, while "women are undoubtedly socialized to think in

---

26 "Award Recognizes 'Digital Democratizer' of History," *National Humanities Center: News and Events,* 13 May 2003.

terms of community," white feminist academics often express this "in ways that can lead to conflict instead of collaboration." As one person has noted, "niceness" is much easier when you don't have to "routinely" deal with "casual racism."[27]

In other words, the community of digital humanities was itself only ever "nice" for some participants. Despite the protestations of inclusivity and openness, digital humanities, at least in the sources I'm using in this essay, deals very little with aspects of identity or power that might challenge our consensus. "Power" appears in only two years of the Twitter corpus (2011 and 2012) and at the same frequency as "plough" if that gives you any indication of how little we discuss power, at least as an abstract concept. Turning to semantic tagging, digital humanities manifestos contain no content for "power, organizing" or "no power." To quote one digital feminisms manifesto: "We can't move forward to new tech with old methodology — this is a recipe to maintain old power."

And yet it wasn't always this way. The first #digitalhumanities tweet containing "community" expressed a desire to "draw" in "women's/gender historians." Yet that need to "draw" in contains the idea that historians who study women and/or gender are not yet involved in #digitalhumanities. While these disparities may mirror those that exist in both higher education and the tech world, because digital humanities describes itself as an open and encompassing community, and at times as even a "model," dedicated to greater "democratization," it sets itself up to be held to a higher standard. Again, this aspect is part of what attracted me to digital humanities in the first place, so I view this as a positive aspect of our beloved community.

How did this happen then, this lack of sufficient attentiveness to power and identity, if aspects of this discourse have been present since the inception of #digitalhumanities? One digital feminist manifesto cuts to the heart of the problem, calling attention to the "exacerbated asymmetries in gendered forms

---

27 Nicole Chung, "What Goes Through Your Mind: On Nice Parties and Casual Racism," *The Toast*, 5 January 2016.

of power and knowledge including or especially where those asymmetries appear to be addressed."[28]

Is the digital humanities one of those sites, presenting itself as addressing asymmetries while exacerbating them? As digital humanities have become increasingly institutionalized through grants, centers, conferences power imbalances have been replicated. Scott Weingart's investigation of gender at digital humanities conferences suggests, however, that power disparities continue to exist, at least for gender, in these face-to-face events. "Women are (nearly but not quite) as likely as men to be accepted by peer reviewers at digital humanities conferences" except that "a lot of the topics women are submitting in aren't getting accepted to digital humanities conferences" including, unsurprisingly, gender studies a field that has almost 70% of its submissions by women, as well as "culture, teaching digital humanities, creative arts & art history, GLAM, institutions." [29]

A sign of the uneasy fit between #digitalhumanities and identity/power is evidenced by the rise of the hashtag #transformDH. #transformDH first emerged as a hashtag for the panel/roundtable "Transformative Mediations: Queer and Ethnic Studies and the Digital Humanities" at the 2011 American Studies Association. Alexis Lothian announced #transformDH as the panel hashtag.[30] Although the panel itself had very low attendance according to live tweeting (six-eight people) and the post conference "call to action" written by one of the panelists, Amanda Phillips, is not contained in the digital manifesto archive, the hashtag clearly captured a felt need as that same day it was picked up as "How to transformDH" appended to a tweet[31] requesting volunteer reviewers for submissions to the Digital Humanities 2012 conference and appended to a tweet about organizing a THATCamp at historically black colleges and uni-

28 Sarah Kember, "Notes Towards a Feminist Futurist Manifesto," *Ada: A Journal of Gender New Media and Technology* 1 (November 2012).

29 Scott Weingart, "Acceptances to Digital Humanities 2015 (part 4)," *The Scottbot Irregular,* 28 June 2015.

30 See Alexis Lothian, Twitter post, 21 October 2011, 10:01 a.m.

31 George H. Williams, Twitter post, 21 October 2011, 11:54 a.m.

versities. The hashtag then gave rise to its own Tumblr and then twitter accounts, both of which continue today.[32]

The oppositional voice implied by the need to "transform DH" gave rise to more focused efforts to organize within digital humanities which were amplified by existing organizations like HASTAC and NITLE (both frequent co-occuring hashtags). In particular, in February of 2012 Adeline Koh tweeted a CFP for a panel on race at the MLA. In July of that year, she invited people "working on a project about race, class, gender & disability in #digitalhumanities" to add themselves to an open Google doc. In March of 2013, Adeline Koh and Roopika Risam founded the DHPoco website. In addition to featuring blog posts about digital humanities and post-colonialism, the site also published hilarious and sometimes controversial comics about digital humanities, hosted a highly successful online summer reading group, and offered an online forum for discussion.[33]

In May of 2013 an open forum about the question "Are the #digitalhumanities a refuge from race/class/gender/sexuality/ disability?" crossed over to Twitter. Sixty-six tweets appeared containing a reference to the "open forum" with the #dhpoco hashtag, but even more remarkably, a robust discussion (one hundred fifty-three comments by thirty-eight unique participants) emerged in the forums. Heather Froehlich and I collaborated on an analysis of the forum in an attempt to answer the rather simplistic question: "Do men and women in the DHPoco thread talk about digital humanities and post-colonialism differently? If so, how?" Conducting the same sort of computational linguistics analysis I've done in this essay on the comments in the forums revealed that women were more likely to comment about *whiteness, color, white, queer, woman,* and *intersecting* as well as all aspects of race (lexical variants of race included *race, racial, racializing, racing, racism, racist*) Men used words like *discourse, technologies, technology, institutional, research, in-*

---

32 See @TransformDH, Twitter feed, https://twitter.com/TransformDH and *#transformDH* (blog), *Tumblr,* http://transformdh.tumblr.com/.

33 See the *DHPoco* homepage at http://dhpoco.org/.

stitutions, and *culture,* more frequently. *Research, technologies, technology, culture* are among the most frequent in used by Day of Digital Humanities definitions as well.

Using these terms as a jumping-off point, I compared their frequency in #transformDH (October 11, 2011 through December 31, 2012) to digital humanities tweets from the calendar year 2012.[34] As expected #transformDH tweets use words that denote identity, *race* (LL 121), *black* (LL 90), *gender* (LL 58), *women* (LL 45), *queer* (LL 33), *feminist* (LL 33), and *power* (LL 12) more frequently than #digitalhumanities tweets. However, when comparing the words central to both definitions of Day of Digital Humanities and used more by men in the DHPoco debate, only *research* and *technologies* are over-used in #digitalhumanities tweets while *institutions, technology,* and *tech* are underutilized as compared to TransformDH. #transformDH then represents a model within digital humanities that not only talks about what the community understands itself as doing, but also attends to identities of practitioners and the transformative potential of the field.

Conclusion

In our enthusiasm for the digital humanities, it seems we may have proselytized too hard and or have been heard as promising too much. The emphasis on digital humanities as "new" or "revolutionary" or "radical" in the current neoliberal context that pervades higher education has created something of a utopian paradox. Whatever our intentions might have been, digital humanities is now positioned as a choice that will lead to greater chances for future success. As a host of authors have noted, some approvingly and others with disdain, digital humanities appears as the answer to everything from the crisis in academic publish-

---

34 TransformDH 1077 tweets from 21 October 2011 through 31 December 2012 as compared to over 5800 in 2012 for #digitalhumanities. Because there was such disparity in volume I kept the extra 2 months of TransformDH to offset the disparity in volume. 1077 compared to # of tweets in 2012. All values here are normalized frequencies per 1000 words.

ing, — "sav[ing] university presses" — to the dismal job market for tenure track professors — "provid[ing] [...] jobs in libraries, institutes, nonprofits" to nothing short of "disciplinary transformation." The inevitable push back, which seems to come along with increasing regularity in long-form writing, also follows an almost inevitable cycle of annual-meeting handwringing. We do ourselves no favors pointing to outsiders foisting the rhetoric of "save the humanities" on us or in denying that it exists. While manifestos are by their nature utopian visions, and might reasonably be expected to contain such a discourse, the more prosaic day of digital humanities definitions also contain words like "hope," "promise," "potential," and "future." The forward-looking tendency is evident in tweets as well, especially in the use of "future." Almost 40% of the occurrences of future are in the pattern *future of* if not predicting then at least debating what will happen to "digital humanities," "humanities," "publishing" books, "libraries," "reading," and even "universities" and "the academy" as a whole.

I love our forward-lookingness and as I've now stated repeatedly, embrace our utopianism. How then do we do this without either positioning ourselves as the answer or unwittingly putting ourselves in the place of primacy? The first, obviously, is to be wary of claiming firsts. Scholars have amply documented how identities other than heteronormative white men have overtaken earlier histories of computing.[35] Origin stories of digital humanities are many and thankfully people are hard at work explicating them for us, but they all trace back to "fathers" of digital humanities. My own roots are closest to those of Rozenzweig who also started as an oral historian of social history. Stephen Robertson outlined this recently in an essay that traces digital humanities back through oral histories and efforts to put

---

35 Jane Margolis and Allan Fisher, *Unlocking the Clubhouse: Women in Computing* (Cambridge: MIT Press, 2003); Kurt W. Beyer, *Grace Hopper and the Invention of the Information Age*, reprint edn. (Cambridge: MIT Press, 2012); Lisa Nakamura, "Indigenous Circuits: Navajo Women and the Racialization of Early Electronic Manufacture," *American Quarterly* 66, no. 4 (2014): 919–41.

historical sources "on the web," but I acknowledge the other roots of our beloved community and embrace those as well.[36] The concordancing software I use may find its starting point in Father Busa, the most cited founding father of #digitalhumanities, but that does not mean we have no mothers, although I have searched in vain for any reference to "mother of the digital humanities."[37] We need more space in our origin narratives for #museums and #archives, #edTech, and Museum Culture Network conferences, people at community colleges, as well as librarians, curators, and archivists, and early, still exemplary projects like Early Modernism in Canada, along with visionaries like Roy Rosenzweig.

We could, and probably should, make a concerted effort to stop talking about the community of digital humanities in utopian terms, dialing back our characterization of collaboration as the definitive aspect of digital humanities, and letting go of our belief in digital humanities as "revolutionary" or "new." As one person in digital humanities has noted, utopianism "contains within itself the seeds of its own undoing" in the "all-but-impossibility" and "promises that are difficult to keep."[38]

I reject the idea that we must let go completely of our utopian leanings. I end therefore with a modest manifesto of my own on how those of us with access to power might start to transform digital humanities.

- Don't do homogenous — panels, books, datasets, anything.
- Throw open the gates — suggest new names for speakers, contributors, reviewers.
- Educate yourself — if you only know people like you, get to know some people not like you.

---

36 Stephen Robertson, "The Differences between Digital History and Digital Humanities," blog post, 23 May 2014.

37 See Steven E. Jones, *Roberto Busa, S.J., and the Emergence of Humanities Computing: The Priest and the Punched Cards* (New York: Routledge, 2016).

38 Neil Freistat, "The Promise(s) of Digital Humanities," Australasia Association of Digital Humanities, Tufts University, 10 April 2014.

- Think harder before writing or speaking — which chronologies, what projects, and who you cite.
- Call out oppression — don't wait for people who are being oppressed to do it.

Bibliography

#transformDH (blog). *Tumblr*. http://transformdh.tumblr.
    com/.

"A Digital Humanities Manifesto." *ex.pecul.ando* (blog).
    6 January 2009. https://expeculando.wordpress.
    com/2009/01/06/a-digital-humanities-manifesto-ucla/.

"About Us." *Editing Modernism in Canada*. 3 May 2010. http://
    editingmodernism.ca/about-us.

"Award Recognizes 'Digital Democratizer' of History."
    National Humanities Center: News and Events. 13 May
    2003. http://nationalhumanitiescenter.org/newsrel2003/
    prlymanaward2003.htm.

Beyer, Kurt W. *Grace Hopper and the Invention of the
    Information Age,* Reprint edition. Cambridge: MIT Press,
    2012.

Brennan, Sheila. "DH Centered in Museums?" *Lot 49*. 16 March
    2015. http://www.lotfortynine.org/2015/03/dh-centered-in-
    museums/.

Chung, Nicole. "What Goes Through Your Mind: On Nice
    Parties and Casual Racism." *The Toast*. 5 January 2016.
    http://the-toast.net/2016/01/05/what-goes-through-your-
    mind-casual-racism/.

DHPoco. http://dhpoco.org/.

The Digital Manifesto Archive. http://digitalmanifesto.omeka.
    net/.

Edwards, Jason A., and David Weiss, eds. *The Rhetoric
    of American Exceptionalism: Critical Essays*. Jefferson:
    McFarland, 2011.

Flanders, Julia. "Collaboration and Dissent: Challenges
    of Collaborative Standards for Digital Humanities." In
    *Collaborative Research in the Digital Humanities: A Volume
    in Honour of Harold Short, on the Occasion of His 65th
    Birthday and His Retirement, September 2010,* edited by
    Marilyn Deegan and Willard McCarty, 67–80. Burlington:
    Ashgate Publishing, 2012.

Freistat, Neil. "The Promise(s) of Digital Humanities." *Australasia Association of Digital Humanities, Tufts University,* 10 April 2014. http://mith.umd.edu/wp-content/uploads/2014/05/The-Promises-of-Digital-HumanitiesaaDH.pdf.

Fyfe, Paul. "Digital Pedagogy Unplugged." *Digital Humanities Quarterly* 5, no. 3 (2011). http://digitalhumanities.org/dhq/vol/5/3/000106/000106.html.

Gold, Matthew K. "The Digital Humanities Moment." In *Debates in the Digital Humanities,* edited by Matthew K. Gold. Minneapolis: University of Minnesota Press, 2012. http://dhdebates.gc.cuny.edu/debates.

Hemmings, Clare. *Why Stories Matter: The Political Grammar of Feminist Theory.* Durham: Duke University Press, 2011.

Jakacki, Diane. "Is There Such a Thing as Digital Exceptionalism ..?" Blog post. 6 October 2013. http://dianejakacki.net/is-there-such-a-thing-as-digital-exceptionalism/.

Jones, Steven E. *Roberto Busa, S.J., and the Emergence of Humanities Computing: The Priest and the Punched Cards.* New York: Routledge, 2016.

Kember, Sarah. "Notes Towards a Feminist Futurist Manifesto." *Ada: A Journal of Gender New Media and Technology* 1 (November 2012). http://adanewmedia.org/2012/11/issue1-kember/.

Kramer, Michael J. "Reviewing Lauren Frederica Klein's Review, 'American Studies after the Internet.'" *Digital American Studies.* 17 January 2013. http://www.michaeljkramer.net/cr/digital-american-studies-2/.

Lothian, Alexis (@alothian). Twitter post. 21 October 2011, 10:01 a.m. https://twitter.com/alothian/status/127383970605305858.

"Manifesto for the Digital Humanities." *THATCamp Paris.* http://tcp.hypotheses.org/411.

Margolis, Jane, and Allan Fisher. *Unlocking the Clubhouse: Women in Computing.* Cambridge: MIT Press, 2003.

Nakamura, Lisa. "Indigenous Circuits: Navajo Women and the Racialization of Early Electronic Manufacture." *American Quarterly* 66, no. 4 (2014): 919–41. DOI: 10.1353/aq.2014.0070.

Nowviskie, Bethany. "Uninvited Guests: Twitter at Invitation-Only Events." In *Hacking the Academy: New Approaches to Scholarship and Teaching from Digital Humanities,* edited by Daniel J. Cohen and Tom Scheinfeldt, 124–31. Michigan: University of Michigan Press, 2013.

Nyhan, Julianne, and Oliver Duke-Williams. "Is Digital Humanities a Collaborative Discipline? Joint-Authorship Publication Patterns Clash with Defining Narrative." *Impact of Social Sciences.* 10 September 2014. http://blogs.lse.ac.uk/impactofsocialsciences/2014/09/10/joint-authorship-digital-humanities-collaboration/.

Pease, Donald E. "Exceptionalism." In *Keywords for American Cultural Studies,* edited by Bruce Burgett and Glenn Hendler, 108–11. New York: NYU Press, 2007. http://keywords.nyupress.org/american-cultural-studies/essay/exceptionalism/.

Robertson, Stephen. "The Differences between Digital History and Digital Humanities." Blog post. 23 May 2014. http://drstephenrobertson.com/blog-post/the-differences-between-digital-history-and-digital-humanities/.

Rosenzweig, Rose. "Collaboration and the Cyberinfrastructure: Academic Collaboration with Museums and Libraries in the Digital Era." *First Monday* 12, no. 7 (2007). http://firstmonday.org/ojs/index.php/fm/article/view/1926/1808.

Sci2. https://sci2.cns.iu.edu/user/index.php.

"Software." Laurence Anthony. http://www.laurenceanthony.net/software.html.

TransformDH (@TransformDH). Twitter feed. https://twitter.com/TransformDH.

Weingart, Scott. "Acceptances to Digital Humanities 2015 (part 4)." *The Scottbot Irregular* (blog). 28 June 2015. http://www.scottbot.net/HIAL/index.html@p=41375.html.

What is Digital Humanities? http://whatisdigitalhumanities. com/.

Williams, George H (@GeorgeOnline). Twitter post. 21 October 2011, 11:54 a.m. https://twitter.com/GeorgeOnline/ status/127412401346838529.

"Wmatrix Corpus Analysis and Comparison Tool." *Lancaster University: University Center for Computer Corpus Research on Language.* http://ucrel.lancs.ac.uk/wmatrix/.

Yeo, Reilly. "Is Creating Community a Primary Function of the Digital Humanities?" *Editing Modernism in Canada.* 18 October 2011. http://editingmodernism.ca/2011/10/ is-creating-community-a-primary-function-of-the-digital-humanities/.

11

The Problem with Prof Hacking

Matt Thomas

> *"Nowhere do you find more enthusiasm for the*
> *god of Technology than among educators."*
> — Neil Postman[1]

To the extent that they think about it at all, when most people hear the word "hacker," they still probably picture someone who uses a computer to gain unauthorized access to information. In most people's minds, hackers, even now, are criminals, and "hacks" such as headline-grabbing data breaches are a regrettable feature of twenty-first century life that one must deal with, not something the average person does. But since at least the mid 2000s, the term "hacking" has increasingly been applied to more and more activities outwardly having nothing to do with computers, and "hacking" has come to mean using ingenuity to improve things. People now routinely talk about hacking their work, their hobbies, their possessions, their bodies, and so on. These people have taken a practice traditionally associated with computers — i.e., "hacking" — and broadened it to encompass literally anything under the sun. Indeed, people now speak of

---

1   Neil Postman, "Virtual Students, Digital Classroom," in *Minutes of the Lead Pencil Club: Pulling the Plug on the Electronic Revolution,* ed. Bill Henderson, 197–215 (Wainscott: Pushcart Press, 1996), 198.

hacking everything from IKEA furniture to democracy like it's the most normal thing in the world.[2] Business and lifestyle publications have rechristened advice from investment how-tos to beauty tips as hacks. I'm not the only one to notice this metaphor creep. People joke about it on social media all the time. Next up, I predict, will be "hacking fatigue."

The application of the term "hacking" presumes that anything and everything is amenable to hacks or clever modifications the same way a computer system is. I maintain that the recent broadening of the term to encompass an endless multiplicity of life's activities suggests the degree to which people in the twenty-first century are increasingly thinking about the world in vaguely computational terms. The broadening of the term "hacking," though it might seem faddish, is thus important to attend to precisely because it reveals how the rhetoric of hacking and the point of view of the hacker have become normalized. I argue this rhetoric and subject position carry with them particular ideas. These ideas have deep roots in Western culture, namely a way of thinking about the world that David Golumbia calls "computationalism," a "belief in the power of computation" that "underwrites and reinforces a surprisingly traditionalist conception of human being, society, and politics."[3] Put differently, in a thoroughly computerized world, hacking becomes a — or perhaps even *the* — preferred "way of seeing," to borrow a phrase from art critic John Berger. What is problematic about this way of seeing, I contend, is that it is in line with long traditions in US culture of self-making and techno-fetishism.

In researching and writing a PhD dissertation on "life hacking," the first of its kind to trace the broadening of the term "hacking" discussed above, I often found myself bouncing back and forth between two opposing positions: eager consumer of "life hacks" on the one hand and vociferous critic of them on the

---

2   Roman Mars, "Hacking IKEA," *99% Invisible,* 19 August 2014, and John Postill, "Democracy in an Age of Viral Reality: A Media Epidemiography of Spain's Indignados Movement," *Ethnography* 15, no. 1 (2014): 51–69.

3   David Golumbia, *The Cultural Logic of Computation* (Cambridge: Harvard University Press, 2009), 2.

other. I started following and reading life hacking blogs more or less when they first appeared in the mid 2000s. Convinced there was something there worth studying, something that might help me illuminate certain contradictions at the heart of American culture, I gave my first paper on life hacking at an academic conference in 2007. As life hacking blogs turned into books, I bought and read them. When Twitter took off, I joined and followed prominent life hackers there. I watched life hacking TV shows. I listened to life hacking podcasts. I went all in. But at the same time, I was suspicious of life hacking's promises of increased productivity via technology. Emboldened by life hacking pioneer Merlin Mann's denunciation of life hacks in 2008 and subsequent critiques of life hacking by critics such as Evgeny Morozov and Nikil Saval, I became increasingly disenchanted with the whole concept.[4] Life hacking's shortcuts, I realized, had become dead ends, the metaphor itself problematic.

As a graduate student, one permutation of life hacking I followed especially closely was "prof hacking," in particular ProfHacker (b. 2009), the *Chronicle of Higher Education* group blog "focused on pedagogy, productivity, and technology, and the various ways these intersect in higher education."[5] I began reading it when I was still an anxious graduate student trying to figure out my place in academe. At the time, ProfHacker's advice, tips, and tutorials felt like they were aimed squarely at me and I devoured them eagerly. My comments here concerning the site are thus informed by the years I've spent reading it and wrestling with its advice as both a student and scholar. It's important to understand that ProfHacker is a specific articulation of a larger cultural phenomenon. That is, prof hacking as a practice marks the linkage of the discourse of life hacking with the location of American higher education. I don't think that linkage is an entirely benign one, as I hope to make clear. Namely,

---

4   See Merlin Mann, "Four Years," *43 Folders* (blog), 8 September 2008; Evgeny Morozov, "Down With Lifehacking!" *Slate,* 29 July 2013, and Nikil Saval, "The Secret History of Life-Hacking," *Pacific Standard,* 22 April 2014.

5   "Welcome to ProfHacker," *ProfHacker* (blog), 19 April 2010.

ProfHacker's application of the life hacking concept to academia ultimately seems congruent to me with the oft-remarked upon neoliberalization of the university insofar as it tends to reiterate discourses of productivity, efficiency, and self-improvement. As such, it is something I find intensely problematic at the same time I personally find it tremendously seductive.

That I read ProfHacker religiously and even contributed a post to it once isn't particularly surprising when one considers that my time in graduate school in the late 2000s coincided with the decisive computerization of academia and the explosion of the digital humanities.[6] Now, at the tail end of the second decade of the twenty-first century, one of the most salient aspects of academic life is the influx of digital technologies. Such technologies are not only central to the basic operations of the contemporary university — admissions, registration, financial aid, administration, record keeping, and so on would be unthinkable without them — but, obviously, to research and teaching as well. They are how professors look up articles, create and share work, prepare lectures, post grades, and so forth. These technologies are all invariably pitched in the same way: as tools that will make the entire process of higher education — from teaching and grading to research and publication to communication and coordination — more efficient and productive. In a word, better. Professors are told, in ways both subtle and obvious, they must not only learn how to use new digital tools, but they must learn how to use them well.

During my time as a PhD student at the University of Iowa, for instance, I saw classrooms go from rooms often consisting of little more than desks and whiteboards (or even blackboards) to "wired" classrooms with computers, digital projectors, dependable WiFi, and the like. What I watched play out at Iowa played out at colleges and universities across the country. Today, if an American college classroom is not yet "wired," you can bet that someone, somewhere (an administrator perhaps, or an educa-

---

6   Here's the post I wrote for it: Matt Thomas, "Managing Twitter Favorites," *ProfHacker* (blog), 19 August 2010.

tional technologist, or maybe a technology company) has plans to fix it, to "rescue" it from its "backwardness" and make it attractive enough to parent and student "consumers" that it can be put in a college brochure or shown off on a campus visit. In the eyes of some, a college classroom that doesn't have at least WiFi can now scarcely even be considered a classroom. Students, after all, have to be able to use their devices to connect to the internet. As the campuses of tech companies resemble more and more the campuses of colleges, the campuses of colleges feature more and more of the products of tech companies. As Michael Bugeja notes, "academe has invested heavily in technology since 1995, funding proliferation with easy student loans, higher tuition, and all manner of technology-related fees."[7] The reason for this? Well, as self-described "ed-tech Cassandra" Audrey Watters has relentlessly documented, the idea that new technologies can improve education is a well-worn — if not well-substantiated — one in American culture.[8] And not only are new technologies billed as tools that will help students, but as tools teachers can use to professionalize themselves. Yet one of the more striking contradictions about pitching technologies as "professionalizing," is that learning to use them, let alone master them, often takes a lot of time. As Margaret Cassidy writes in her 2004 book *Bookends: The Changing Media Environment of American Classrooms*:

> Advocates are once again offering the argument that new technology will professionalize teaching — for example, by bringing to the teaching profession the kind of productivity that has purportedly come to professionals in other fields through their use of technology. However, many teachers are experiencing something different when they start to use new technology. The amount of time required to produce appro-

---

7   Michael Bugeja, "E-Tymology of Inefficiency: How the Business World Colonized Academe," in *The Culture of Efficiency: Technology in Everyday Life,* ed. Sharon Kleinman, 173–90 (New York: Peter Lang, 2009), 175.

8   To learn more about Watters's work, see her website: http://hackeducation. com. Her use of the word "hack" here is not lost on me.

priate and valuable curricular materials is staggering. On top of that is the time involved in locating useful Web sites and software programs, evaluating possible software and hardware purchases.[9]

Precisely because they demand so much time of instructors, rather than unambiguously aiding them in their individualized aspirations for prestige through productivity, Cassidy sees these technologies as *de-skilling teachers* by 1) forcing them to rely on other people's content and software and 2) cutting into their time to do things such as research and talk to other teachers, once ordinary parts of the job they now no longer have time for because they are too busy simply trying to stay up to date with the latest software application they are being told is the solution to all their troubles. It is, as one might imagine, a vicious cycle. Here's Cassidy again: "Although using technology might appear, even to teachers, to professionalize their work, it may merely add additional tasks onto an already difficult workload, thus creating a work speed-up that leaves teachers looking for short-cuts and ready-made solutions to their problems."[10] Although Cassidy's focus is on public K–12 education, her overall argument about the hype and hope invested in new technologies is applicable to higher education as well. As media scholar Harold Innis quipped over sixty years ago, "The blight of mechanization spreads from the high schools to the universities."[11]

Here one might reasonably ask: Shouldn't professors make time to learn how to use new computer technologies? Isn't that a good thing? My response to that is to ask a different question: Isn't the pressure to do exactly that overwhelming, even irresistible, in higher education, where digital technologies have almost wholly infiltrated how we communicate, research, write, publish, and teach? Does one even have a choice not to use such

9   Margaret Cassidy, *Bookends: The Changing Media Environment of American Classrooms* (Cresskill: Hampton Press, 2004), 250–1.

10 Ibid., 268.

11 Harold Innis, *The Bias of Communication* (Toronto: University of Toronto Press, 1951), 207.

technologies? For the few holdouts left, there exist institutional structures whose *raison d'être* is getting professors to use the latest tech: educational technologists; administrative initiatives; training sessions; for-profit companies coming into schools and foisting, via incentives or mandates, such technologies on professors. You have to work now not to use the whole assemblage of digital technologies that have taken over academia. Would it be possible, for instance, to teach a college class without using email? To submit grades without going online? To design assignments that don't require computers? Would you even want to? Is this something you even think about? I find that computer access and use is taken as such a given that it is only when my computer is on the fritz or a student tells me they don't have reliable internet access at home that I realize how much I and the classes I teach are dependent on them.

Cassidy is worried about the focus on "issues of implementation and execution,"[12] wherein the teacher is re-imagined more as "a technician that simply helps students use the technology, not a person who selects technology (or some other resource) as a way to help students learn."[13] I worry about that too. But are "prof hackers" similarly worried? For what Cassidy, with an eye to the past, sees as *de-skilling,* they, with an eye to the future, seem to see as skill-enhancing. What might explain this divide?

Since its inception in 2009, ProfHacker has published technology-centric tips and tricks aimed at educators in colleges and universities, especially professors in the humanities and social sciences, i.e., those most often thought of as "backwards," recalcitrant parts in a system in dire need of speeding up and optimizing in the eyes of many administrators, boards, politicians, and critics. Hosted on the *Chronicle of Higher Education*'s website since 2010, and thus benefiting from the *Chronicle*'s imprimatur, ProfHacker is a group blog that, even though it periodically takes pains to assert that it's not only about technology, to date has mainly celebrated and sought to help academics inte-

---

12   Cassidy, *Bookends,* 265.
13   Ibid., 267.

grate the latest technologies into their professional and personal lives. As its name suggests, it is the chief prof hacking blog on the internet, but it is far from the only site applying hacking to academia. Similar contemporaneous blogs include GradHacker (housed at the *Chronicle of Higher Education*'s rival website *Inside Higher Education* since 2014, a fact which suggests how higher education trade publications feel the need to have a hacking "vertical" for the purposes of branding), and HackCollege (b. 2006). If ProfHacker is aimed mostly at professors, GradHacker is aimed mostly at graduate students, and HackCollege is aimed mostly at undergraduates. All, however, are similar in that they apply the metaphor of "hacking" to academic life, and their tips, though aimed at a more specific audience than the tips proffered by general interest life hacking blogs, nevertheless feel of a piece with those blogs. Collectively, they constitute a popular but under-analyzed discursive formation.

In my darker moods, despite its periodic posts acknowledging the structural problems plaguing contemporary academe, I worry ProfHacker works to shift attention from such problems to smaller — some might say trivial — technical matters. In other words, it addresses itself to those looking to change themselves more than it marshals those looking to change the system. Notably, it is when I felt at my most vulnerable as a graduate student that I turned to it and blogs like it for solace in the form of easy-to-follow advice, seeking to change myself so that I might better fit into academe, not changing academe so that it might better fit me. So even though I once contributed to ProfHacker, and ProfHacker has linked to my writing elsewhere, and even though I used to read it regularly, am friendly with several of its contributors, and have found some of its advice useful, I think it's time to pose some neglected questions about it. In so doing, I am not trying to take potshots at individual contributors but, in the spirit of good-natured provocation, offer a critique of ProfHacker as an articulation of a larger propensity. Is the best way to improve academic life to hack it? More importantly, regardless of how useful much of ProfHacker's advice is at the individual level, what are the social and political consequences of propagating it? In whose interest does ProfHacker dispense these tips? And to what end?

As I have already noted, academics, particularly those in the humanities and social sciences, are increasingly pressured to use digital technologies so as not to become "obsolete." I would argue that ProfHacker both responds to and perpetuates this pressure with the largely tech-centric advice it has offered to date. Take, for instance, seemingly harmless posts like "Using Google Forms for In-class Polling" (4/10/2016), "Create a New Habit with an App" (2/25/2016), "Managing References with Paperpile" (1/27/2016), "Tune In to Focus at Will" (12/3/2015), "Preparing Lectures for Large Online Classes" (9/14/2015), "Dropbox's File Request Eases Receiving Files and Assignments" (9/8/2015) and "How (and Why) to Generate a Static Website Using Jekyll, Part I" (8/31/2015).[14] Though their particulars differ, all of these posts are essentially about how already put-upon professors can do more via technology. "Prof hacking" writ large might thus be seen as a response to changing socioeconomic and technological conditions that's congruent with those conditions themselves. And though it's invariably presented in a can-do way, the advice ProfHacker dispenses is in line with the so-called neoliberalization of the university insofar as it addresses the individual, professional academic looking to improve him or herself, generally via technology, more than the larger structural circumstance academics now find themselves embedded in, even if it doesn't deny that those structural circumstances exist.

Part of ProfHacker's mission seems to be to help what might be dubbed the "hack-curious" academic transform

14 See Amy Cavender, "Using Google Forms for In-class Polling," *ProfHacker* (blog), 10 March 2016; Natalie Houston, "Create a New Habit with an App," *ProfHacker* (blog), 25 February 2016; Amy Cavender, "Managing References with Paperpile," *ProfHacker* (blog), 27 January 2016; Natalie Houston, "Tune In to Focus at Will," *ProfHacker* (blog), 3 December 2015; Anastasia Salter, "Preparing Lectures for Large Online Classes," *ProfHacker* (blog), 14 September 2015; Jason B. Jones, "Dropbox's File Request Eases Receiving Files and Assignments," *ProfHacker* (blog), 8 September 2015; and "How (and Why) to Generate a Static Website Using Jekyll, Part I," *ProfHacker* (blog), 31 August 2015.

into a full-fledged "hackacademic," the result of which is supposedly marked gains in productivity. Though billed as non-
disciplinary,[15] ProfHacker, in the time I've been reading it, seems
to be stealthily addressing the aforementioned academics in the
humanities and social sciences whose curiosity about technology is rooted in a fear of being outpaced and obsolesced. In
short, ProfHacker is not for professors who are already hackers,
so much as it is, one post at a time, trying to turn professors into
hackers by sharing with them a series of hacker-esque productivity tips.

Conspicuously absent from ProfHacker's posts as of this writing are extended criticisms of technology. This is not to suggest
that those who write for ProfHacker are not themselves reflec-
tive about the technology they let into their lives, or that none of
its posts grapple with technology's Faustian bargains (a January
2012 post, for instance, admits that it would be impossible to
"try out *all* ProfHacker recommendations and still maintain careers and families"[16]). Rather, that from its inception ProfHacker
has hailed a particular kind of reader, one who uses and cheers,
and teaches others to use and cheer, digital technology, not one
who asks pesky questions about its deeper rhetorical sources
and structures. In other words, its attitude is more "Digital technologies are shiny new toys that are fun to play with" than "Digital technologies are perhaps something we need to be wary of."

But this, according to cultural critic and educator Neil Postman, whose words serve as an epigraph to this essay, gets it
backwards. From his point of view, the sort of technical tips put
forward by ProfHacker are essentially trivial. What we need to
know about technologies, he argues, "is not how to use them
but how they use us." My sympathies here are with Postman. He
uses the example of cars. "In the case of cars," he writes, "what
we needed to think about in the early twentieth century was
not how to drive them but what they would do to our air, our

---

15 Jason B. Jones, "Welcome to ProfHacker.com (Open Thread Wednesday),"
ProfHacker (blog), 9 September 2009.

16 Ryan Cordell, "Avoiding Tool Takeover," *ProfHacker* (blog), 16 January 2012.

landscape, our social relations, our family life and our cities." "I am talking here," he concludes, "about making technology itself an object of inquiry."[17] I feel like Postman, if he were still alive, would subscribe to, as I do, Thomas Haigh's (and others') up-ended version of digital humanities: a digital humanities that seeks to "apply the tools and methods of the humanities to the subject of computing" rather than vice versa.[18] For Postman was "more interested in asking questions about the computer than getting answers from it."[19] ProfHacker's interest, by contrast, lies more in the other direction: with getting answers from computers not asking questions about them. Tellingly, as of this writing, the most popular category on the blog is "Software."

While ProfHacker occasionally features a post about one structural problem in academe or another, what it never quite gets to is a more complicated and troubling question: Are digital technologies partially responsible for, or at least congruent with, the labor situation in which professors are increasingly asked to use digital technologies? And that's fine. If it doesn't want to ask this question, it doesn't have to. But ProfHacker does seem concerned about academic exploitation and imagines its tips as empowering. Its overall approach, however, suggests a reluctance to think through how technological and economic changes are linked, which is a question I am preoccupied with and feel is important to advance.

Perhaps the closest ProfHacker has gotten to this since I've been a reader is a June 11, 2010 post that asked "To what extent are edupunks, DIY faculty, and, heck, we ProfHackers, useful idiots in the destruction of higher education?"[20] The post links to a

---

17 Postman, "Virtual Students," 206–7.

18 Thomas Haigh, "We Have Never Been Digital," *Communications of the* ACM 57, no. 9 (2014).

19 Postman, "Virtual Students," 207. See also the six questions Postman thinks should be asked about any new technology in his *Building a Bridge to the 18th Century* (New York: Vintage, 2000), 42–53.

20 Jason B. Jones, "Weekend Reading: State of Higher Education edition," *ProfHacker* (blog), 11 June 2010.

post by Jim Groom on his personal blog from the day before.[21] In his post, Groom is responding to an article by blogger and law professor Glenn Reynolds in which Reynolds expresses enthusiasm for "edupunks" because they're doing things like teaching students how to code — i.e., in his estimation, giving them skills that might lead to gainful employment. Groom, the coiner of the term "edupunk," is put off by Reynolds's careerist thinking. Groom writes: "What we are seeing is the gentrification of higher ed as an impulse to razing public education though the liberatory rhetoric of innovation and efficiency — only to have the process devoured by the wolves of the free market."[22] But what seems to irk Groom the most is how Reynolds sees the work of "edupunks." He writes: "Reynolds understands the 'edupunks' as the useful idiots who very well may help bring the public education system down." Groom, while wanting the term "edupunk" to remain polysemic, takes issue with Reynolds's interpretation: "an EDUPUNK that devastates public education in service to the unregulated promise of free markets and capital is possibly the worst vision one can imagine." Basically, Groom is worrying out loud about how a term he coined is being co-opted. "To what degree," he wonders, "is the dream vision of DIY U a means of further gutting the salaries, rights, and benefits of educational professionals?" By linking to Groom's piece, ProfHacker is, by extension, asking this question as well, albeit indirectly. ProfHacker has a tendency to link to articles and blog posts that raise similarly thorny questions, but then not answer them. Indeed, ProfHacker does not answer Groom's question, either in the post that links to it or anywhere else I know of. Nevertheless, this represents a time ProfHacker raised the question of its role in the "destruction of higher education," and as such, it is important to note.

Frankly, it's the sort of question I wish ProfHacker would grapple with more. Higher education, especially in public uni-

---

21 Jim Groom, "EDUPUNK or, on Becoming a Useful Idiot," *bravatuesdays* (blog), 10 June 2010.

22 Ibid.

versities, has been in an increasingly obvious and remarked upon state of crisis since the Great Recession of the late 2000s. This crisis is marked by things such as budget cuts at public universities; the miserable job market for newly minted, debt-ridden PhDs; the adjunctification of faculty; and the erosion of tenure. Collectively, these things are often referred to as the cor-poratization or neoliberalization of the university. Myriad and growing books, articles, blog posts, and symposia have regis-tered and discussed this shift.

The "collapse" of academe has been marked at the same time by, as I started to sketch out above, a huge investment in tech-nology. We don't have the money to take care of professors, but we do have the money to take care of computers. How can this be? Or as historian of technology David Noble asks in his 2001 book *Digital Diploma Mills,* "What drives this headlong rush to implement new technology with so little regard for deliberation of the pedagogical and economic costs and at the risk of stu-dent and faculty alienation and opposition?" "A short answer," he writes, "might be the fear of getting left behind, the incessant pressures of 'progress.'"[23] Yet Noble's answer here, as much as it would seem to explain the "hack curious" hailed by a blog like ProfHacker, is incomplete. It is not simply a fear of being left behind that explains why ProfHacker continues as a fixture of the *Chronicle of Higher Education,* and similar blogs like Grad-Hacker and HackCollege continue to pump out tip after tip day after day, week after week, year after year. As Noble writes, "For the universities [in the 1990s] were not simply undergo-ing a technological transformation. Beneath that change, and camouflaged by it, lies another: the commercialization of higher education. For here as elsewhere technology is but a vehicle and a disarming disguise."[24]

Noble's treatment of this issue, though from 2001, is useful precisely because it is historical. His insight is to historicize the

23 David Noble, *Digital Diploma Mills: The Automation of Higher Education* (New York: Monthly Review Press, 2001), 26.
24 Ibid., 26.

corporatization/neoliberalization — or what he calls the "commercialization" — of academe and connect it back to technology. For him, the crises confronting academe and the technologization of academe are linked. According to Noble, the commercialization of academia happened in two separate but overlapping stages: first research, then instruction. Put differently, first research became something one could profit from, then teaching became something one could profit from. Noble's framing here gives us, perhaps, a clue as to why so many of Prof-Hacker's tips and tricks relate to teaching. The commercial potential, it is probably fair to say, of a lot work in the humanities and social sciences is low, but professors in the humanities and social sciences can participate in the commodification of the education function of the university through technology without much difficulty. They can, for instance, offer tech-centric tips on the *Chronicle of Higher Education*'s website to fellow academics looking to "upgrade" themselves. The *Chronicle of Higher Education,* in turn, gets more hits on its website. And it's a vicious cycle because the tips keep coming. There is always another "hack" to learn. As Noble explains, pointing a finger at ed-tech boosters directly:

> Ignoring the true sources of the financial debacle — an expensive and low-yielding commercial infrastructure and greatly expanded administrative costs — the champions of computer-based instruction focus their attention rather upon increasing the efficiencies of already overextended teachers. And they ignore as well the fact that their high-tech remedies are bound only to compound the problem, increasing further, rather than reducing, the costs of higher education.[25]

Noble was writing nearly a decade before ProfHacker launched, but is not ProfHacker in many ways an attempt to use various technologies to increase the efficiencies of already overextended academics? What Noble realized that ProfHacker seems loath

---

25   Ibid., 28.

to discuss is that by acquiescing to the high-tech, higher-ed paradigm, one becomes weirdly complicit in one's own obsolescence. ProfHacker is arguably giving prestige to the technologies that are undermining professors. And what is especially ironic is that it does so at precisely the same time professors are trying to avoid being obsolesced. Perhaps the problem is not so much that academics are inadequately hacking themselves; perhaps the problem is with hacks themselves, specifically the various structural pressures that make them seem like the answer, which is how I tended to see them as a graduate student.

It has been said that any discussion of the role of technology in higher education is also a discussion about labor relations. Suffice it to say, ProfHacker does not present its tips as being dictated from on high. Just the opposite. Its tips are presented as stratagems contributors have chosen to employ on their own for the benefit of themselves and their students. ProfHacker does not ask, however, why so many stratagems have to be employed in the first place, or why constant self-improvement along technological lines is put forward as the only reliable insurance against the changes happening in academe. Ultimately, one of the more troubling things to me about ProfHacker is how it has internalized and repackaged for consumption what might be dubbed managerial values as tips and tricks coming from the "bottom up." Not only are academics being asked to do it themselves, they are being asked to do it *to* themselves. As Jim Groom's post suggests, many ProfHacker-adjacent academics — those in digital humanities, advocates of open access publishing, and so on — see themselves on the vanguard of a revolution in higher education. What I am suggesting, however, is that the embrace of technology as a category for solving problems inevitably and regrettably aligns one with neoliberal forces.

Academe as a whole would be better served, in my view, by a more philosophical and skeptical attitude toward technology, one that attempts to think through how structural issues and digital technologies might be connected à la Noble. Insofar as ProfHacker pays lip service to collective action but then turns around and encourages professors, like other professional workers, to see the changes wrought by neoliberalization as problems they have to work out themselves by better managing themselves through computer technology, it undercuts itself. Even if, at some level, everyone acknowledges that much of what we experience as personal problems are in fact deeply rooted and broad-ranging sociocultural ones, individualist, technocratic solutions are still the order of the day. This frustrates me to no end. Constant self-improvement via technology — often coded as "personal/professional development" — is the solution people keep coming back to. ProfHacker can't help but contribute to this state of affairs with its tips. It presents the needy, desperate, overstressed, and overburdened contemporary academic subject to neoliberal pressures with the apparent opportunity to do things more quickly and more easily. It says, in effect, "Hey, you've got all this work you have to do now. Let me show you how to manage it all more easily using this new piece of software." Now, I will be the first to admit that I used to find this sort of stuff incredibly attractive, and to some extent still do. If I read, for instance, that I should be automating my writing with TextExpander scripts or writing in Markdown or using Zotero as a reference manager, to use three actual examples from Prof-Hacker that I've incorporated into my "academic workflow," I will think, "Yes, of course I should be doing these things."[26] Every post makes me ask, "Am I being as productive and efficient as I can be?" But while I'm questioning myself, these kind of posts don't question why academics might be feeling like their tools are inadequate and how such feelings might be related to changing conditions of academic labor, which are mostly structural, not individual, in nature. In this way, a site like ProfHacker contributes to the problems it purports to help solve. By turning the focus inward, away from larger systemic issues and to one's own habits, it encourages a form of academic navel-gazing

---

26 See, respectively, Jason B. Jones, "Automating Writing with TextExpander Scripts," *ProfHacker* (blog), 8 October 2015; Lincoln Mullen, "Markdown: The Syntax You (Probably) Already Know," *ProfHacker* (blog), 12 August 2011; and Amy Cavender, "ProfHacker 101: Getting Started with Zotero," *ProfHacker* (blog), 13 November 2009.

that eclipses more collective worries. After all, who has time to remake academe when they're busy playing around with their computers all day?

Let me be clear: I'm not saying we should go back to a pre-computerized academia (though that's an interesting thought experiment), or that we should leave professors looking for advice about how to use computers hanging, or even that Prof-Hacker should close shop, merely that we need to step back from the incessant tips and tricks and ask ourselves the more meta question of why such tips and tricks seem to be so appealing to us. The answer to that question might lead us into a conversation less about how we can use technology better, and into a more Postman-esque one about how technology might be using us, and how we might be able to resist it.

Bibliography

Bugeja, Michael. "E-Tymology of Inefficiency: How the Business World Colonized Academe." In *The Culture of Efficiency: Technology in Everyday Life,* edited by Sharon Kleinman, 173–90. New York: Peter Lang, 2009.

Cassidy, Margaret. *Bookends: The Changing Media Environment of American Classrooms.* Cresskill: Hampton Press, 2004.

Cavender, Amy. "ProfHacker 101: Getting Started with Zotero." *ProfHacker* (blog). 13 November 2009. http://chronicle. com/blogs/profhacker/profhacker-101-getting-started-with-zotero/22829.

———. "Managing References with Paperpile." *ProfHacker* (blog). 27 January 2016. http://chronicle.com/blogs/ profhacker/managing-references-with-paperpile/61592.

———. "Using Google Forms for In-class Polling." *ProfHacker* (blog). 10 March 2016. http://chronicle.com/blogs/ profhacker/using-google-forms-for-in-class-polling/61823.

Cordell, Ryan. "Avoiding Tool Takeover." *ProfHacker* (blog). 16 January 2012. http://chronicle.com/blogs/profhacker/ avoiding-tool-takeover/37948.

Golumbia, David. *The Cultural Logic of Computation.* Cambridge: Harvard University Press, 2009.

Groom, Jim. "EDUPUNK or, on Becoming a Useful Idiot." *bravatuesdays* (blog). 10 June 2010. http://bavatuesdays. com/edupunk-or-on-becoming-a-useful-idiot/.

Haigh, Thomas. "We Have Never Been Digital." *Communications of the ACM* 57, no. 9 (2014). http://cacm. acm.org/magazines/2014/9/177930-we-have-never-been-digital/abstract.

Houston, Natalie. "Tune In to Focus at Will." *ProfHacker* (blog). 03 December 2015. http://chronicle.com/blogs/ profhacker/tune-in-to-focus-at-will/61367.

———. "Create a New Habit with an App." *ProfHacker* (blog). 25 February 2016. http://chronicle.com/blogs/profhacker/ create-a-new-habit-with-an-app/61784.

Innis, Harold. *The Bias of Communication.* Toronto: University of Toronto Press, 1951.

Jones, Jason B. "Welcome to ProfHacker.com (Open Thread Wednesday)." *ProfHacker* (blog). 9 September 2009. http://chronicle.com/blogs/profhacker/welcome-to-profhackercom-open-thread-wednesday/22689.

————. "Weekend Reading: State of Higher Education edition." *ProfHacker* (blog). 11 June 2010. http://chronicle.com/blogs/profhacker/weekend-reading-state-of-higher-education-edition/24703.

————. "Dropbox's File Request Eases Receiving Files and Assignments." *ProfHacker* (blog). 8 September 2015. http://chronicle.com/blogs/profhacker/dropboxs-file-request-eases-receiving-files-and-assignments/61071.

————. "Automating Writing with TextExpander Scripts." *ProfHacker* (blog). 8 October 2015. http://chronicle.com/blogs/profhacker/automating-writing-with-textexpander-scripts/61163.

Mann, Merlin. "Four Years." *43 Folders* (blog). 8 September 2008. http://www.43folders.com/2008/09/08/four-years.

Mars, Roman. "Hacking IKEA." *99% Invisible.* 19 August 2014. http://99percentinvisible.org/episode/hacking-ikea/.

Morozov, Evgeny. "Down With Lifehacking!" *Slate.* 29 July 2013. http://www.slate.com/articles/technology/future_tense/2013/07/lifehacking_is_just_another_way_to_make_us_work_more.html.

Mullen, Lincoln. "Markdown: The Syntax You (Probably) Already Know." *ProfHacker* (blog). 12 August 2011. http://chronicle.com/blogs/profhacker/markdown-the-syntax-you-probably-already-know/35295.

Noble, David. *Digital Diploma Mills: The Automation of Higher Education.* New York: Monthly Review Press, 2001.

Postill, John. "Democracy in an Age of Viral Reality: A Media Epidemiography of Spain's Indignados Movement." *Ethnography* 15, no. 1 (2014): 51–69. DOI: 10.1177/1466138113502513.

Postman, Neil. "Virtual Students, Digital Classroom." In *Minutes of the Lead Pencil Club: Pulling the Plug on the Electronic Revolution,* edited by Bill Henderson, 197–215. Wainscott: Pushcart Press, 1996.

———. *Building a Bridge to the 18th Century.* New York: Vintage, 2000.

ProfHacker (blog). "Welcome to ProfHacker." 19 April 2010. http://chronicle.com/blogs/profhacker/welcome-to-profhacker/23211.

———. "How (and Why) to Generate a Static Website Using Jekyll, Part I." 31 August 2015. http://chronicle.com/blogs/profhacker/jekyll1/60913.

Thomas, Matt. "Managing Twitter Favorites." *ProfHacker* (blog). 19 August 2010. http://chronicle.com/blogs/profhacker/managing-twitter-favorites/26119.

Salter, Anastasia. "Preparing Lectures for Large Online Classes." *ProfHacker* (blog). 14 September 2015. http://chronicle.com/blogs/profhacker/prep-lectures-large-online/61087.

Saval, Nikil. "The Secret History of Life-Hacking." *Pacific Standard.* 22 April 2014. http://www.psmag.com/business-economics/the-secret-history-of-life-hacking-self-optimization-78748.

12

Digital Humanities and the Erosion of Inquiry

Sean Michael Morris[1]

There's something die-hard about academics. When you think about it, there's nothing more unlikely than leading a life devoted to knowledge, to discovery and inquiry, to curiosity. It's not practical. Not many parents want their kids to grow up to be scholars. We'd rather our kids have security, stability — an education, yes, and a job and a house and a family, yes. But few of us say "I want you to spend the rest of your life in a library carrel, carrying on conversations arcane to other people." Academics hoe a row that may or may not yield a crop. It's risky, doing this work. And yet there's nothing else we'd like to be doing.

There's a lot of talk about the liberal arts and the humanities being in peril, or that scholarship itself is in profound crisis. To this, I have said that "If higher education is ailing, it is only because its many doctors have not applied themselves to its resuscitation." There is no better solution to the problems scholarship faces than its professors, adjuncts, and students. People are the solution to whatever ails the profession. Each of us has agency

---

1 Parts of this piece are adapted from a piece originally published on *Hybrid Pedagogy.* See Sean Michael Morris, "Digital Humanities and the Erosion of Inquiry," *Hybrid Pedagogy,* 12 February 2016.

we can apply — not only to create our own success, but to ensure the success of those who come after us.

Agency is tricky, though, especially within the reputational economy of higher education. Too often it seems that in order to enjoy reputational gains — cool publishing opportunities, speaking gigs, keynotes, better and better project opportunities, grants — agency must be set aside. The institution itself, as well as the bodies that fund it, delight in setting up obstacle courses, hoops for jumping through, rhetorics to echo, expectations which must be met that have nothing or little to do with our agency, with our passions and excitement. Projects that we love become dreams we must negotiate or defer until the system which holds purview is satisfied.

Or, as Simon Ensor writes, until we learn to speak dog.[2]

In kindergarten we dream about the agency afforded the sixth grader. In the sixth grade, we dream of the agency of high schoolers to drive cars and eat lunch off campus. In high school, we dream — well, we dream about sex… but we also dream about the agency promised us in the offing: leaving home and going to college. And it goes on. We work for our PhD only to have to further graduate from being junior faculty, only to have to work for tenure and the title of professor. And even then — if we're not exhausted — there are bodies that judge and govern and peer-review our work until it can seem like we never left kindergarten.

And it's not a kind governance we encounter.

The Tumblr site "Shit My Reviewers Say" offers an example. Reality so absurd that we don't even need parody.[3] The site invites "real sentences from reviews you received […] the harsh, the weird, the passive aggressive, the active aggressive and the downright mean."[4] For example, "No new insights, no impor-

---

2   Simon Ensor, "My Lawyer is a Dog," *touches of sense…* (blog), 27 January 2016.

3   *Shit My Reviewers Say* (blog), *Tumblr,* http://shitmyreviewerssay.tumblr.com/.

4   "About/Submit," *Shit My Reviewers Say,* http://shitmyreviewerssay.tumblr.com/submit.

tant question addressed, no problem solved."[5] A single blunt sentence fragment. Others are more overtly cruel.

This is not what scholarship is for. This is not the risk we were meant to take. Scholarship should be expressive, experimental, and liberatory. And it should land within a community of support rather than a pit of critics.

If it's not clear, what I'm saying is that the systems of rigor that we've created, and that we submit to, and which purport to elevate us — in fact oppress. The academy, through some trick of mass hypnosis, makes us dependent upon its reputational economy. For marginalized people — women, people of color, queer, or trans people — this struggle is even more poignant. Women forego pregnancy to achieve tenure. Families are split in order to secure a "good" job. We strive harder and harder to meet the expectations of the academy, but rarely receive praise. The reputational economy is unforgiving. From kindergarten through the writing of a dissertation, we wait upon the satisfaction of others, a nod, and permission to speak. Indeed, the final step in our long life of study is not to present a dissertation, but to *defend one.* A dissertation should be met with applause, not with a defense.

Years and years and years go by and the most consistent message we get from the academy is to sit down, and shut up.

And that's why *Hybrid Pedagogy* was founded. To say instead, no. Stand up, and speak.

Both Jesse Stommel and I have watched teachers and students be silenced, cowed into conformity, broken by the need to please their peer reviewers, their instructors, and their administrations. *Hybrid Pedagogy* is an effort — however small, however emergent — to provide a space where academic voices can be heard in important, authentic ways.

The journal's collaborative peer review process provides authors with a supportive editorial team, partners in an effort to amplify their voices — amplify, rather than reduce — and to

---

5   Link to the post: http://shitmyreviewerssay.tumblr.com/post/138673489984/no-new-insights-no-important-question-addressed.

broadcast them to the widest possible audience. And the journal unabashedly promotes each of its articles in order to give authors a broad stage from which to address that audience.

Taking that stage isn't always easy. The academic stage is most usually crowded with experts, collaborators, citations and references. Scholars aren't used to an open spotlight, broad and bright, one that is specifically their own. Moreover, they aren't accustomed to being told they are *good* writers. Some of them have a very hard time breaking free from the style of writing that's been branded onto their skin — the peculiar rigor that requires more *absence* of the writer's own voice and perspective and insists on the *presence* of the voices of esteemed others. The removal of the first person pronoun from our work is a violence that leaves us utterly unsure of our own expertise, our own genius.

What happens when we put the "I" back into our work is really nothing short of the return of the human to that work. We forget to oppress ourselves. And the next step after that is to begin to forget to oppress others. Because in truth, we are all always almost about to oppress.

So much of academic work aims at conformity. Even as we push against the oppression of the academy, we recycle and reuse that oppression in our relationships with others. As we work with one another, we frame relationships with expectations. We install and enforce — even unknowingly, even unwillingly — standards for participation in the community.

This is especially prevalent in the project of the digital humanities. Not only does funding require conformity, not only does prestige rely upon it, but we keep the gates of our relationships by those standards. We align ourselves with the "right" people, we collaborate on the "right" projects. We do not spend our sabbaticals breaking molds, but building them up. Risking otherwise leads to criticism at best, excommunication from our communities at worst. We don't just peer review the work of the field, we peer review its people.

For example, can we look at Adeline's Koh's *Sabbatical Beauty* project as DH?[6] Well known for her work as a Digital Humanist, Adeline spent her sabbatical developing a line of beauty products aimed specifically at the academic who has little time for self-care. It's a decidedly feminist project, not at all divorced from the politics of identity, and a project for which she has had to rely on the skills she's developed as a humanities and technology scholar. But are we tempted to look down our nose at her inventiveness? Are we tempted to peer review her life as we would a research article from her? She says that

> Academic culture asks you to champion some ways of thinking over others (in the humanities: capitalism/neoliberalism = bad! not getting a tenure-track job at a research institution=failure), in ways which are often completely uncritical, but imperative for one to fit into the culture.[7]

The institutionalization of the digital humanities has made it largely inaccessible to those who remain outliers to the institution. As DH has grown in prominence, as it has become what William Pannapacker once called "the first 'next big thing'"; it's also become all too discriminating about what and whom the field may include. And as the academy is wont, it has forgotten those upon whose backs the digital humanities was built. Twitter activist @so_treu, responding to "Hybrid Pedagogy, Digital Humanities, and the Future of Academic Publishing," an article by Jesse and me, wrote:

> Academia with this digital humanities push is rushing to catch up with centuries old practices of marginalized wmn / & really, academia made itself via the exclusion/delegitimizing of these kinda open grassroots scholarship practices.

---

6   *Sabbatical Beauty,* https://sabbaticalbeauty.com/.

7   Adeline Koh, "Academia, You Don't Own Me Any Longer: Or, Why I Started a Small Business While on Sabbatical," *Medium,* 9 February 2016.

/ Now it's ready to acknowledge them, but only via the approved bodies & positionalities.[8]

The perfect unfairness of this is that when we limit another's voice, we create an ecosystem reliant upon limitation. And that's the ecosystem in which we now swim. We can't ask for freedom to take risks, to follow our own curiosity — to be scholars — if we don't offer that freedom to others.

Where DH grew out of positions of deep and necessary inquiry — deep and necessary especially in that its early advocates had to form communities of practice, interest, and support beyond the pale of traditional academic communities — today that inquiry has eroded into gratuitous and massively funded career-building projects. Not only has digital humanities exhaled its sense of urgency, but in doing so it has lost its soul, its spirit, and its ecstatic necessariness.

Interestingly, it's actually the shackles of rigor and rules of participation — the burdened infrastructure that we build — out of which can come something vital. When people are oppressed, expression rises up. Sometimes violently, sometimes rudely, sometimes in quiet ways that catch us by surprise. When they are unrepresented, when they are oppressed and voiceless, humans find new ways to speak, new forms, new words. Few of us think about scholarship as expression created under duress, but the best of it is. The best of scholarship rises from our need to speak, and to have our observations heard.

Just as a journal run by teachers is — or should necessarily be — a classroom, so must scholarly fields be rich with dialogue. We must open our infrastructure to let in the unexpected and the curious if we want our own curiosity to thrive. The digital humanities will not survive without its collaborators.

8   @so_treu, Twitter post, 3 January 2015, 9:22 a.m.

Turmoil as Process

Paulo Freire writes of how the oppressed — or in the case of the digital humanities, the outlier — must undo their own oppression. The outlier, she whose work is not seen, not heard, not viewed, who is precarious precisely because of her location outside prescribed bounds — of acceptability, of rigor, of form, of content, of perspective, of political orientation, of gender, of color, of credential, of imagination or invention, of syntax, of mediocrity — she must make her own voice heard.[9] In some ways, this is how the Digital Humanities got its start. As Jesse Stommel and Dorothy Kim say in their introduction to the *Disrupting the Digital Humanities* project,

> Many scholars originally were drawn to the Digital Humanities because we felt like outcasts, because we had been marginalized within the academic community. We gathered together because our work collectively disrupted the hegemony and insularity of the "traditional" humanities.[10]

This is where I see the value of digital humanities: in its knack — indeed, its proclivity — for tearing apart, breaking down, for parsing all the way to the marrow, the code behind the text and image. There is something organically extra-institutional, even anti-institutional, about the quizzical approach of digital humanities. It is as much a practice of inspection as it is of invention. The hack is its pedagogy. At its best, it is playful and original, daring but not decadent, irreverent even of its own tenets. It is its own pain in the ass. And when it's not, it fails to be noteworthy.

I would say that the pedagogy of digital humanities must be comfortable with its own discomfiting processes. This is

---

9   See Kathi Inman Berens, "Want to 'Save the Humanities'? Pay Adjuncts to Learn Digital Tools," *Disrupting the Digital Humanities: Digital Edition,* 5 January 2015.

10 Dorothy Kim and Jesse Stommel, "Introduction: Disrupting the Digital Humanities," *Disrupting the Digital Humanities: Digital Edition.*

my common soapbox, an aspirational rebellion of intellect and heart that reforms education upon a post-digital landscape. And when speaking about the nature of disruption it is all too tempting to glide into superlative anxiety and deliver the stump speech about change and transformation... and disruption. To some extent, more stump speeches are necessary, more manifestos are due; to assume that digital humanities has shaped itself, clean and done, is to presume there is no contention in the field at all. But it is the field's contentious nature that provides its most fertile manure. Resistance to the awkwardness of argument and dissent is a whitewashing maneuver designed to keep only certain voices dominant.

There must never be dominant voices in digital humanities. We must walk always along the precipice of radical inclusion — not so that we may include others in the conversations we're having, but so that we may have the chance to be included in their conversations.

Inclusion is disruptive precisely because it does not level the playing field; rather, it points out how uneven that field is, and also that the game we're playing may not be the right game at all. Part of the effort of digital humanities, if its aim truly is to flourish, must be outreach. To focus on individual projects, to exhaust our muscles on the work of just a few, is to sacrifice the field itself on the sword of individual reputation and career-building. Instead, we must absent ourselves, willingly, intentionally leaving space for others to speak. For the field to develop at all, it must look outwardly; and it must not cower before the havoc which that may wreak.

What I find valuable about digital humanities is not its products, not its notaries, but its processes — the dialogues, the bickering, the fastidious attention to the discourse as it struggles with issues of race and gender, equality and oppression, multilingualism and ethnicity, play and seriousness.

There is very little point in stoic civility. A commitment to disruption is a commitment to breaking our own selves upon the matter of our work — humorously, dangerously, compassionately, bravely, fiercely. We must not cow to the standards of the academic hegemony, but must rail against them. When digital humanists eschew performance and bare instead their knuckles, I am reminded what academia is for, and what it can do for me.

My hope, however naive, is that digital humanities might attune us more completely to our humanity, finding the muscle and smile, the delicacy and humor, the beauty and the viscera in our digital presences — to focus upon how the humanities plays against the backdrop of the digital, yes, but also upon how the digital makes us ever increasingly human.

Bibliography

Berens, Kathi Inman. "Want to 'Save the Humanities'? Pay Adjuncts to Learn Digital Tools," *Disrupting the Digital Humanities: Digital Edition*. 5 January 2015. http://www. disruptingdh.com/want-to-save-the-humanities-pay-adjuncts-to-learn-digital-tools/.

Ensor, Simon. "My Lawyer is a Dog." *touches of sense…* (blog). 27 January 2016. http://tachesdesens.blogspot.com/2016/01/my-lawyer-is-dog.html.

Jones, Juju (@so_treu). Twitter feed. https://twitter.com/so_treu

Kim, Dorothy, and Jesse Stommel. "Introduction: Disrupting the Digital Humanities." *Disrupting the Digital Humanities: Digital Edition*. http://www.disruptingdh.com/.

Koh, Adeline. "Academia, You Don't Own Me Any Longer: Or, Why I Started a Small Business While on Sabbatical." *Medium*. 9 February 2016. https://medium.com/@adelinekoh/academia-you-don-t-own-me-any-longer-or-why-i-started-a-small-business-while-on-sabbatical-5605f7087d66#.veocyvdwg.

Morris, Sean Michael. "Digital Humanities and the Erosion of Inquiry." *Hybrid Pedagogy*. 12 February 2016. http://www.digitalpedagogylab.com/hybridped/digital-humanities-erosion-of-inquiry/.

Sabbatical Beauty. https://sabbaticalbeauty.com/.

Shit My Reviewers Say (blog). *Tumblr*. http://shitmyreviewerssay.tumblr.com/.

#transform(ing) DH Writing and Research: An Autoethnography of Digital Humanities and Feminist Ethics

Moya Bailey

As I began to work on my book, *Transforming Misogynoir: Black Women's Digital Resistance in US Culture,* I prioritized transparency. I used my own blog and the Digital Humanities blog at Northeastern University to provide readers with insights into the academic publishing process and my efforts to shape this process into a more just experience for my research collaborators.[1] My research highlights the networks contemporary Black trans women create through the production of digital media and in this article I make the emotional and uncompensated labor of this community visible. I provide an added level of insight into my research process as a way to mirror the access I was granted by these collaborators. I use Digital Humanist Mark Sample's concept of *collaborative connections* to demonstrate my own efforts to enact a transformative feminist process of writing and researching in the Digital Humanities (DH) while highlighting

---

1   Moya Bailey, "I'm Back!" blog post, 21 October 2014.

the ways in which the communities I follow are doing the same in their spheres of influence.[2]

The networks built through digital media production are significant attempts to redress the lack of care that Black Trans women receive from the healthcare community and society. I argue that these processes of digital media production produce more than just redefined representations but connections that can be understood as a form of health care praxis themselves. To reach these conclusions I have charted a new methodology that incorporates theoretical perspectives from Black queer theory, digital humanities, and feminist theory as well as transform my relationship to the people producing these digital representations.

Media and cultural studies scholars have long understood the epistemological and pedagogical significance of popular media.[3] Similarly, marginalized groups have often used media production to challenge dominant scripts within mainstream outlets and the rise of digital platforms makes this task even easier. Black trans women's use of Twitter, an existing digital media platform, creates new and alternate representations as a practice of health promotion, self-care, and wellness that challenge the ways they are depicted in popular culture. I focus on a digital project by Black trans women that involves the collaborative creation of images, links, and other digital media that trouble problematic representations through a curation process that also works to enrich the lives of those participating. I look at trans advocate Janet Mock's twitter hashtag #girlslikeus and discuss the many issues of trans women's collective survival signaled via the tweets marked by the tag.

2   Mark Sample, "Building and Sharing (When You're Supposed to Be Teaching)," *Journal of Digital Humanities* 1, no. 1 (Winter 2011).

3   Stuart Hall, "What Is This 'Black' in Black Popular Culture?" *Social Justice* 20, nos. 1–2 (1993): 104–14; Amelia Jones, ed., *The Feminism and Visual Culture Reader* (London and New York: Routledge, 2003); Chris Jenks, ed., *Visual Culture* (London and New York: Routledge, 1995); Joy Sperling, "Reframing the Study of American Visual Culture: From National Studies to Transnational Digital Networks," *The Journal of American Culture* 34, no. 1 (2011): 26–35; and Myra Macdonald, *Representing Women: Myths of Femininity in the Popular Media* (London and New York: E. Arnold, 1995).

I focus on Mock's use of the hashtag because I sought and achieved her permission to work on the project. I parse my process for achieving informed consent and how it differs from the paternalism of the Institutional Review Board (IRB) process. I acknowledge the negotiated terms under which this project is discussed to signal my own queer feminist praxis in conducting this research. I build towards an understanding of what I call *digital alchemy* as health praxis designed to create better representations for those most marginalized in the Western biomedical industrial complex through the implementation of networks of care beyond its boundaries.

Alchemy is the "science" of turning regular metals into gold. When I discuss *digital alchemy* I am thinking of the ways that women of color, Black women in particular, transform everyday digital media into valuable social justice media magic that recodes failed dominant scripts. Digital alchemy shifts our collective attention from biomedical interventions to the redefinition of representations that provide another way of viewing Black queer and trans women. I argue that this process of redefining representation challenges the normative standards of bodily representation and health presented in popular and medical culture.

Connection through #girlslikeus

Black trans women negotiate unique threats to life and health as those multiply marginalized by gender, race, sexuality, and the disproportionate amounts of violence their communities face. On December 5, 2014, Deshawnda Sanchez, a Los Angeles area Black trans woman, was killed while trying to escape her murderer. Her death was one of 226 murders of trans women in 2014.[4] The LGBTQ magazine *The Advocate* reported that in 2015, one trans woman has been murdered every week and the major-

---

4   "Transgender Europe: TDOR 2014," *Transrespect versus Transphobia,* 15 November 2014.

ity of these deaths are women of color.[5] These statistics do not reflect the frequent harassment, abuse, and harm that trans women of color survive. The deadly violence that Black trans women negotiate is often perpetrated by intimate partners, many of whom when asked claim they were "duped" or "tricked" by their partners, and then justifiably enraged to the point of murder.[6] The popular media trope of trans identity as a form of deception is iconized in the 1992 film *The Crying Game,* a film that shows a man vomit when he realizes that Dill, the Black woman he loves is trans.[7] Laverne Cox, a Black trans woman portraying a trans woman on the Netflix original series *Orange is the New Black,* notwithstanding, the generally portrayal of trans women and trans women of color in popular culture is one of hypersexual tricksters who deserve to be victimized for deceiving cis men.[8] It is again in the area of representation, in visual assessment, that trans women of color, are judged and then responsible for the resulting violence it illicit from the people closest to them.[9]

These acts of violence also coincide with increasing visibility and advocacy by trans women of color, particularly through digital media outlets and in online media. In 2012, Mock was moved to become a more outspoken trans activist because of the rise in the number of murders and suicides of queer and trans youth. She has used her platform as a former web editor for *Marie Claire* as well as digital tools like videos and hashtags to

5   Mitch Kellaway, "Miami: Seventh Trans Woman Murdered in U.S. in 2015," *Advocate,* 20 February 2015.
6   Kat Callahan, "'Gay' or 'Trans Panic' Defenses Are Crap and Now Illegal in California," ROYGBIV, 30 September 2014.
7   Neil Jordan, *The Crying Game,* film, directed by Neil Jordan (1993; Palace Pictures and Channel Four Films), DVD.
8   "Victims or Villains: Examining Ten Years of Transgender Images on Television," GLAAD, 20 November 2012.
9   Cynthia Lee, "The Gay Panic Defense," *UC Davis Law Review* 42 (2008): 471–566; Cynthia Lee and Peter Kar Yu Kwan, "The Trans Panic Defense: Heteronormativity, and the Murder of Transgender Women," *Hastings Law Journal* 66 (2014): 77–132; and Victoria L. Steinberg, "A Heat of Passion Offense: Emotions and Bias in 'Trans Panic' Mitigation Claims," *Boston College Third World Law Journal* 25, no. 2 (2005): 499–524.

reach out to other trans women in society. In explaining the origin of her hashtag #girlslikeus, Mock describes her support of Jenna Talackova, a contestant disqualified for the Miss Universe Pageant for in the words of the pageant officiates, "not being a natural born female."[10] Mock's desire to help Talackova achieve her dream lead to the creation of the hashtag. On her personal blog Mock wrote,

> So I shared Jenna's petition on Twitter, and said: Please sign & share this women's rights petition in support of transgender beauty queen Jenna Talackova & #girlslikeus: ow.ly/9TYc6b 27 Mar 12 And that was the online birth of #girlslikeus. I didn't think it over, it wasn't a major push, but #girlslikeus felt right. Remarkably a few more women — some well-known, others not — shared the petition and began sharing their stories of being deemed un-real, being called out, working it, fighting for what's right, wanting to transition, dreaming to do this, accomplishing that…. #girlslikeus soon grew beyond me… my dream came true: #girlslikeus was used on its own without my @janetmock handle in it. It had a life of its own.[11]

Other trans women have embraced the tag, including Laverne Cox; they use it to discuss everything from the desire to transition and the violence of being outed in unsafe situations as well as the banality of everyday living and dreams of job success.

Computational Scientist Alan Mislove created a database that collects a random one percent of all tweets, tweeted since Twitter began.[12] As scholars in sociology and DH have noted, a one percent sample of such a large database can provide statical-

---

10 Alexis L. Loinaz, "Miss Universe Pageant Allows Transgender Women to Compete," *E! Online,* 10 April 2012.

11 Janet Mock, "Why I Started #GirlsLikeUs Twitter Hashtag For Trans Women," blog post, 28 May 2012.

12 Alan Mislove, Sune Lehmann, Yong-Yeol Ahn, Jukka-Pekka Onnela, and J. Niels Rosenquist. "Understanding the Demographics of Twitter Users," *5th International AAAI Conference on Web and Social Media,* last modified 5 July 2011.

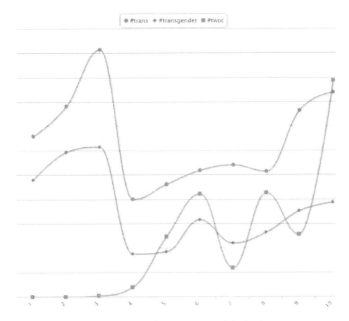

Fig. 1. Voyant Screen capture of the instance of the hashtags #trans, #transgender, and #twoc.

ly significant information about the representative population.[13] I examined all publically accessible tweets using #girlslikeus between Mock's first uses in March 2012 until October 2014. With the help of computer science graduate student Devin Gaffney, we created a script that gathered all instances of the hashtag within this database. With over 11,000 tweets stored that used the hashtag I could begin to assess significance. I used Voyant to mine the texts of the collected tweets. Voyant is a web-based textual analysis tool. It can generate word visualizations and can

13 See Carolin Gerlitz and Bernhard Rieder, "Mining One Percent of Twitter: Collections, Baselines, Sampling." *M/C Journal* 16, no. 2 (2013); "Twitter Usage Statistics," Internet Live Stats, http://www.internetlivestats.com/twitter-statistics/; and Axel Bruns and Stefan Stieglitz, "Towards More Systematic Twitter Analysis: Metrics for Tweeting Activities," *International Journal of Social Research Methodology* 16, no. 2 (2013): 91–108.

measure the frequency and occurrence of words in a corpus. The most popular hashtags used with the hashtag were other words of identity affirmation, including, #trans and #transgender, #twoc (trans women of color).

Words related to representation were very popular, ultimately representing the content of nearly every tenth tweet. Variations of the words *picture, selfie,* as well as types of media and news outlets drove this significance often pointing to pictures, and links that affirmed trans women in ways that counter their portrayal in mainstream news stories. For example, Janet Mock in May of 2013 attended the celebration of Laverne Cox's Time magazine cover, captioning their photo together, "Celebrating Laverne Cox's time cover tonight in NYC. Sisterhood in action! #girlslikeus [#transwomenofcolor]."[14]

Though celebrating a friend's appearance on the cover of time magazine is not something that we can necessarily relate too, celebrating friends and their success through posting a picture on Twitter is very common. I won't provide visual referents of the types of images that these daily life photos subvert but think about the trauma and violence that befalls Dil in *The Crying Game* or the way trans women are often made to account for their genitalia in popular culture by journalists in order to be invited guests. Tweets like this one, which show the daily and collective joys that trans women experience, challenge the proliferation of media representations that focus solely on their anatomy and victimization. Despite this tendency to over represent trans women in salacious ways, mainstream media continues to under report and misgender trans women in the news.[15]

---

14 James Nichols, "Laverne Cox Responds To Controversy Regarding Katie Couric's Invasive Comments," *The Huffington Post,* 2 February 2016.

15 Jay Michaelson, "Media Ignores Rash of Assaults on Transgender Women," *The Daily Beast,* 6 June 2012.

Transforming methodology through collaborative consent

Given the frequent attacks on trans women, particularly trans women of color, I wanted to be sure that my scholarly inquiries about the hashtag were welcome and did not bring undue negative attention to the community. I reached out to Mock via Twitter to see if she was interested in my researching the tag. I told her I was interested in how it has been used and what sorts of actions have developed through its use. She said she would be excited for me to work on the project. I also asked her what sort of information would be most useful to her to know about the hashtag. She was specifically interested in the most popular retweets and users along with gathering a sense of where the hashtag has shown up in mainstream media.

Though Mock welcomed my research on the hashtag, I wanted to give her the opportunity to say no to the project. My own interest was not enough justification for pursuing this type of potentially risky research that could expose an already vulnerable community to more vitriol and negative visibility. Trans women of color are not understood as one of the vulnerable communities identified by the Institutional Review Board that assesses the potential harms of academic research on those researched. The intentions of the IRB are quite noble and necessary. IRBs are charged with ensuring that human subject's research conducted at institutions that receive federal resources is ethical and does not cause harm to participants.[16] In response to human rights abuses of marginalized groups within the US, the IRB was established to ensure that human subjects research did not cause undo harm to those participating in studies.[17] This function of the IRB tends to result in a paternalistic orientation

---

16 Won Oak Kim, "Institutional Review Board (IRB) and Ethical Issues in Clinical Research," *Korean Journal of Anesthesiology* 62, no. 1 (2012): 3–12.

17 See Peter C. Williams, "Success in Spite of Failure: Why IRBs Falter in Reviewing Risks and Benefits," *IRB: Ethics and Human Research* 6, no. 3 (1984): 1–4 and Matt Bradley, "Silenced for Their Own Protection: How the IRB Marginalizes Those It Feigns to Protect," *ACME: An International E-Journal for Critical Geographies* 6, no. 3 (2007): 339–49.

towards research subjects.[18] The IRB evaluates the proposed research to ensure that investigators have thought through the potential impacts of the research on the people being researched. This paternalism may be a valid position for the researcher to take given the type of research to be conducted but in the digital landscape, a more nuanced and fluid understanding of the way power flows between researcher and researched is needed.

Previous studies have attempted to address IRB paternalism and its impact on research subjects in several ways. In the social sciences, there have been efforts to shift research from a top down orientation to one that is side to side or bottom up.[19] These methodologies attempt to undo the power imbalance by shifting people's relationships to power. Participation observation or becoming an observing participant are other ways that this conversation is framed, relying on the researchers shifting location into the research as the mechanism which disrupts the traditional flow of power.[20] My experience in trying to shift my relationship to power involved collaboration even before developing my research methodology.

Social media users are not the traditionally infantilized research subject that the IRB assumes. As people who are actively

18 Franklin G. Miller and Alan Wertheimer, "Facing Up to Paternalism in Research Ethics," *Hastings Center Report* 37, no. 3 (2007): 24–34.

19 See Paul A. Sabatier, "Top-down and Bottom-up Approaches to Implementation Research: A Critical Analysis and Suggested Synthesis," *Journal of Public Policy* 6, no. 1 (1986): 21–48; Ursula Plesner, "Studying Sideways: Displacing the Problem of Power in Research Interviews With Sociologists and Journalists," *Qualitative Inquiry* 17, no. 6 (2011): 471–82; and Frederick Erickson, "Studying Side by Side: Collaborative Action Ethnography in Educational Research," in *Innovations in Educational Ethnography: Theory, Methods, and Results,* eds. George Spindler and Lorie Hammond, 235–58 (reprint; Mahwah: Psychology Press, 2010).

20 See Patricia Maguire, "Doing Participatory Research: A Feminist Approach" (1987), http://www.popline.org/node/369992; George J. McCall and Jerry Laird Simmons, eds., *Issues in Participant Observation: A Text and Reader* (Reading: Addison-Wesley Publishing Co., 1969); and Lisa H. Weasel, "Conducting Research from the Ground Up Using Feminist Participatory Methodologies to Inform the Natural Sciences," *International Review of Qualitative Research* 4, no. 4 (2011): 417–29.

engaged in the ongoing generation of their digital content, so-cial media users require a level of forethought that extends be-yond the purview of the IRB. While IRB review might be able to forestall some of the harms that are involved in academic research, issues of consent beyond an initial yes to participate in a research project or plans to address issues that may not be anticipated by the IRB are obfuscated.

By speaking with Mock, before even designing my research plan, I ensured that the findings are useful to her as the creator of the hashtag. In addition to speaking with Mock directly, I spoke with my online community of friends and activists to create an advisory panel. I reached out to Black women and queer thought leaders within the arenas of Twitter and Tumblr, which included a diverse group of activist, artists, and academics, who have cre-ated some of the most popular hashtags related to social justice issues. Twitter and Tumblr users like Sydette Harry and Jamie Nesbitt Golden have been central to the creation and popularity of many social justice related hashtags. Many of them have direct experience with journalists and scholars using their social media posts without their consent. These incidents range from annoy-ing to outright dangerous. Many amongst this group of twelve have had their work stolen by journalists, interpreted by scholars all without attempting to contact them. They have experienced rape and death threats for simply voicing their opinions online. Given these histories, I did not want to repeat a pattern of begin-ning work without asking for permission first nor giving them the opportunity to ask their own research questions.

I plan to ask within the networks using #girlslikeus what us-ers think about the project and what questions they want an-swered about how it moves through cyberspace and the world. Even though tweets are understood as public, I err on the side of privacy and anonymity for Twitter users of the hashtag. I will not use the handles of users without first attempting to contact them. By allowing Twitter users to shape the project and their participation at multiple points, I create a more flexible process, one that is dynamic and shifting. But the collaboration doesn't end there.

My experience with issues of ongoing consent suggests that the way we do digital humanities needs to shift. Digital Humanists interested in conducting research that is ethical and feminist must go beyond the simple politics of citation, as citation itself may be the thing that creates the harm to the community. As noted in the #TwitterEthics Manifesto, "both academics and journalists should ask each individual user on Twitter for consent. They should explain the context and the usage of their tweets."[21] This allows a Twitter user to prepare for a potential onslaught or simply say no. Consent is a form of collaboration, a collaborative process which means that a *no* may come later in the course of the collaboration. Similarly, the process of doing this research necessitates an ongoing commitment to understanding dynamics that emerge through engagement in the medium. To that end, the advisory panel proved to be more dynamic than even I could have anticipated.

Collaborative construction: What transforms in the doing

I sent the potential members of my advisory committee an email asking if they'd like to participate. Those who agreed were invited to join a listserv where I post questions and solicit help as needed. While this group was initially created to ensure that my work was ethical and useful, it has already been repurposed for needs I could not anticipate. One member of the group is an internationally renowned journalist. She queried the collaborators, asking for their thoughts on a story she was writing about the ways that the media has represented Black people, particularly Black women in coverage of the Ebola outbreak. Many in the group responded to her request, providing important quotes that shaped her article in really helpful ways. When I envisioned the group, I imagined what I could do for them and what they could do for me. Even in creating a collaborative space it didn't occur to me that there might be things that we could collectively

---

21 Dorothy Kim and Eunsong Kim, "The #TwitterEthics Manifesto," *Model View Culture*, 7 April 2014.

do for another member of the circle. I knew what I wanted from the group and assumed that I could anticipate the group's needs from me. However, this emergent need and purpose for the group was possible through the creation of the network itself. Even before I could use the network in the way I envisioned, my collaborators were able to leverage it to meet their own needs.

Stressing a non-hierarchal circular collaboration is what helped make this possible. I had transformed the traditional top down approach of my research but I had only imagined a relationship that was side by side, a horizontal relationship with me on one side and my collaborators on the other. This moment allowed for the realization that we had co-created a three dimensional space with multiple directions of flow. What I appreciate about the language of *collaborators* as opposed to *research subjects* is that it provides the potential for multiple levels of relationship between those participating in the research; the structure is dynamic.

Those of us who participated in interviews for the journal article were not compensated, and there were a few minor errors in the article that I believe in another context would have been hurtful. However, the initial transparent ask, the openness to feedback, the knowledge that we were all participating in a process that extends beyond the published article and even my book, created a level of trust that allowed for forgiveness and correction among peers. Many of these collaborators have experienced the unsolicited appropriation of their words by people they did not know. By developing a new community, albeit a short lived one, the power differentials among the members are mitigated. While I still maintain epistemic privilege as the gatherer and convener of this circle, I am interested in using that privilege for the collective good.[22]

Because academic books make very little money, I endeavored to find meaningful ways to compensate this group for their

---

22 Bat-Ami Bar On and Bat Ami, "Marginality and Epistemic Privilege," in *Feminist Epistemologies*, eds. Linda Alcoff and Elizabeth Potter, 83–100 (London: Routledge: 1993).

work and time. This includes using university research money to help convene gatherings for existing digital networks that may have trouble connecting in another capacity. Creating a fund from my honoraria received in relationship to the work in addition to coauthoring articles with interested collaborators are a few of the ways I am exploring transforming my relationship to my collaborators. I've been able to leverage my position within academic communities to write grants that include budget opportunities for these collaborators and more. As the project continues I am creating more opportunities that challenge the ways that researchers have traditionally compensated and shared in the benefits that come from doing research.

One of the important aspects of DH work is the emphasis on collaboration. Scholars in the humanities are still primarily rewarded for single-author texts. Tenure and promotion committees regard books and articles that have one author more favorably than multi-authored texts. One of the ways that DH is creating a different methodological practice is by supporting connection through collaboration across multiple aspects of digital projects. Digital Humanist Mark Sample discusses this idea in relationship to student work through his concept *collaborative construction*. He writes:

> [B]y collaborative construction, I mean a collective effort to build something new, in which each student's contribution works in dialogue with every other student's contribution. A key point of collaborative construction is that the students are not merely making something for themselves or for their professor. They are making it for each other, and, in the best scenarios, for the outside world.[23]

I adapt this idea of collaborative construction such that it has import outside the realm of the classroom; it is critical to the way that I shape my research project. Consequently, a transformation of the goals of my academic work was necessary, as well

---

23 Sample, "Building and Sharing."

as my relationship to the usually distinct categories of research subject, researcher, and audience.

My work has multiple audiences. There are scholars in fields like women's, gender, and sexuality studies, ethnic studies, and digital humanities who might be interested understanding the networks contemporary Black trans women create through the production of digital media. However, I am interested in creating a book that also and perhaps most importantly, is useful to the communities on which my examples draw. More than just the world outside the academy, I want to create a resource for the communities with whom I am collaborating on the research. This practice involves a more intentional form of collaboration than I have attempted before.

I am creating a new way of practicing the relationships I am developing through my the advisory panel, transforming a researcher/researched relationship into one of collaboration, thereby shifting out of the position of researcher into a more equal role. This process also includes developing new models of expressing the value of everyone's contributions. For me, the process is the product, meaning that the process itself is productive, creative, and transformative of the conditions we are seeking to understand through the research.[24] We are building collaboratively in ways that build community and shift existing dynamics. We are actively shaping the project of collaborating through our collective participation so that an end product, while potentially very useful, is not the only thing created by this collaborative investigation.

The example of trans women of color's digital activism demonstrate the power of digital media to redefine representations of marginalized groups and their ability to impact a white supremacist, heterosexist, and trans misogynist media culture without that being their primary goal. The practices of reclaiming the screens of our computers and phones with content is not simply one of creating new representations, but is a practice of self-preservation and health promotion through the networks

---

24 Kim and Kim, "The #TwitterEthics Manifesto."

of digital media. While often celebrated for the rehabilitated images, this media is not often interrogated as processes that support the development of community and individual health strategies. Trans women of color aren't simply naming the violence they experience but are building networks of support and recognition for their work that helps them have safer environments in which to live.

I understand trans women of color's production of media as an act of self-preservation and one of health praxis that is not centered on appeals to a majority audience. The creation of media by minoritarian subjects about themselves and for themselves can be a liberatory act. These acts of image redefinition actually engender different outcomes for marginalized groups and the processes by which they are created build networks of resilience that far outlives the relevant content. Black women and queer and trans folks reconstruct representations through *digital alchemy.*

#Girlslikeus rejects the assimilationist invisibility of another potential hashtag like #imagirltoo, in favor of a declaration that makes trans women the undeniable center of their own project, where they are their own referents, not cis women. #Girlslikeus signals a conversation that is for, by, and about trans women and not their proximity to another group of relative power. #Girlslikeus exemplifies the magic of digital alchemy through this practice of shifting from margin to center utilizing established mediums to create literally transformative realities.

The added benefit of creating this community online is that it is visible to those outside the identity and does the work of humanizing inadvertently and without draining energy from the more important work of supporting each other. Digital media is creating and supporting a network of connection among communities that have traditionally had trouble finding each other let alone reaching a larger audience. By doing the work of community building online, groups are leveraging both visibility and education at once. Trans women of color are telling their own stories but in the process are forcing more recognition for their identities in mainstream publics.

Putting process into practice

As a member of the Allied Media Projects community, I have been shaped by the organization's values and principles. Allied Media Projects mission is to "cultivate media strategies for a more just, creative, and collaborative world."[25] The annual Allied Media Conference that highlights work from activists, artists, and organizers in service of this mission, highlights the words create, connect, and transform in their advertising for the event. I find these verbs particularly useful in marking the different components of this project that trouble more traditional methodologies in my fields of study. I see these three components of connection, creation, and transformation as the template for the types of questions we should be asking about our digital research. I have identified the questions these verbs raise for me in my own research which may be a good starting place for others who are interested in the same.

Connect

1. Who are your collaborators?
   - What community is your research accountable to beyond your academic community?
   - How will you demonstrate your desire to be accountable to them?
   - Are there people you can talk to about the impact of your research beyond the IRB?
2. How does everyone benefit from the research?
   - What questions does the community want answered?
   - Can people be compensated in ways that honor their time and skills?

---

25 *Allied Media Projects,* http://amc.alliedmedia.org/.

Create

1. What tools and or methods encourage multidirectional collaboration?
   - What mechanism of accountability can you create?
   - Are there ways that collaborators can use the research process to their own ends?
2. What kind of process can you create for your research?
   - Is there room for collaborators to give and rescind consent at different times during the research process?
   - Does the pace of the project meet your needs and your collaborators needs?

Transform

1. How will you take care of yourself in the research process?
   - What do you and your collaborators need to stay sustained while conducting the research?
   - What happens after the research product is complete?
2. How will you be transformed?
   - Will the research strengthen your connection to your collaborators?
   - Did you and your collaborators come to new understandings?

I offer these questions as a starting place for conducting digital research within a feminist ethical frame. I am reminded of Octavia Butler's aphorism "all you touch you change; all you change, changes you."[26] With this tenant in mind, I think that the important take away here is that the very process of conduct research creates shifts in the landscape. These shifts have incredible potential to be both helpful and harmful, depending

---

26 Patricia Melzer, "'All That You Touch You Change': Utopian Desire and the Concept of Change in Octavia Butler's Parable of the Sower and Parable of the Talents," *Femspec* 3, no. 2 (2002): 31–52.

on how you frame your project and interactions with your collaborators.

The Research Justice collective at the AMC frames their work with the question "is this *just* research or just research?"[27] I want my research to be just and I had to set my parameters for what that means. I realized that I can't answer these questions by myself. I am close but not embedded in these digital networks I study. I've tried to shape the process of my investigation in a way that honors the principles set out in the Allied Media Project mission. My project is collaborative and builds connection; it's creative in that I'm generating something useful in the media; and it's transformative in that my collaborators and I will be changed by the process of doing the research.

27   "Call for Session Proposals," *Allied Media Projects*, 12 February 2013.

Bibliography

Allied Media Projects. http://amc.alliedmedia.org/about/amc-mission.

Bailey, Moya. "I'm Back!" Blog post. 21 October 2014. http://moyabailey.com/2014/10/21/im-back/.

Bradley, Matt. "Silenced for Their Own Protection: How the IRB Marginalizes Those It Feigns to Protect." *ACME: An International E-Journal for Critical Geographies* 6, no. 3 (2007): 339–49.

Bruns, Axel, and Stefan Stieglitz. "Towards More Systematic Twitter Analysis: Metrics for Tweeting Activities." *International Journal of Social Research Methodology* 16, no. 2 (2013): 91–108. DOI: 10.1080/13645579.2012.756095.

Callahan, Kat. "'Gay' or 'Trans Panic' Defenses Are Crap and Now Illegal in California." *ROYGBIV.* 30 September 2014. http://roygbiv.jezebel.com/gay-or-trans-panic-defenses-are-crap-and-now-illegal-in-1641055202.

"Call for Session Proposals." *Allied Media Projects.* 12 February 2013. https://www.alliedmedia.org/news/2013/02/12/call-session-proposals-research-justice.

Erickson, Frederick. "Studying Side by Side: Collaborative Action Ethnography in Educational Research." In *Innovations in Educational Ethnography: Theory, Methods, and Results,* edited by George Spindler and Lorie Hammond, 235–58. Reprint. Mahwah: Psychology Press, 2010.

Gerlitz, Carolin, and Bernhard Rieder. "Mining One Percent of Twitter: Collections, Baselines, Sampling." *M/C Journal* 16, no. 2 (2013). http://journal.media-culture.org.au/index.php/mcjournal/article/view/620.

Hall, Stuart. "What Is This 'Black' in Black Popular Culture?" *Social Justice* 20, nos. 1–2 (1993): 104–14. http://www.jstor.org/stable/29766735.

Jenks, Chris, ed. *Visual Culture.* London and New York: Routledge, 1995.

Jones, Amelia, ed. *The Feminism and Visual Culture Reader.* London and New York: Routledge, 2003.

Jordan, Neil. *The Crying Game.* Film. Directed by Neil Jordan. 1993; Palace Pictures and Channel Four Films. DVD.

Kellaway, Mitch. "Miami: Seventh Trans Woman Murdered in U.S. in 2015." *Advocate.* 20 February 2015. http://www.advocate.com/politics/transgender/2015/02/20/miami-seventh-trans-woman-murdered-us-2015.

Kim, Dorothy, and Eunsong Kim. "The #TwitterEthics Manifesto." *Model View Culture.* 7 April 2014. https://modelviewculture.com/pieces/the-twitterethics-manifesto.

Kim, Won Oak. "Institutional Review Board (IRB) and Ethical Issues in Clinical Research." *Korean Journal of Anesthesiology* 62, no. 1 (2012): 3–12.

Nichols, James. "Laverne Cox Responds To Controversy Regarding Katie Couric's Invasive Comments." *The Huffington Post.* 2 February 2016. http://www.huffingtonpost.com/2014/01/09/laverne-cox-katie-couric_n_4568495.html.

Lee, Cynthia. "The Gay Panic Defense." *UC Davis Law Review* 42 (2008): 471–566. http://papers.ssrn.com/sol3/papers.cfm?abstract_id=1141875.

———, and Peter Kar Yu Kwan. "The Trans Panic Defense: Heteronormativity, and the Murder of Transgender Women." *Hastings Law Journal* 66 (2014): 77–132. http://papers.ssrn.com/sol3/papers.cfm?abstract_id=2430390.

Loinaz, Alexis. L. "Miss Universe Pageant Allows Transgender Women to Compete." *E! Online.* 10 April 2012. http://www.eonline.com/news/307600/miss-universe-pageant-allows-transgender-women-to-compete.

Macdonald, Myra. *Representing Women: Myths of Femininity in the Popular Media.* London and New York: E. Arnold, 1995.

Maguire, Patricia. "Doing Participatory Research: A Feminist Approach." 1987. http://www.popline.org/node/369992.

McCall, George J., and Jerry Laird Simmons. *Issues in Participant Observation: A Text and Reader.* Reading: Addison-Wesley Publishing Co., 1969.

Melzer, Patricia. "'All That You Touch You Change': Utopian Desire and the Concept of Change in Octavia Butler's Parable of the Sower and Parable of the Talents." *Femspec* 3, no. 2 (2002): 31–52.

Michaelson, Jay. "Media Ignores Rash of Assaults on Transgender Women." *The Daily Beast.* 6 June 2012. http://www.thedailybeast.com/articles/2012/06/06/media-ignores-rash-of-assaults-on-transgender-women.html.

Miller, Franklin G., and Alan Wertheimer. "Facing Up to Paternalism in Research Ethics." *Hastings Center Report* 37, no. 3 (2007): 24–34. DOI: 10.1353/hcr.2007.0044.

Mislove, Alan, Sune Lehmann, Yong-Yeol Ahn, Jukka-Pekka Onnela, and J. Niels Rosenquist. "Understanding the Demographics of Twitter Users." *5th International AAAI Conference on Web and Social Media.* Last modified 5 July 2011. http://www.aaai.org/ocs/index.php/ICWSM/ICWSM11/paper/view/2816.

Mock, Janet. "Why I Started #GirlsLikeUs Twitter Hashtag For Trans Women." Blog post. 28 May 2012. http://janetmock.com/2012/05/28/twitter-girlslikeus-campaign-for-trans-women/.

On, Bat-Ami Bar, and Bat Ami, "Marginality and Epistemic Privilege." In *Feminist Epistemologies,* eds. Linda Alcoff and Elizabeth Potter, 83–100. London: Routledge: 1993.

Plesner, Ursula. "Studying Sideways: Displacing the Problem of Power in Research Interviews With Sociologists and Journalists." *Qualitative Inquiry* 17, no. 6 (2011): 471–82. DOI: 10.1177/1077800411409871.

Sabatier, Paul A. "Top-down and Bottom-up Approaches to Implementation Research: A Critical Analysis and Suggested Synthesis." *Journal of Public Policy* 6, no. 1 (1986): 21–48. DOI: 10.1017/S0143814X00003846.

Sample, Mark. "Building and Sharing (When You're Supposed to Be Teaching)." *Journal of Digital Humanities* 1, no. 1 (Winter 2011). http://journalofdigitalhumanities.org/1-1/building-and-sharing-when-youre-supposed-to-be-teaching-by-mark-sample/.

Sperling, Joy. "Reframing the Study of American Visual Culture: From National Studies to Transnational Digital Networks." *The Journal of American Culture* 34, no. 1 (2011): 26–35.

Steinberg, Victoria L. "A Heat of Passion Offense: Emotions and Bias in 'Trans Panic' Mitigation Claims." *Boston College Third World Law Journal* 25, no. 2 (2005): 499–524.

"Transgender Europe: TDOR 2014." *Transrespect versus Transphobia*. 15 November 2014. http://transrespect.org/en/transgender-europe-tdor-2014/.

"Twitter Usage Statistics." *Internet Live Stats*. http://www.internetlivestats.com/twitter-statistics/.

"Victims or Villains: Examining Ten Years of Transgender Images on Television." GLAAD. 20 November 2012. http://www.glaad.org/publications/victims-or-villains-examining-ten-years-transgender-images-television.

Weasel, Lisa H. "Conducting Research from the Ground Up Using Feminist Participatory Methodologies to Inform the Natural Sciences." *International Review of Qualitative Research* 4, no. 4 (2011): 417–29. http://www.jstor.org/stable/10.1525/irqr.2011.4.4.417.

Williams, Peter C. "Success in Spite of Failure: Why IRBs Falter in Reviewing Risks and Benefits." *IRB: Ethics and Human Research* 6, no. 3 (1984): 1–4.

14

DH and Adjuncts: Putting the Human Back Into the Humanities

Kathi Inman Berens & Laura Sanders

Note: Our voices blend in this essay. Because of our uneven workloads and compensation for work, Kathi wrote much of the essay, but the essay is borne of conversation between us and represents our shared enterprise to talk about adjuncting and digital humanities. To differentiate, Laura is *italicized*. We are happy to share our title with the terrific essay "Putting the Human Back in the Digital Humanities: Feminism, Generosity and Mess," by Elizabeth Losh, Jacqueline Wernimont, Laura Wexler, and Hong-An Wu. Their essay appears in *Debates in Digital Humanities 2016,* edited by Matthew K. Gold and Lauren F. Klein, and was published as ours was going to press.

For MLA 2015, I wrote a piece called "Want to Save the Humanities? Teach Adjuncts to Learn Digital Tools."[1]

At the time, I was an adjunct. Fall 2015 I started my first year as a tenure-track professor after teaching nine years as an NTT and four years as an adjunct. I am a mid-career scholar who

---

1   Kathi Inman Berens, "Want to Save the Humanities? Teach Adjuncts to Learn Digital Tools," *Disrupting the Digital Humanities: Digital Edition,* 5 January 2015.

is also legitimately "early career," writing a proposal for my first book contract, preparing lectures for a slate of new course preps, looking for grants. I'm happy to be "junior," even if the nomenclature of our profession is an atavism from when the field was more homogenous and marked rites of passage for mostly young white men from "assistant" to "associate." Many of the "junior" faculty I know today are my age, and have walked what would have seemed a twisty path twenty years ago. In 1969, tenured or TT faculty comprised 80% of the professoriate. Today that ratio has flipped.[2] But the field's nomenclature is rooted in a time when, as a colleague once told me of his own childhood, an English professor could buy a house and raise four children on a single income.

"Adjunctification" is the graceless word that describes our field's slide into temporary employment. Part-time and non-tenure track faculty comprise 76% of the humanities professoriate, according to a 2013 report from the American Association of University Professors.[3] *That means three of four humanities college instructors might be income-insecure. They occupy a liminal space between professional and hourly temp worker. While contract workers in other fields might be released from specific professional obligations, freeway flyers may hold conferences in their cars or write letters of recommendations during breaks. They may be expected to attend in-services or other departmental meetings and trainings. Sometimes they will be paid for participating in such events; sometimes they will not.*

Adjuncts are also the fastest-growing class of the professoriate. Hundreds of thousands of students are taught by part-time

---

2   Adrianna Kezar and Daniel Maxey, "The Changing Faculty and Student Success: National Trends for Faculty Composition over Time" (Los Angeles: University of Southern California, 2012), 1. See also "Table 303.40: Total Fall Enrollment in Degree-Granting Postsecondary Institutions, By Attendance Status, Sex, and Age: Selected Years, 1970 through 2024," *Institute of Education Sciences National Center for Education Statistics*.

3   "Here's the News: The Annual Report on the Economic Status of the Profession, 2012–13," *American Association of University Professors* (March–April 2013).

faculty. These faculty aren't paid to advise students or offer the kind of hands-on mentoring that allows digital humanist inquiry to flourish. "Notably," observes Katina Rogers, "provosts expect equal or greater reliance on adjunct labor in coming years, and demonstrated little faith in a continued tenure system."[4]

"There are no adjunct administrators"

Fundraising, marketing, student life, facilities management: every administrative department is rewarded with job security except teaching, the one performing the institution's core function. There are no adjunct administrators. Further, institutions may carve traditional instructor duties so thinly that staff members with income-secure positions in coordinating, advising, and mentoring may have never taught a college course. In this structure, adjunct faculty expertise is tapped purely for classroom labor as though it were separate from these functions. Through this process, the people who may have the keenest understanding of students and what they need to be successful can be excluded from key conversations on campus. Faculty are no longer the professional group that drives campus decision making, despite instruction being the core mission of most colleges and universities.

Tenure-track academics accept overwork as a condition of our lives, but overwork among adjuncts isn't an investment in tenure. It's just the condition of keeping your gig. During MLA 2016, Ian Bogost tweeted disdainfully of adjuncts, wondering why anybody would persist teaching part-time when intelligent people can earn a living in more remunerative, less demeaning ways.[5] When TT professors work weekends, defer sleep and rest, catch up on grading during spring break, and make "vacations" out of conference travel, it's on the mutual understanding that this is the agreed-upon path toward professional promotion. For adjuncts, there is no path. Instead, it's a stick dangling a car-

---

4   Katina Rogers, "Humanities Unbound: Supporting Careers and Scholarship Beyond the Tenure Track," *Digital Humanities Quarterly* 9, no. 1 (2015): par. 4.

5   Ian Bogost (@ibogost), Twitter post, 9 January 2016, 9:38 p.m.

rot just out of grasp. The allure of full-time work, income stability, health insurance, a long-term contract: feeling *this close* to those things keeps people adjuncting. I did it, and every adjunct I know does it. You say yes to every request because any one of those requests might yield a switchback to full-time work. In my case, miraculously, it did. I landed a full-time, TT job the year after I won a Fulbright. This is a byproduct of hard work and privilege: my partner earned the family's main income while I built my research profile. Now it's flipped, and I'm the main earner while he builds his profile as a futurist, heading up research projects for the USC Annenberg Center for the Digital Future.

The first months my life on the TT awakened how much I'd missed collegiality when I was adjuncting, how collegiality is a social bond that grows our capacity in a thousand small ways.

I love that the DH community shares abundantly and makes many tools and trainings freely available. I have benefitted from that sharing ethos and contributed to it. But I worry that, absent a shared communal enterprise, DH training for adjuncts risks raising a yet higher and less attainable bar. Could DH training for adjuncts make adjuncts as a class *more vulnerable* in a precarious job market? Could DH become a hiring expectation in various institutions' rush to adopt DH, before digital pedagogy experts provide guidelines or best practices articulated specifically for adjunct working conditions?

Rather than "disrupting DH," this essay aims to awaken in DH practitioners a will to disrupt adjunctification. It seeks to restore the human in a humanities that, three times out of four, views people from the vantage of the bottom line.

The next section of the essay traces a genealogy of care in digital humanities. The final section examines practical solutions to the problem of how to teach DH in environments where neither students nor professor have adequate time to "fail." An iterative, heuristic method is not ideally suited to adjunct practice because failure is a luxury. There has to be a fund of trust that failure can draw upon, and adjuncts lack that fund. As Lisa Marie Rhody observed in her MLA 2016 talk "Building Trust: Designing Digital Projects for the Public through Care and Repair," if you were writing on a deadline and Zotero were suddenly to strip your bibliography of diacritical marks, your response would not be: "oh how charming! I celebrate this moment of failure!" Not all DH "fails" are productive. Their impact has to be contained and provide an opportunity for critical reflection.

A DH genealogy of care

Tara McPherson's March 2015 lecture "DH by Design: Alternative Origin Stories for the Digital Humanities" sketches a genealogy that locates screen-cultures and audio-visual aesthetics as a parallel development to "Father Busa as patron saint of a certain version of textually focused computing." Tracing an origin in visual design, film, and screen installation art, McPherson highlighted the expressive capacities of screen cultures and the critical traditions they fostered, such as feminist film studies, queer studies, and critical race studies. In this screen cultures origin of DH, liveness, intimacy, and affect figure both in the art and analytical appraisal. A different commitment to experimentation and political engagement manifests itself. In this origin story of DH, human experience, critical agency, and the body's sensorium is never at far remove.

It is this DH that has the critical tools and sensibility to recognize in adjuncting a systemic erosion of the collective humanist endeavor. These practitioners might respond by asking whether, or under what conditions, DH training might responsibly serve adjunct faculty and the hundreds of thousands of students who take classes with them.

Not all DH research scholars will see in a DH customized for the most disadvantaged learners and teachers a method or approach consonant with their own definition of the field. But as I suggest in "Sharing Precarity," an essay in the DH *Debates 2017* collection, the modular logics of higher education funding make adjuncts and DH centers surprisingly aligned in their vulnerability.[6] DH is a critical method and interdisciplinary practice, yes, but it is also a cultural totem. In a recent talk at UC Irvine, "Money and Time: Some Hard Truths About Institutional Support for Digital Humanities," Miriam Posner debunked the myth that DH funding is "flush."[7] She observed that funding for DH centers is "boom and bust" and that many people charged with this work are overworked and exhausted. In the rush to establish digital humanities centers, some institutions have hired "post-docs with a laptop," a synecdoche for the scant infrastructural support that would nurture growth, faculty integration and project sustainability. Adjuncts and DH centers are more aligned in their function at today's university than they are distinct.

Adjuncting is lonely. It's isolating. I missed having my name and email address in the university directory. Only my own students could find me. I was invisible unless I took steps in social media to work against invisibility. I missed having a mailbox in the bank of faculty boxes. Most keenly I missed the shared collective enterprise, the good faith that we're all in this together.

Collegiality is at the core of what professors do and why we do it. Couldn't every single teacher have an office with her name on the door, even if it's a shared space with many other names on the door? Couldn't the nameplate, which signals belonging and respect, be something more durable than a scrap of printed paper in Times New Roman 12-point font? Why wouldn't faculty who perennially teach the same class at the same institution be recognized as functionally permanent and treated as such by their colleagues? My own university, Portland State, has approved just such a move. To me, the acknowledgement that part-time is in some cases permanent removes the dangling carrot. It doesn't extort free labor on the implied promise of first consideration when a full-time position becomes available.

---

6   Kathi Inman Berens, "Sharing Precarity: Adjuncts and Digital Humanities," in *Debates in the Digital Humanities 2017*, eds. Matthew K. Gold and Lauren F. Klein (Minneapolis: University of Minnesota Press, forthcoming).

7   Miriam Posner posted the talk to her blog.

"Expertise in taking risks and failing"

DH is being *disseminated widely*. It is no longer the exclusive provenance of top-tier research universities [R-1s]. In such conditions, *we cannot once again ask "part-time" employees to donate even more time to develop skills that may benefit their students tremendously but have no financial reward whatsoever for themselves.*

For DH, *public humanities, and maybe even liberal education in general, we need to strategize for wider range of college students, those whose undergrad experiences may be nothing like those of their instructors. According to Institute of Education Sciences National Center for Education Statistics, over 40 percent of college students in 2013 were 25 years old or older. Over 25 percent were 30 years old or older. In the United States, most college students are not at private institutions. Most are taught by adjuncts.*

On my [Laura's] campus, about 55 percent of credit students do not self-identify as white. Some of my students are former foster children, recovering addicts, formerly/currently homeless, and ex-offenders. My students have life experience that is rarely reflected in popular media coverage of college students, such as the student who told me she lived under an overpass for years until she "got pregnant again," and the online student who revealed to the class that she chose her paper topic because she was a survivor of sex trafficking. These students have deep expertise in taking risks and failing and learning and trying again.

Originally, I had doubted that DH *could offer my students more than just another set of barriers to academic success. However, after I attended the National Endowment for the Humanities Office of Digital Humanities Summer Institute, "Beyond Pockets of Innovation, Toward a Community of Practice," I started to realize the incredible possibilities* DH *has for traditionally underserved students who enroll at community colleges. My students rarely see themselves reflected in dominant narratives, but as they become digital humanists, they will no longer have to wait to be nominated to be part of the conversation. Digital tools will allow them to represent themselves and tell stories from their own perspectives.*

My goal for the NEH institute was to develop a pedagogy of digital citizenship for students who often see themselves as outsiders at institutions of higher learning and strangers to civic engagement. I wanted to connect DH practices with my work in community-based learning (CBL). Both DH and CBL require instructors to forfeit some control (and maybe some ego) to empower their students through a "hands-on, learn as you go, fail harder" approach. Discomfort is necessary for learning because it prompts reflection on how knowledge itself is produced.

At community colleges, we discuss concepts like "growth mindset" and "grit" as strategies for helping students from communities without a college-going tradition to persist in higher education. Dovetailing into these conversations, I encourage my colleagues to experiment with DH and fail, so they can "model resiliency" for their students and possibly teach them valuable lessons that will serve them for years after the class ends.

While my original goals for pursuing DH approaches were rooted in my desire to add social justice content to my composition courses, I now believe that DH itself is a social justice practice. DH has the potential to level the playing field by giving underrepresented communities a voice and the opportunity to offer their own narratives. As my students develop digital confidence as well as the ability to interrogate how knowledge itself is constructed, they also develop the tools they need to authorize themselves to be part of larger conversations.

"Yes, I will": What "part-time" looks like

Teaching time may no longer be the accurate measure of adjunct labor. Here's what "part-time" teaching work entails among the adjuncts we spoke with.

Yes, I'll head up Community-Based Learning. I'll serve as the liaison between nonprofit partners and instructors while helping teachers use community work to achieve academic outcomes required by our institution.

Yes, I will mentor new distance learning faculty. I'll teach them the LMS ropes as well as translate their teaching persona to an online classroom.

Yes, I will edit the accreditation report. I'll work with colleagues across campus to refine institutional responses that demonstrate how we meet regional accreditation standards in preparation for a high-stakes campus visit.

Yes, I will meet to discuss current composition theory in our reading group, so my teaching can be informed by the latest theories of skill transfer and competency grading.

Yes, I will coach various departments at various levels of knowledge and enthusiasm about creating, executing, and reporting on their required assessment projects.

Yes, I'll write a competitive grant proposal. I'll help colleagues across campus articulate their goals in the proposal.

Yes, I will speak at the in-service on any number of projects and campus initiatives where I have developed expertise over the past few years.

Yes, I'll advise students informally with their personal, professional, and academic challenges. I'll refer them to a variety of student resources I have come to know through my years on campus.

Yes, I'll join that search committee and event planning committee, while I also attend meetings about the new campus learning garden.

Yes, I will present on DH at campus meetings, department meetings, and our annual professional development conference.

Yes, I will attend that important meeting about placement exams, completion rates, retention strategies, supporting veteran students, financial aid policy, FERPA, copyright law, Title IX, sexual harassment, disability as cultural competency, making my courses culturally relevant, stereotype threat, fire drills, earthquake drills, active shooter drills, assessment, new core outcomes, new general education requirements, and leading without power.

For some of this work (such as grant writing, assessment coaching and online training), a part-timer in Portland makes about $30/hour. But note that the workload is much more diversified than that of a typical full-time faculty member. Part-timers touch every aspect of work done in English departments and indeed, stretch into other units such as library, events planning, student life, and assessment.

Invisible labor

Yes, I will learn how to use the online ordering system for the bookstore,
the add/drop system for students,
the email system,
the voice mail,
the photocopy code,
the security code.
the method for ordering classroom keys,
the paperwork I need to complete to get paid,
the podium,
the projector,
the LMS,
the rules for my syllabi,
withdrawal deadlines,
major academic policies,
organization charts, and
office hour policies
for every school where I work.

I will meet or call students at random times to accommodate their schedules. I will learn how to accommodate learning differences and support success for all students but still help them reach institution-specific outcomes that can be assessed by a third party.

Most likely, I will not receive dedicated office space, the promise of a job beyond this term, health and retirement benefits equal to non-casual employees, guaranteed funding for conferences, or sabbatical eligibility offered to academic professionals who have worked at the institution fewer years than I have.

Given these significant demands, many adjuncts simply lack the bandwidth to take on additional responsibilities without remuneration.

"Right sizing" DH methods for adjuncts

Digital humanists make many of their resources open, public, and freely available. DHers' moral commitment to public good is ethical, and it bonds the community together in shared, iterative enterprise. Generosity is an efficient way to diversify the field. It's evolutionarily advantageous.

But how many adjunct faculty can afford to take up DHers' offer of free tools and training? What if "time" and most adjuncts' inability to defend against its appropriation prevents teaching-only faculty from integrating digital methods into their work? Free tools might be a necessary but insufficient condition to provide access.

Here's our *Disrupting DH* provocation: Don't just make tools freely available. Pay adjuncts for their *time* to learn them.[8] Let's right-size some DH assignments and methods so that they can be used by faculty who teach the majority of students in English and writing classes, and accept that the minimization of DH technique doesn't tarnish the field's credibility or give students a false impression of it. We don't say of English instruction at community colleges: "that's not English" because it is taught differently at the CC than at the R-1, but we risk doing that with the digital humanities. I see tremendous goodwill for identifying DH methods appropriate to community college learners and their precariously employed professors, but we have to do the work to figure out what those community-specific best practices might be.

The National Endowment for the Humanities Office of Digital Humanities [NEH ODH] has taken steps toward building a DH commons for community college faculty; however, the large majority of participants in these efforts is full-time. ODH funded a 2014 survey of community college instructors to gauge interest in bringing DH methods into the classroom, and a week-long 2015 summer institute (which Laura attended, and about which

---

8   Kathi Inman Berens, "Wanna Save the Humanities? Pay Adjuncts to Learn Digital Tools."

she wrote above), where twenty-nine CC instructors learned digital tools and DH methods. "The goal of the institute was to get like-minded folks together who were ready to be part of a movement, to embed DH in their courses and to act as ambassadors for DH at their institutions," observes Prof. Ann McGrail, the Lane Community College professor who won the ODH Start-up Grant to spearhead these endeavors. McGrail reached out to part-timers via Twitter and received a signal boost from Adjunct Nation "and other handles that seemed to pass the word along." I saw McGrail's tweet and sent it to Laura. Of the 100 faculty who filled out the survey and the 29 who participated in the Institute, the "overwhelming response [was] from full-timers and very few part-timers." McGrail made an effort to reach adjuncts, and the Institute application "didn't inquire" about full or part-time status. Still, McGrail acknowledges institutional obstacles to including part-timers: messages circulate through listservs and professional organizations to which adjuncts may not have access. Even open access platforms like Twitter are chancy, because people have to know which hashtags and people to follow. And they must have the time and surplus energy to track it.

McGrail's vision for a "DH Commons" is a solid start in the work of opening DH pedagogy to adjuncts, but it's only a first foray into work that requires consultation and collaboration across the spectrum of institutions from CC to R-1.

Such consultation is urgent. Textbook companies are poised to sweep into writing classrooms with pre-fab, start-to-finish learning modules that would usurp the dynamic interrogative space of digital humanities. Gabriel Kahn's *Slate* article "College in a Box" details the apparent financial attraction of massive scale in these templatized (and for-profit) forms of online learning that outsource precisely the situated, local and emergent properties of learning that digital humanities instruction fosters.[9] When universities subcontract teaching to for-profit organizations, they strain the dynamic bonds between faculty, students, and local community.

---

9   Gabriel Kahn, "College in a Box," *Slate,* 4 September 2014.

We want culturally sensitive, empowering lessons for students who may have trouble seeing themselves as college students, much less as curators of their own knowledge. But to create customized DH, *we need to put the humans back into humanities. We need to authorize* PT *faculty (as well as students) to fail and survive. However, it is hard to ask people to "model resiliency" when certain types of failure can fray nerves about job security.*

Financial commitments from the Modern Language Association, the National Council of Teachers of English, the American Historical Association, the Allied Digital Humanities Organizations and other professional bodies could earmark funds to pay adjuncts for their time to get digital training. Such organizations, mostly run by tenured faculty, have historically not seen alignment between adjuncts' needs and their own.

What would "paying" an adjunct to learn digital tools look like? Buy a contingent faculty member out of one class for one term and stipulate that the time be spent in DIY, on-demand digital tools training. Pay adjuncts to attend in-person training institutes like Digital Humanities Summer Institute or the Digital Pedagogy Lab.[10] Follow up to see how such training changes classroom practice, whether or not it creates conditions of increased job security, or augments job satisfaction.

Pedagogical resources are abundant, but time is scarce. Diane Jakacki's soul-searching post "Disrupt DH?" ends with a be-the-change challenge: "Is the ultimate act of disruption modeling the kind of behavior that you want others to adopt?"[11] Several full-time professors and digital technologists are making and sharing "DH 101" coursework. But the field has yet to discern best practices for community college learners and their professors, the majority of whom are adjunct. Katherine Harris, like Jakacki a longtime digital humanities pedagogy innovator, routinely publishes her DH assignments and students' results on

---

10 See "Course Offerings," *The Digital Humanities Summer Institute,* http:// dhsi.org/courses.php, and "Institute," *Digital Pedagogy Lab,* http://www. digitalpedagogylab.com/institute/.

11 Diane Jakacki, "Disrupt DH?" blog post, 7 June 2015.

her scholarly blog Triproftri.org. Her students published in the peer-reviewed *Journal of Interactive Technology and Pedagogy* an account of their two-year Beard-Stair project.[12] Harris's new project is a "Bootstrap DH" method to explore topic modeling, genre, and periodization in her small data project working with the Forget Me Not Hypertext Archive (2005).[13] Harris has consistently generated digital pedagogy work that might adapt well to plug-and-play needs of adjuncts. Diane Jakacki has co-taught the DHSI Digital Pedagogy course for years, and has amassed some excellent tools and methods which she shares via Creative-Commons licensed course materials.[14] Howard Rheingold, a pioneer of virtual learning and environments since 1964, runs 5-week online courses for "co-learners" aimed at the five social media literacies he's identified.[15] Jesse Stommel and Sean Michael Morris's Digital Pedagogy Lab offers a slate of online courses that are steeply discounted for adjuncts.[16]

The DH at the CC Commons is just now being populated with assignments and commentary, another useful starting point when imagining how to right-size DH projects for adjunct labor conditions. I would hope that DH at the CC would explicitly flag

---

12 David Coad, Kelly Curtis, Jonathan Cook, and Katherine Harris. "Beard-Stair: A Student-Run Digital Humanities Project History, Fall 2011 to 16 May 2013." *Journal of Interactive Technology and Pedagogy* 4 (2013).

13 Katherine Harris, "Using Bootstrap Digital Humanities to Explore Topic Modeling Ghosts, Haunted Houses, and Heroines in 19th-Century Literature (UC Irvine Talk)," *triproftri* (blog), last modified 9 February 2016, https://triproftri.wordpress.com/2016/02/05/uc-irvine-talk/. See also *Forget Me Not: A Hypertextual Archive of Ackermann's 19th-Century Literary Annual,* last modified 28 January 2007, http://www.orgs.miamioh.edu/anthologies/fmn/Index.htm.

14 Diane Jakacki and Mary Galvin, "Digital Pedagogy Integration in the Curriculum," shared document, https://docs.google.com/document/d/1-xuSSYVco9zLZDINCO-WolPEQP8znQ2A_hyYNHxdlU/edit#heading=h.g9r-6777j2vxt.

15 Howard Rheingold, interviewed by Justin Ellis, "Howard Rheingold on how the five web literacies are becoming essential survival skills," *Nieman Lab,* 11 May 2012.

16 "Online Courses," *Digital Pedagogy Lab,* http://www.digitalpedagogylab.com/courses/.

or foreground assignments suited to adjuncts who aren't em-powered to be "ambassadors" at their institutions. Jesse Stom-mel's "Free College; Free Training for College Teachers" declares that "[i]f college is ever to be 'free' in any broad or expansive sense of the word, we must start by fostering pedagogical work as an ethic" that meets the needs of the majority of teachers in humanities classrooms today.[17] Stommel and Sean Michael Morris argue that pedagogical instruction is marginalized and undervalued in graduate programs despite more than 75% of new job postings seeking teachers.

"A feminist ethic of care — like many a DH research agenda or platform for large-scale visualization and analysis," Bethany Nowviskie notes, "seeks to illuminate the relationships of small components, one to another, within great systems."[18] In aggre-gate adjuncts are massive, but adjuncts are treated in university systems individually as "small components" — alienable labor, with specific privileges added on only as necessary in order to complete the work. At one point when I was adjuncting, my email address could access mail servers but not the research li-brary. The rushed, protean work conditions of adjuncts keeps them isolated. If DHers were to turn their critical knowledge of systems to visualize, interpret and improve the work conditions of three-quarters of the professoriate, it might supply evidence for a more holistic, humane, and sustainable working condi-tions for students and all members of our profession.

---

17 Jesse Stommel, "Free College; Free Training for College Teachers," blog post, 28 February 2016.
18 Bethany Nowviskie, "Capacity through Care," blog post, 20 February 2016.

Bibliography

Berens, Kathi Inman. "Sharing Precarity: Adjuncts and Digital Humanities." In *Debates in the Digital Humanities 2017*, edited by Matthew K. Gold and Lauren F. Klein. Minneapolis: Minnesota University Press, forthcoming.

———. "Want to Save the Humanities? Teach Adjuncts to Learn Digital Tools," *Disrupting the Humanities: Digital Edition*, 5 January 2015, http://www.disruptingdh.com/ want-to-save-the-humanities-pay-adjuncts-to-learn-digital-tools/.

Bogost, Ian (@ibogost). Twitter post. 9 January 2016, 9:38 p.m. https://twitter.com/ibogost/status/686014212657692672.

Coad, David, Kelly Curtis, Jonathan Cook and Katherine Harris. "BeardStair: A Student-Run Digital Humanities Project History, Fall 2011 to 16 May 2013." *Journal of Interactive Technology and Pedagogy* 4 (2013). https://jitp. commons.gc.cuny.edu/beardstair-a-student-run-digital-humanities-project-history-fall-2011-to-may-16-2013/.

Davis, Rebecca Frost, Matthew Gold, Katherine Harris, and Jentery Sayers. *Digital Pedagogy in the Humanities. Forthcoming.* New York: Modern Language Association Press.

The Digital Humanities Summer Institute. http://dhsi.org/ courses.php.

Digital Pedagogy in the Humanities: Concepts, Models, and Experiments. https://digitalpedagogy.commons.mla.org/.

Digital Pedagogy Lab. http://www.digitalpedagogylab.com/.

Forget Me Not: A Hypertextual Archive of Ackermann's 19th-Century Literary Annual. Last modified 28 January 2007. http://www.orgs.miamioh.edu/anthologies/fmn/Index.htm.

Harris, Katherine. "Using Bootstrap Digital Humanities to Explore Topic Modeling Ghosts, Haunted Houses, and Heroines in 19th-Century Literature (UC Irvine Talk)." *triproftri* (blog). Last modified 9 February 2016. https:// triproftri.wordpress.com/2016/02/05/uc-irvine-talk/.

"Here's the News: The Annual Report on the Economic Status of the Profession, 2012–13." *American Association of University Professors* (March–April 2013). https://www.aaup. org/report/heres-news-annual-report-economic-status-profession-2012–13.

Jakacki, Diane. "Disrupt DH?" Blog post. 7 June 2015. http:// dianejakacki.net/disrupt/.

——— and Mary Galvin. "Digital Pedagogy Integration in the Curriculum." Shared document. https://docs.google.com/ document/d/1-xuSSYV-co9zLZDINCO-WolPEQP8znQ2A_ hyYNHxdlU/edit#heading=h.g9r6777j2vxt.

Kahn, Gabriel. "College In a Box: Textbook Giants Are Now Teaching Classes." *Slate*. 4 September 2014. http://www. slate.com/articles/life/education/2014/09/online_college_ classes_textbook_companies_offer_courses_with_minimal_ university.html.

Kezar, Adrianna, and Daniel Maxey. "The Changing Faculty and Student Success: National Trends for Faculty Composition over Time." Los Angeles: University of Southern California, 2012. http://eric.ed.gov/?id=ED532269.

Losh, Elizabeth. "Respect, Niceness and Generosity." *Virtualpolitik*. http://virtualpolitik.blogspot.no/2014/08/ respect-niceness-and-generosity.html.

McGrail, Ann. "Doing DH at the CC." Blog post. 19 October 2015. https://blogs.lanecc.edu/dhatthecc/.

McPherson, Tara. "DH by Design: Alternative Origin Stories for the Digital Humanities," lecture at Washington State University Pullman, 3 March 2016.

Nowviskie, Bethany. "Capacity Through Care." In *Debates in the Digital Humanities 2017*, edited by Matthew K. Gold and Lauren Klein. Minneapolis: Minnesota University Press, forthcoming. Cross-posted to Nowviskie's scholarly blog. 20 February 2016. http://nowviskie.org/2016/capacity-through-care/.

Posner, Miriam. "Money and Time: Some Hard Truths about Institutional Support for Digital Humanities," talk at UC Irvine's "Data Science and Digital Humanities Symposium.

5 February 2016. Posted to Posner's blog. 14 March 2016. http://miriamposner.com/blog/money-and-time/.

Rheingold, Howard. Interviewed by Justin Ellis. "Howard Rheingold on how the five web literacies are becoming essential survival skills." *Nieman Lab.* 11 May 2012. http://www.niemanlab.org/2012/05/howard-rheingold-on-how-the-five-web-literacies-are-becoming-essential-survival-skills/.

Rhody, Lisa Marie. "Building Trust: Designing Digital Projects for the Public through Care and Repair," on the panel "Care and Repair: Designing Digital Scholarship." Modern Language Association Convention, 9 January 2016. Austin.

Rogers, Katina. "Humanities Unbound: Supporting Careers and Scholarship Beyond the Tenure Track." *Digital Humanities Quarterly* 9, no. 1 (2015). http://www.digitalhumanities.org/dhq/vol/9/1/000198/000198.html.

Stommel, Jesse. "Free College; Free College Training for Teachers." Blog post. http://jessestommel.com/free-college-free-training-for-college-teachers/.

"Table 303.40: Total Fall Enrollment in Degree-Granting Postsecondary Institutions, By Attendance Status, Sex, and Age: Selected Years, 1970 through 2024." *Institute of Education Sciences National Center for Education Statistics.* https://nces.ed.gov/programs/digest/d14/tables/dt14_303.40.asp.

15

Not Seen, Not Heard

Liana Silva

I am a brown Puerto Rican female, first-generation graduate with a PhD. Until recently, I worked within academia, and now I am an independent scholar and editor for an online publication on sound studies. Despite my scholarly history, I hesitate to call myself a digital humanist. Twitter was my introduction to digital humanities, very much like Annemarie Pérez states in her contribution "Lowriding through the Digital Humanities."[1] I joined Twitter in 2009 out of curiosity but didn't become an avid user until mid 2010, when I moved to Kansas City and started a new job while still having a dissertation to finish. I found an academic and professional community on Twitter that filled a void I had in my offline life. On Twitter I felt seen and heard in academic circles in a way I hadn't before.

By virtue of being a humanist and the academics I followed on Twitter, I quickly learned that there was a thing called "digital humanities." I had never heard the term in my college classroom or from my professors or college peers. I came across the short-hand DH and the hashtags #DH, #digitalhumanities and later #TransformDH. The medium, which lended itself to casual conversations as well as bite-sized insights, made me feel com-

---

1 Annemarie Pérez, "Lowriding through the Digital Humanities," *Disrupting the Digital Humanities: Digital Edition,* 6 January 2016.

fortable enough to follow along with these conversations and occasionally even chime in on the latest DH topic of the week.

In the meantime, as I waded through the Twitter waters of DH, I switched career tracks and eventually finished a PhD. I also became Managing Editor of the sound studies blog *Sounding Out!* and started asking questions about what publishing about sound in a digital medium means.[2] Questions such as "why a blog and not a journal?" would send us down a rabbit hole of what each word entailed and the heft of our editorial choices. Editing for the blog enabled me to learn about the field of sound studies through practice: by engaging with the ideas of our writers and pressing questions about how power affects the way we interpret sound. As I thought about the conversations I saw on Twitter among digital humanists and under the hashtags, I often thought, "we are doing digital humanities... I think."

I struggled to articulate our connection to DH. I could sense it... but whenever it came time for me to explain why we "did" digital humanities, I drew a blank. Imposter syndrome quickly flared up when the best response I could come up with in Twitter, Facebook, and face-to-face conversations was "we're a blog, all of our content is available digitally" and I had "established" digital humanists tell me "making a blog is not digital humanities." I focused on the *digital* aspect instead of the humanities aspect. For me, the digital in *Sounding Out!* was what made us relevant to digital humanities, and all I could see was the wordpress.com blog where *Sounding Out!* has lived for over six years. By digital humanities standards, our site wasn't anything new. I took it personally.

I hesitate to call myself a digital humanist because of the nagging feeling that I am always in disguise, always playing at being something I am not. The circular conversations about what is digital humanities (as well as who is a digital humanist, like Tressie McMillan Cottom pointed out) feed this feeling that somehow, I will never know enough or that I am not well versed enough in the theory of digital humanities in order to practice

---

2   *Sounding Out!,* https://soundstudiesblog.com/.

it.[3] In the rhetoric of "hack versus yack," yack (or lack thereof) can be used as a way to isolate and exclude those who want to participate, even if they are hacking.[4]

In reality, I have been hacking, for some time. Although I heard time and time again that a blog did not constitute "doing digital humanities," the blog has allowed us to push against the limits of digital writing about sound. In academic humanities, conversations about sound often happen on paper and in person at conferences and in classrooms. As a Managing Editor, I work side by side (virtually) with my writers, let their arguments sit in my head as they do in the heads of the writers, and engage with their point's one on one.

However, trying to figure out what side of the hack & yack coin *Sounding Out!* was on took me away from one of *Sounding Out!*'s major contributions: we have rendered visible and audible to a broad, intellectual audience conversations about sound and sound studies. As a member of the editorial committee, I have worked for years in thinking and rethinking what academic scholarship looks like and how to give voice (no pun intended) to people who write about sound. In the process, we have helped our writers/bloggers think about what it means to practice sound studies and what are the questions sound studies is thinking about. On the other hand, our bloggers have also opened up our views and pushed our thinking about what "counts" as sound studies and who is doing sound studies.

But I know that my work as a humanist (in the digital?) should not remain only in writing, only from the sidelines of editorial work. I know that my mere presence in the academy makes a statement. The editorial committee I am a part of (comprised by all minorities: women, freelancer, junior faculty, once-upon-a-time grad students, Latina, African American) makes the statement that difference matters, and that pretending not to

---

3 Tressie Cottom, "Digital Humanities: Egalitarian or Just A New Elite?" blog post, 1 November 2012.

4 Bethany Nowviskie, "On the Origin of 'hack' and 'yack,'" blog post, 8 January 2014.

see difference or operating from the premise that difference will make the whole weaker is just another way to silence difference.

True story that I rarely tell: as a college instructor in Upstate New York, I remember teaching in classrooms where I was the only person of color. I remember one time when I gasped quietly while my students worked. I have always wondered, how did my presence in the classroom affect them? And that leads me to think: how could my presence in the classroom, nay, in the academy, affect those who are just like me: first-generation Latinas? Like Annemarie Pérez in "Textual Communities: Writing, Editing, and Generation in Chicana Feminism," I wonder how being a Latina editor helps me step into academic conversations I otherwise would find myself excluded from.[5]

And that is why I decided to be a part of a collection of essays on digital humanities, even though I still struggle to fully articulate what about my work makes me a digital humanist. People need to at least hear my voice. More importantly, other Latinas who have considered what the humanities can offer them must see me and hear me. They must know they are not alone, and that their experience, skills, and knowledge are valued.

When Tara McPherson asks "why are the digital humanities so white?"[6] I want to respond with "we're not all white. We're just looking for a place to stand." I offer this solution to readers today: when you approach your next digital humanities meeting or next digital humanities project, ask yourself, who is missing here? Who do I not see? More importantly, ask yourself, who do I not hear? Start with questions. It's what we humanists do best.

---

5 Annemarie Pérez, "Textual Communities: Writing, Editing, and Generation in Chicana Feminism," *Hybrid Pedagogy*, 15 October 2015.

6 Tara McPherson, "Why Are the Digital Humanities So White? or Thinking the Histories of Race and Computation," in *Debates in the Digital Humanities*, ed. Matthew K. Gold, 139–60 (Minneapolis: University of Minnesota Press, 2012).

Bibliography

Cottom, Tressie McMillam. "Digital Humanities: Egalitarian or Just A New Elite?" Blog post. 1 November 2012. http://tressiemc.com/2012/11/01/digital-humanities-egalitarian-or-just-a-new-elite/.

McPherson, Tara. "Why Are the Digital Humanities So White? or Thinking the Histories of Race and Computation." In *Debates in the Digital Humanities,* edited by Matthew K. Gold, 139–60. Minneapolis: University of Minnesota Press, 2012. http://dhdebates.gc.cuny.edu/debates/text/29.

Nowviskie, Bethany. "On the Origin of 'hack' and 'yack.'" Blog post. 8 January 2014. http://nowviskie.org/2014/on-the-origin-of-hack-and-yack/.

Pérez, Annemarie. "Lowriding through the Digital Humanities." *Disrupting the Digital Humanities: Digital Edition.* 6 January 2016. http://www.disruptingdh.com/lowriding-through-the-digital-humanities/.

———. "Textual Communities: Writing, Editing, and Generation in Chicana Feminism." *Hybrid Pedagogy.* 15 October 2015. http://www.digitalpedagogylab.com/hybridped/writing-editing-generation-in-chicana-feminism/.

Sounding Out! https://soundstudiesblog.com/.

16

Disrupting Labor in Digital Humanities; or, The Classroom Is Not Your Crowd

Spencer D.C. Keralis

"Stick to the boat, Pip, or by the Lord, I won't pick you up if you jump; mind that. We can't afford to lose whales by the likes of you; a whale would sell for thirty times what you would, Pip, in Alabama."
— Herman Melville, *Moby-Dick; or, The Whale*[1]

Digital humanists have a labor problem, but it's not what you might think. It's not about humanities faculty getting credit in tenure and promotion for digital informatics work. It's not about the adjunctification of teaching labor. It's not about the devaluation of humanities PhDs, the contraction of the faculty job market, the rise of #alt-ac, nor the para-professionalization of academic libraries. And while I care deeply about these problems, at least three of which affect me personally almost every day, what I am concerned with here is a type of labor that the collective preoccupation with these issues effaces: the use of student labor on digital humanities projects in and out of the classroom.

---

1   Herman Melville, *Moby Dick, or The White Wale* (Boston: C.H. Simonds Co., 1922), 390.

Without student labor, the academy as a whole would grind to a halt. From the office of the Registrar to the digital library, student labor keeps the wheels of the neoliberal university spinning. These students, even those in skilled technical jobs, are generally paid at or just above minimum wage. Right or wrong, this is a truth of the university that is well established, and many digital humanities projects with the funding to do so participate in the student labor economy. What isn't so widely recognized is that, in the absence of funding for student wages, some faculty use the classroom as a locus for exploiting student labor. The aim of this chapter is not to point fingers or name names, but to adumbrate a trend in disciplines I value that I, and others, find deeply troubling. In what follows, I will describe the problem of student labor in digital humanities as I see it, and examine some of the structural issues that drive the use of student labor. I will place the labor economy of digital humanities projects within the broader context of the innovation economy writ large to demonstrate how labor within the academy cannot operate under the same system of consensual participation which informs movements like crowdsourcing and crowdfunding. And in conclusion, I will offer suggestions for how ethically managed student labor in the classroom can empower students to demonstrate both CV-ready skills and humanistic knowledge in durable products for which they receive full credit.

On digital humanities panels at conferences ranging from the Modern Language Association, to the Digital Library Federation Forum, to the Alliance of Digital Humanities Organizations' annual Digital Humanities, to C19, to the Texas Conference on Digital Libraries, I've been struck again and again by how glibly panelists, upon describing their project, will declare something to the effect of: "[A]nd we incorporate the grunt work into a syllabus and have students do it as part of a class." Under the rubric of "skills building," these comments are usually met with nods of knowing approval by attendees. During the Q&A of the Feminist DH panel at Digital Humanities 2013 when a few audience members questioned the legitimacy of using student labor in the classroom, they were piously dismissed both in the room and on social media, with Tweeters wielding hashtags like #pedagogicalvalue and #computationalthinking as though they expressed some ineffable, self-evident good.[2] This sort of dismissal is the rhetorical equivalent of #NotAllMen or #AllLivesMatter — a sleight of hand leveraged by a vocal and influential clique of DH true believers to efface the legitimacy of claims of student labor exploitation.[3]

This circling of the wagons is reflexive and unreflective and has more, I believe, to do with a sort of siege mentality on the part of a DH in-crowd (in particular those who have attempted to focus the conversation on recognition of digital work for tenure and promotion), than it does with those individuals' actual convictions about the use of student labor.[4] The desire on

---

2   Kathryn Tomasek (@KathrynTomasek): "@4Hum Hope someone is noting that not all student labor is exploitation. #pedagogicalvalue #computation-althinking #DH2013," Twitter post, 19 July 2013, 12:02 p.m..

3   #notallmen is used by "men's rights" advocates to dismiss arguments about rape culture (i.e., #notallmen are violent against women). For a useful examination of how #notallmen is an impediment to serious conversation about rape, see Phil Plait, "#YesAllWomen," *Slate*, 27 May 2014. Similarly, #AllLivesMatter is used in response to #BlackLivesMatter to "take race out of the equation" and "[turn] our eyes away from acknowledging America's racist past, functioning as a form of dismissal or denial." David Bedrick. "What's the Matter with 'All Lives Matter?'" *Huffington Post*, 24 August 2015.

4   The notion of a "DH in-crowd" isn't new. In *Debates in the Digital Humanities*, William Pannapacker notes that "the field, as a whole, seems to be developing an in-group, out-group dynamic that threatens to replicate the culture of Big Theory back in the 80s and 90s, which was so alienating to so many people." Pannapacker notes, as I do, that this cliquishness is notable on social media, and observes that DH seems "more exclusive, more cliquish," each year. William Pannapacker, "Digital Humanities Triumphant?" in *Debates in the Digital Humanities*, ed. Matthew Gold, 233–34 (Minneapolis: University of Minnesota Press, 2012). In the same volume, Lisa Spiro resists this idea, suggesting that in-crowd is "an ironic label for a group of people who have long felt like misfits. Lisa Spiro, "'This Is Why We Fight': Defining the Values of the Digital Humanities," in *Debates in the Digital Humanities*, 16–35, at 16. The two notions are not, however, mutually exclusive, since the misfit status Spiro describes is precisely what produces the smug, cliquish "cool kids' table" mentality that I and Pannapacker experience. They are different sides of the same coin with social media as their echo chamber.

the part of those attempting to deflect and dismiss criticism of practices to which they have so fiercely allied themselves comes instead from a collective defensiveness driven in part by a very real desire to ensure that digital work within the humanities is valued and recognized. Meg Worley asserts elsewhere in this volume that imbalances in power "are rarely attributable to individual action or an intent to oppress."[5] But decisions about student labor are often individual decisions made from a range of alternatives within institutional contexts. At some point all digital humanities practitioners choose, to a greater or lesser degree depending on institutional realities and individual values, to willingly capitulate to the logic of what Richard Grusin describes as "bottom-line economics and the need for higher education to train students for jobs[,] not to read literature or study culture."[6] Digital projects have the potential to allow faculty to have their neoliberal cake and teach literature and history too, and any criticism of the practices that support digital projects is rejected out of hand.

But the economic motivation goes further than simple pandering to shifts in administrative priority away from producing thoughtful citizens and toward making corporate minions. As Melville's Stubb reminds us "man is a money-making animal, which propensity too often interferes with his benevolence,"[7] and digital humanities has woven money into the social fabric of humanities pedagogy in unprecedented ways. The misperception of university administrators that digital humanities will bring in money has unfortunately been encouraged by the same community of practitioners who are most invested in defending the maker culture fostered by the National Endowment for Humanities Office of Digital Humanities (NEH-ODH). The ability to bring in grant money has been a key point for those fighting for recognition of informatics and computer science projects for

---

5   Meg Worley, "The Rhetoric of Disruption: What Are We Doing Here?," this volume, 64.
6   Richard Grusin, "The Dark Side of the Digital Humanities — Part 2." *Thinking C21*, 9 January 2013.
7   Melville, *Moby Dick*, 390.

tenure in humanities departments, and the fact that there really isn't much money out there to support these projects creates a culture of scarcity that fosters, among other inequities, the move of student labor from campus employment or work study into the classroom.

Let's break this down a little. NEH-ODH Digital Humanities Start up Grant Level 1 Awards, which for years were the gold standard for funding early stage humanities informatics projects, max out at $40,000. Considering that the average indirect costs charged against grants often exceed 50%, there's very little room to fund the sort of skilled labor necessary to produce the technical innovation prioritized by the ODH. Given that hourly rates for freelancers with experience in R, Python, or other programming languages can exceed $100/hour,[8] the pittance remaining after universities take their facilities and administration cut might cover less than 200 hours of skilled labor, with nothing left over for hardware, hosting, travel, or other research-related costs.

The deficit internship

The solution for some scholars is to shift this work away from paid professionals, or even paid apprentice labor like graduate research assistants, and into the classroom. They provide just enough training in code, content management, and style sheets for students to contribute some basic programming, write content for blogs and wikis, transcribe manuscripts and primary source documents, or develop visualizations and design. Students that come to the classroom with skill in computing, design, or even statistics can face an undue burden compared to their peers both in terms of supporting and mentoring their less technically savvy classmates and in terms of what the instructor

---

8   A search for R programmers on UpWork, a web-based service that matches clients with freelancers, generates a list of hourly rates ranging from $35 to $250 per hour. See *UpWork,* https://www.upwork.com/o/profiles/browse/c/web-mobile-software-dev/.

expects them to contribute to the project. Even when students are given credit for their work, they often end up building portfolios for fields they'll never crack, or which don't help them in their chosen major (this is especially true of students outside the major in which these courses are offered, who are also sometimes the more technically skilled of the students). [9] Under the rationale of promoting skills building and in-class collaboration, the faculty essentially gets the benefit of free labor on their projects. Free, that is, to the faculty. Students still pay tuition for these courses, making them not just unpaid internships, but deficit internships subsidized in no small part by student loan debt accrued by the students. If faculty can't get federal money to support their research, this is a back door to getting its equivalent, and students foot the bill in both their labor and their future debt burden.

In the culture of perpetual lack that is humanities research funding, there has been very little scrutiny of these practices. They are, in fact, difficult to identify unless faculty come right out and discuss it at conferences or in other venues, which happens not infrequently. In an environmental scan of 129 syllabi gathered online we found no instances of instructors explicitly stating in their syllabi that student work would contribute to a faculty project (individual assignments, however, are generally not visible). The practice of using student labor in the classroom is naturalized into the fabric of digital pedagogy, and some large scale collaborative projects actively provide mechanisms for the effacement of student labor.

One example of this is the *History Engine*, "an educational tool that gives students the opportunity to learn history by doing the work — researching, writing, and publishing — of a historian."[10] A collaborative project of the University of Richmond and a number of liberal arts colleges, the Engine is spon-

---

9   Spencer Roberts (@robertssw87): "@digiwonk: 'you're building a portfolio for a field you'll never break into.' how not to help students. #DH2013 fem and DH panel," Twitter post, 19 July 2013, 11:57 a.m.

10   *The History Engine,* http://historyengine.richmond.edu/.

sored by the University of Richmond Digital Scholarship Lab, the Virginia Center for Digital History, and the National Institute for Technology in Liberal Education (NITLE). The site has received press in *Inside Higher Ed, Academic Commons,* and the *Chronicle of Higher Education*; was awarded NITLE's 2009 Community Contribution Award; and was written up by faculty and project staff in three essays in Perspectives on History in 2009.

The *Engine* is based around a database of student authored "Episodes" describing moments in history. These episodes are assigned as part of courses at participating institutions, and the *Engine* provides sample assignments, lesson plans, and style guides for completing the essays in accordance with the site's standards. As an example, let's look at the episode describing the Keating-Owen Child Labor Act of 1916. [11] The episode is tagged with metadata including the date of the episode, location, topic tags, and the course and institution which produced the essay. There is no metadata field for author, and author is not a searchable term in the site's advanced search function. In the process of producing work for the site, work which students are "fully aware that future classrooms will engage with and critique," [12] the student author is erased and anonymized. While the site claims it is providing students with the experience of writing and publishing as an historian, it is in fact structured to ensure that students' contributions are unidentifiable.

What this amounts to is an undergraduate student paying for the privilege of contributing his work anonymously to the project. Students at US institutions participating in the *History Engine* pay an average of $954 per credit hour, and as much as $2200 per credit hour to contribute, without credit, to the database. Whatever the pedagogic value of these small episode essays may be, one lesson the students must certainly internalize is that their work does not belong to them, and can be subsumed silently by a larger entity. This is great preparation for the corporate world, but it seems we should be having a more nuanced

---

11 "The Keating-Owen Child Labor Act of 1916," *The History Engine.*
12 "What is the History Engine?" *The History Engine.*

conversation about intellectual property with students we hope to cultivate as future scholars. While the *Engine* purports to help students "learn history by doing the work [...] of an historian," the way the site treats the products of that work complicates the relationship between labor and pedagogy. The Engine remains in use in classrooms and continues effacing the labor of its student contributors, with episodes from courses at Marist College, Richard Bland College, University of Richmond, University of Toronto Scarborough, Widener University, and Indiana University-Purdue University, Indianapolis appearing in 2017.[13]

The networked machine

It is sometimes argued that the use of student labor in the classroom operates as a form of crowdsourcing (that is certainly the discursive angle taken in the *History Engine* documentation), and crowdsourcing has been a popular if unevenly successful method of doing some kinds of digital humanities work. But crowdsourcing operates under specific conditions of informed consent and volunteerism which labor in the classroom cannot support.

Along with crowd-funding, crowdsourcing has emerged as one the twin pillars of the neoliberal entrepreneurship economy. It is broadly accepted that the term was first coined in 2006 by *Wired* columnist Jeff Howe to describe "The new pool of cheap labor: everyday people using their spare cycles to create content, solve problems, even do corporate R&D."[14] Crowdsourcing relies on low- to no-cost labor to produce a wide variety of products, from computer code to photography, and deploys an instrumentalist ethic toward those contributing their labor — note the word "cycles" in Howe's description, a term describing the fundamental steps a CPU performs to execute commands. Crowdsourcing dehumanizes individual contributors, reducing them effectively and affectively to anonymous components in a networked machine.

---

13   "Schools Using the History Engine," *The History Engine.*
14   Jeff Howe, "The Rise of Crowdsourcing," *Wired,* 1 June 2006.

Of course not all advocates of crowdsourcing are so transparently mechanistic in describing their labor pool. Clay Shirky describes the pool of skilled individuals willing to spend their spare time working on projects as offering a "cognitive surplus" which takes advantage of the networked social sphere and a collective spirit of generosity.[15] As Shirky describes it, crowdsourcing relies on a particular social contract between labor and producer, encouraging a spirit of volunteerism to produce collaborative projects at scale. If there's a product that can be developed collaboratively, using small slices of time contributed by people with a particular skill and interest, the internet economy has found a way to wrangle those people together to produce that product. These products range from those with potentially significant social effects — Shirky offers Wikipedia and Ushahidi, a platform that aggregates citizen reports of ethnic violence in Kenya, as examples — to the merely entertaining, like meme factory ICanHasCheezburger.com.

This model can provide a significant return for the companies leveraging this diffuse labor force. Aside from being "incredibly cheap,"[16] the company benefits from greater intellectual diversity than any one workforce could support. The benefits to workers are less tangible. In those rare instances where the labor is paid, the pay is minimal. Contributions to crowdsourced projects have a similarly minimal impact as cv padding, and are designed to rely on the experience of workers rather than providing experience for them. While one study finds that offering financial incentive is one of the most reliable means of soliciting participants in crowdsourced projects, interest in the topic, ease of participation, altruism, and the desire to share knowledge are also motivators. Interestingly, the perceived sincerity of the project organizers is also highly valued by participants, and projects that

---

15   Clay Shirky, *Cognitive Surplus: Creativity and Generosity in a Connected Age* (New York: Penguin Books, 2010).
16   Howe, "The Rise of Crowdsourcing."

"appreciate and celebrate their community" are more likely to be perceived as an "honest beneficiary" of crowdsourced labor.[17]

Despite the benefits companies can derive from crowd-sourced labor, one of the essential assumptions of crowdsourcing from Howe's first elaboration of the term, is that "The crowd produces mostly crap." As Howe describes it, "Any open call for submissions — whether for scientific solutions, new product designs, or funny home videos — will elicit mostly junk. Smart companies install cheap, effective filters to separate the wheat from the chaff."[18] Paradoxically, one of those filters is the crowd itself, as "a networked community [...] ferrets out the best material and corrects errors. Wikipedia enthusiasts quickly fix inaccuracies in the online content."[19]

Both situations are certainly true of humanities data generated by crowdsourcing. Begun in 2001, the New York Public Library's "What's on the Menu?" project invites users to help transcribe historical restaurant menus.[20] User-entered information is recorded in large open datasets that can be accessed through the website. In terms of engagement, the project has been incredibly successful, with 1,331,934 dishes transcribed from 17,545 menus as of this writing. However, these transcriptions are notoriously messy. The Digital Humanities Data Curation Institute instructors Trevor Muñoz and Dorothea Salo used the NYPL data sets as object lessons in how to curate and clean up crowdsourced data using tools like OpenRefine.[21] In an elegant response to the economies of scale inherent to the project, the "cheap, effective filter" the NYPL deploys to fix this messy crowd-produced data

17 Piotr Organisciak, "Why Bother? Examining the Motivations of Users in Large-Scale Crowd-Powered Online Initiatives" (MA Thesis, University of Alberta, 2010).

18 Jeff Howe, "5 Rules of the New Labor Pool," *Wired,* 1 June 2006.

19 Ibid.

20 "What's on the Menu?" *New York Public Library,* http://menus.nypl.org/.

21 Digital Humanities Data Curation Institute, http://www.dhcuration.org/institute/ (accessed 21 June 2016). For Muñoz's exploratory work using the data, see Trevor Muñoz, "What IS on the Menu? More Work on NYPL's Open Data — Part 1, blog post, 8 August 2013 and "Refining the Problem — More Work with NYPL's Open Data — Part 2," blog post, 19 August 2013.

is the crowd itself. They invite users to help review transcribed menus invites to "fix misspellings, fill in missing data."[22] There's something a little utopian about the notion of crowdsourcing being a sort of self-healing system in which the crowd fixes errors the crowd produced, but on some level the benefit received by the project or organization outweighs the crap generated by the user base, otherwise the practice would long since have died out.

Even under apparently ideal conditions of pure volunteerism as in the NYPL's case, or modestly compensated contributions to Human Intelligence Tasks (HITs) on Amazon's Mechanical Turk marketplace,[23] crowdsourcing is not without its ethical pitfalls. I've alluded to one above in describing how contributors are dehumanized as part of an anonymous labor network, and Jonathan Zintan, co-founder of the Berkman Center for Internet and Society at Harvard, argues that there is a "Tom Sawyer syndrome" involved in crowdsourcing labor "in which people are suckered into doing work thinking that it's something to be enjoyed," and criticizes the gamification aspect of crowdsourcing in which contributors are given points or badges for recognition within the volunteer community in lieu of compensation.[24] Other critics focus on issues of data privacy and data integrity that crowdsourcing input and analysis of research data may involve. In terms of issues in crowdsourcing in the humanities, Julie McDonough Dolmaya argues that the use of crowdsourced labor for translation devalues the work of translation, and lowers the "occupational status" of professional translators.[25] This critique offers an analogue for the devaluation of labor in the humanities at large: does, for example, the *History Engine* devalue the work of historians by shifting the labor of content production onto

---

22 "What's on the Menu."

23 *Amazon Mechanical Turk,* https://www.mturk.com/mturk/welcome.

24 Aminda, "Crowdsourcing: an ethics dilemma?" *ideaconnection,* 23 April 2011.

25 Julie McDonough Dolmaya, "The Ethics of Crowdsourcing," *Linguisitica Antverpiensia: New Series — Themes in Translation Studies* 10 (2011): 97–111. See also Mia Ridge, ed., *Crowdsourcing Our Cultural Heritage* (Burlington: Ashgate, 2014).

anonymous student authors? The staff and teachers involved in the Engine would likely argue that it does not, but what if these student-authored texts were cited instead of other scholarly works? How much authority do they want the entries to accrue, and how much does this anxiety contribute to the decision to keep the texts anonymous?

Amanda Fucking Palmer: A cautionary tale

Even under what appear to be the most clearly voluntary of circumstances, when the social contract of a crowdsourcing engagement seems obvious to the participants, the practice is not exempt from the criticism that crowdsourcing devalues professional practice. One of the most visible examples of this is the controversy that erupted in 2012 around Amanda Palmer's invitation to musicians in towns visited by her tour to play onstage for free. In August 2012, Palmer posted a call on her blog: "Wanted: Horny-y and String-y Volunteers for the Grand Theft Orchestra Tour!!!!" The post asked for "professional-ish horns and strings for EVERY CITY to hop up on stage with us for a couple of tunes" and in return, "we will feed you beer, hug/ high-five you up and down (pick your poison), give you merch, and thank you mightily for adding to the big noise we are planning to make."[26] This had been Palmer's practice for years. Her punk cabaret act The Dresden Dolls relied in part on volunteer musicians when touring. She toured Australia in 2008 with The Danger Ensemble, four performance artists and a violinist who traveled with her for room and board, and they passed the hat at each gig. Palmer espouses an ethic of sharing and giving-what-you-will developed in her years busking as the 8-Foot Bride in Harvard Square, or playing her ukulele for change, and elaborated in a 2013 TED talk and her 2014 book *The Art of Asking*.

What was different about the Grand Theft Orchestra Tour was that Palmer had just completed a wildly successful Kick-

---

26 Amanda Palmer, "Wanted: Horny-y and String-y Volunteers for the Grand Theft Orchestra Tour!!!!" blog post, 21 August 2012.

starter campaign to support the recording of the LP *Theatre Is Evil* and the tour to promote the record. The campaign was supported by 24,883 backers and raised $1,192,793 — the highest-grossing musical campaign on the platform at the time. These funds were intended to cover recording and distribution costs for the record (Palmer has self-released her work since her acrimonious split from Roadrunner Records in 2010), and to pay a salary to the core band that would accompany her on tour. Professional musicians were outraged and blasted her website with comments decrying her use of volunteer players. Industry heavy hitters like Raymond Hair Jr., president of the American Federation of Musicians weighed in, and producer Steve Albini published a particularly vitriolic post on the message board for his studio Electrical Audio, which was reproduced on *Pitchfork* and subsequently went viral.[27] The crux of the arguments against Palmer, aside from those that just called her an idiot or worse in a downward spiral of grotesque misogyny, was that asking musicians to play for free, when she had the resources to pay them, devalued professional musicianship. Palmer found herself on the defensive, explaining the request to the *New York Times,* and claiming, "If you could see the enthusiasm of these people, the argument would become invalid [...]. They're all incredibly happy to be here."[28] Palmer wrote on her blog that none of the volunteer musicians dropped out, but she ultimately moved money from the Kickstarter campaign around and paid those who played with her on the tour.

For Palmer and her proponents, including if not especially the musicians who stuck with her, volunteerism, community, and informed consent were more important than the perceptions of their critics. For the critics, the threat to professionalization and the devaluing of the labor of musicians trumped the social contract between Palmer and her community. The par-

---

27 Carrie Battan, "Steve Albini Slams Amanda Palmer for Asking Fans to Play in Her Band for Free," *Pitchfork,* 13 September 2012.

28 Daniel J. Wakin, "Rockers Playing for Beer: Fair Play?" *Arts Beat; The New York Times,* 12 September 2012.

allels between the Palmer controversy and what I'm critiquing in the DH classroom should, I think, be fairly obvious. In both cases a practitioner relies on unpaid labor to complete a project for which funding is available to compensate that labor. What may not be so obvious is why I empathize with Amanda Palmer, and reject out of hand the professors who use student labor in the classroom.

Palmer and the musicians who chose to play with her were operating within a social contract in which both perceived a benefit to themselves and agreed to participate under conditions of informed consent. The musicians knew in advance the situation they were entering into and did so willingly, eyes open. Despite the economic disparity between Palmer and the musicians who volunteered for her (the fact that Palmer had accounted for the entirety of the Kickstarter funds — which were also given willingly under conditions of informed consent — for the operation of her business and her brand notwithstanding), Palmer had absolutely no power to coerce or compel labor from these musicians, and articulated no expectations beyond those in the original call: show up early, practice a little, play your hearts out, get some beer and hugs.

Conversely, student labor in the classroom is never not coerced. Other critics of student labor in the classroom suggest that alternate assignments could be offered in lieu of project-oriented or public-facing work. While this may be possible if students are doing work as individual contributors for the assignment, I believe that under circumstances where students are expected to work on a professor's project, even if an alternative assignment is offered, students will feel coerced to participate in the professor's project, or that students choosing the alternative project will be penalized for not contributing. The power dynamic of the classroom is such that student choice in this situation cannot be unequivocal, and that faculty objectivity will always be suspect. Miriam Posner in collaboration with her students at UCLA recently developed "A Student Collaborators' Bill of Rights" which articulates these principles quite clearly: "It's important [...] to recognize that students and more senior scholars don't operate from positions of equal power in the academic hierarchy. In particular, students' DH mentors may be the same people who give them grades, recommend them for jobs, and hold other kinds of power over their futures." [29]

The social contract of the classroom

The social contract of the professor-student relationship only allows for limited roles in which the two parties may operate ethically: teacher-student, mentor-mentee, and sometimes employer–employee. In the teacher–student relationship, the professor is responsible for imparting information, knowledge, and skills as defined by a syllabus and course description, and evaluating student work according to an established rubric. The student is responsible for attending class, completing reading and other assignments as described in the syllabus, and demonstrating subject mastery in exams or assignments to meet the requirements defined in the grading rubric. The student (or their proxy in the form of scholarships, grants, or other financial aid) is paying to participate in the course, and while I strenuously resist the neoliberal notion that students are customers engaging in a classroom-based market transaction, the fact that students are paying at least implies that their labor in the classroom, including intellectual property for the work they produce, should belong to them at the end of the day.

The mentor–mentee role involves the professor or other advisor supporting the student in their professional development, providing opportunities to build expertise and gain professional exposure, and supporting their psychic welfare as they progress toward their occupation. The mentee is responsible for articulating their needs, evaluating and implementing their mentor's advice, and taking necessary steps to advance in the profession.

The employer–employee relationship has arguably the strongest delineation, in which the employer supervises the

---

29 Miriam Posner et al., "A Student Collaborators' Bill of Rights," UCLA *Digital Humanities*, 8 June 2015.

work of the employee as defined by a job description, provides training as needed for the employee to perform their job, and pays the employee for their labor in accordance with an agreed-upon wage and schedule. The employee must be present and punctual, represent their skills accurately in order to perform the job, learn what they need to do their duties, and complete their duties as assigned and in a timely manner. This labor may be in the service of a professor's research or project development, funded either with departmental or grant money.

These roles may overlap in that a professor may be teacher, mentor, and employer for a given student, but under distinct circumstances. For example a student may be in a seminar taught by the professor for whom they TA in another class, who is also their dissertation director. But these roles must remain distinct in order for the professor to adequately fulfill their responsibilities to their student, their mentee, and their employee. The social contract of each of these roles allows for distinct expectations for credit and compensation. In the teacher–student relationship, the student has the right to expect that their work is evaluated fairly, that they retain intellectual property, and will receive attribution for the work they produce. If these expectations cannot be met, then the social contract of the classroom has been violated. A grade is neither credit nor compensation. The mentee should not be expected to contribute to the professor's research or project in exchange for their mentorship, unless other arrangements for compensation and credit are made. And even if the employee is paid, they have a right to receive credit for the labor they perform on a project.

The neoliberal university is an easy straw man on which to blame inequities in the treatment of student labor, since it is the values of the neoliberal university that drive both the culture of lack and the shift from a pedagogical to a consumer model. But it is individual faculty who are responsible for the content of their courses and their conduct toward their students, and those most able to report on violations of the social contract of the classroom are also those most liable to be subject to these depredations. As Posner and her student colleagues note, "Students may not feel entirely comfortable raising objections to certain practices if they feel these objections could endanger their academic or career prospects;" an understatement if ever there was one.[30]

Therefore it is up to the community of digital humanities practitioners to acknowledge and engage constructively with this problem. In a positive sense, as a community we can adopt and endorse the principles outlined in the Student Collaborator's Bill of Rights and work to socialize them throughout our institutions, much as many of us have striven to advocate for the principles of open access, or promoted the guidelines for professional collaboration outlined in the Collaborators' Bill of Rights.[31] We can develop and share resources for constructively encouraging students to produce durable public work in the classroom, and for engaging student labor in digital projects in a way that is meaningful to students, as well as to the faculty.

One outstanding example of this is the *Perseus Project* which incorporates student-translated texts into its database. The Perseids platform "offers students an opportunity to produce original scholarly work, which they can then list on their resumes in the context of a job search or when seeking admission to graduate school." Student translators are credited by name, and the site provides durable URIs to student work which can be incorporated into CVs or e-portfolios.[32] The *Perseus Project* offers a model of digital pedagogy that combines academic rigor with technical innovation, allowing students to produce durable products demonstrating their skills and to receive equally durable credit for their labor.[33]

---

30 Ibid.

31 "Collaborators' Bill of Rights," *Off the Tracks: Laying New Lines for Digital Humanities Scholars,* http://mcpress.media-commons.org/offthetracks/part-one-models-for-collaboration-career-paths-acquiring-institutional-support-and-transformation-in-the-field/a-collaboration/collaborators%E2%80%99-bill-of-rights/.

32 Bridget Almas and Marie-Claire Beaulieu, "Developing a New Integrated Editing Platform for Source Documents in Classics," *Literary and Linguistic Computing* 28, no. 4 (2013): 493–503, at 502.

33 It must be noted that there are Family Educational Rights and Privacy Act of 1974 (FERPA) implications for requiring students to produce publicly

But positive methods are unlikely to have a universal impact on the misuse of student labor in the DH classroom. Regardless of the adoption of principled declarations like the Student Collaborators' Bill of Rights by organizations or institutions, there will always be faculty who can justify using student labor in the classroom. In those cases, negative remedies may be necessary. The Collaborators' Bill of Rights includes the provision that "Funders should take an aggressive stance on unfair institutional policies that undermine the principles of this bill of rights."[34] A similar approach should be taken in fostering the ethical use of student labor (which is not addressed in the Collaborators' Bill of Rights). But if we as a community have to wait for funders, particularly those most invested in promoting the maker culture that has enabled these practices, to intervene, we're already lost. Faculty members, librarians, administrators and staff should actively promote the principles of ethical student engagement described by Posner and her collaborators, going so far as to establish Provost-level policies governing this behavior with serious implications for tenure, promotion, and eligibility for Principle Investigator status for violations. This may seem extreme, but students learn what's permissible in the academy and in society from how they are treated in the classroom. Students who experience the anonymization and devaluation of their labor in the classroom will be well equipped to justify labor alienation in their careers as leaders in business, industry, and the academy. This is not a future I want to see and am eager to resist, though it may well be already inevitable.

---

visible work in the classroom — from translations to participation in class blogs, wikis, and Twitter discussions. The University of Oregon Libraries have developed models for informed consent releases that allow students to participate and still protect their privacy. See "For Instructors: Student Privacy and FERPA Compliance," *University of Oregon Libraries,* http://library. uoregon.edu/cmet/blogprivacy.html. Thanks to Charlotte Nunes for bringing these resources to my attention. Other models for FERPA release forms have been developed and are in use elsewhere.

34 "Collaborators' Bill of Rights," *Off the Tracks.*

If, as the DH true believers contend, digital scholarship is the future of the humanities and the academy, we as a community have a responsibility to our students and ourselves to ensure the future DH produces is one we all can live with.[35]

---

35 This chapter is derived and expanded from a talk I gave as part of the Mini-Symposium on DH and Collaboration at THATCamp DHCollaborate at Texas A&M University on May 16, 2014. I was respondent for a roundtable discussion on "Digital Humanities in the Classroom: Students as Collaborators" with Amy Earhart and Toniesha Taylor. My remarks were inspired by the lively Q&A that followed Digital Humanities 2012 session #PS08 "Excavating Feminisms: Digital Humanities and Feminist Scholarship" with Katherine D. Harris (whose paper was read by George Williams), Jacqueline Wernimont, Kathi Inman Berens, and Dene Grigar.

I am grateful to Amy, Toniesha, Sarah Potvin, Liz Grumbach, and Ann Hawkins for their responses to my comments at Texas A&M and subsequent conversations on this topic. The conversation from THATCamp DHCollaborate is storied here: https://storify.com/trueXstory/thatcamp-dh-collaborate-2014#publicize.

Miriam Posner is a collegial and generous interlocutor, and was kind enough to share a draft of the "Student Collaborators' Bill of Rights" she developed with her students. The final document is here: http://cdh.ucla.edu/news/a-student-collaborators-bill-of-rights/

This work would be impossible without the labor of my talented and chaos-tolerant student assistants, past and present. Jeanette Laredo in particular was invaluable to developing the data set for the ongoing research inspired by this topic. Braden Weinmann has provided moral support and a fresh eye as we brought this project over the finish line. My student workers are collaborators in the truest sense, and I'm honored that they choose to work with me and proud to credit them for their efforts. I'm also grateful that my institution enables me to pay them, though not nearly as much as they are worth.

Bibliography

Almas, Bridget, and Marie-Claire Beaulieu. "Developing a New Integrated Editing Platform for Source Documents in Classics." *Literary and Linguistic Computing* 28, no. 4 (2013): 493–503. DOI: 10.1093/llc/fqt046.

Amazon Mechanical Turk. https://www.mturk.com/mturk/welcome.

Aminda. "Crowdsourcing: an ethics dilemma?" *ideaconnection.* 23 April 2011. http://www.ideaconnection.com/blog/2011/04/crowdsourcing-an-ethics-dilemma/.

Battan, Carrie. "Steve Albini Slams Amanda Palmer for Asking Fans to Play in Her Band for Free." *Pitchfork.* 13 September 2012. http://pitchfork.com/news/47845-steve-albini-slams-amanda-palmer-for-asking-fans-to-play-in-her-band-for-free/.

Bedrick, David. "What's the Matter with 'All Lives Matter?'" *Huffington Post.* 24 August 2015. http://www.huffingtonpost.com/david-bedrick/whats-the-matter-with-all-lives-matter_b_7922482.html.

"Collaborators' Bill of Rights." *Off the Tracks: Laying New Lines for Digital Humanities Scholars.* 2011. http://mcpress.media-commons.org/offthetracks/part-one-models-for-collaboration-career-paths-acquiring-institutional-support-and-transformation-in-the-field/a-collaboration/collaborators%E2%80%99-bill-of-rights/.

Digital Humanities Data Curation Institute. http://www.dhcuration.org/institute/.

Dolmaya, Julie McDonough. "The Ethics of Crowdsourcing." *Linguisitica Antverpiensia: New Series — Themes in Translation Studies* 10 (2011): 97–111.

"For Instructors: Student Privacy and FERPA Compliance." University of Oregon Libraries. http://library.uoregon.edu/cmet/blogprivacy.html.

Grusin, Richard. "The Dark Side of the Digital Humanities — Part 2." *Thinking C21.* 9 January 2013. http://

www.c21uwm.com/2013/01/09/dark-side-of-the-digital-humanities-part-2/.

The History Engine. http://historyengine.richmond.edu/.

Howe, Jeff. "The Rise of Crowdsourcing." *Wired.* 1 June 2006. http://archive.wired.com/wired/archive/14.06/crowds.html.

———. "5 Rules of the New Labor Pool." *Wired.* 1 June 2006. http://archive.wired.com/wired/archive/14.06/labor.html.

"The Keating-Owen Child Labor Act of 1916." *The History Engine.* https://historyengine.richmond.edu/episodes/view/5309.

Melville, Herman. *Moby Dick, or The White Wale.* Boston: C.H. Simonds Co., 1922.

Moby Dick. Princeton University. http://etcweb.princeton.edu/batke/moby/.

Muñoz, Trevor. "What IS on the Menu? More Work on NYPL's Open Data — Part 1." Blog post. 8 August 2013. http://www.trevormunoz.com/notebook/2013/08/08/what-is-on-the-menu-more-work-with-nypl-open-data-part-one.html.

———. "Refining the Problem — More Work with NYPL's Open Data — Part 2." Blog post. 19 August 2013. http://www.trevormunoz.com/notebook/2013/08/19/refining-the-problem-more-work-with-nypl-open-data-part-two.html.

Organisciak, Piotr. "Why Bother? Examining the Motivations of Users in Large Scale Crowd-Powered Online Initiatives." MA Thesis, University of Alberta, 2010. https://era.library.ualberta.ca/downloads/c821gk82p.

Palmer, Amanda. "Wanted: Horny-y and String-y Volunteers for the Grand Theft Orchestra Tour!!!!" Blog post. 21 August 2012. http://blog.amandapalmer.net/20120821/.

Pannapacker, William. "Digital Humanities Triumphant?" In *Debates in the Digital Humanities,* edited by Matthew Gold, 233–34. Minneapolis: University of Minnesota Press, 2012. http://dhdebates.gc.cuny.edu/debates/text/46.

Plait, Phil. "#YesAllWomen." *Slate.* 27 May 2014. http://www.slate.com/blogs/bad_astronomy/2014/05/27/not_all_men_how_discussing_women_s_issues_gets_derailed.html.

Posner, Miriam, Haley Di Pressi, Stephanie Gorman, Raphael Sasayama, and Tori Schmitt. "A Student Collaborators' Bill of Rights." UCLA *Digital Humanities.* 8 June 2015. http://www.cdh.ucla.edu/news-events/a-student-collaborators-bill-of-rights/.

Ridge, Mia, ed. *Crowdsourcing Our Cultural Heritage.* Burlington: Ashgate, 2014.

Roberts, Spencer (@robertssw87). Twitter post. 19 July 2013, 11:57 a.m. https://twitter.com/robertssw87/status/358269395551789056.

Shirky, Clay. *Cognitive Surplus: Creativity and Generosity in a Connected Age.* New York: Penguin Books, 2010.

Spiro, Lisa. "'This Is Why We Fight': Defining the Values of the Digital Humanities." In *Debates in the Digital Humanities,* edited by Matthew Gold, 16–35. Minneapolis: University of Minnesota Press, 2012. http://dhdebates.gc.cuny.edu/debates/text/13.

Tomasek, Kathryn (@KathrynTomasek). Twitter post. 19 Jul 2013, 12:02 p.m. https://twitter.com/KathrynTomasek/status/358270653708763137.

UpWork. https://www.upwork.com/o/profiles/browse/c/web-mobile-software-dev/.

Wakin, Daniel J. "Rockers Playing for Beer: Fair Play?" *Arts Beat; The New York Times.* 12 September 2012. http://artsbeat.blogs.nytimes.com/2012/09/12/rockers-playing-for-beer-fair-play/.

"What Is the History Engine?" *The History Engine.* https://historyengine.richmond.edu/pages/about/what_is_the_history_engine.

"What's on the Menu?" *New York Public Library.* http://menus.nypl.org/.

17

The "Unbearable" Exclusion of the Digital

Maha Bali

Digital tools are largely Western products, dominated by American and Western European interests; as such, they can somewhat colonize the spaces and networks depending on them, including by making the "other" invisible or tokenized, if not silenced or oppressed.

This chapter begins with some critiques about the illusions of inclusion in digital spaces, adapted from a two-part article that I wrote with Shyam Sharma for *Hybrid Pedagogy* using a postcolonial perspective, and then moves on to a more focused account of possibilities of creating more open and inclusive spaces and networks, subverting existing digital power structures based on some open online work one or both of us have been involved in.[1] Then I offer some practical directions while remaining critical of the potentials of colonizing treatments of non-Western participants in digital spaces. This work is an attempt to use my grasp of Western digital discourse, as partial insider, to metaphorically "subvert the master's house using the master's tools," something Audre Lorde suggested was impos-

---

1   Maha Bali and Shyam Sharma, "Bonds of Difference: Illusions of Inclusion," *Digital Pedagogy*, 4 April 2014.

sible (and may yet be).[2] The intention is to highlight the importance of creating more equitable opportunities for marginalized or excluded voices from the global peripheries and the margins of geopolitical centers themselves in digital spaces. I argue that this needs to be done by outsiders, insiders, and semi-insiders alike because it is just and beneficial to everyone to diversify communities and enrich learning and sharing for everyone in increasingly globalized digital spaces.

The title of this chapter echoes a blog post I wrote previously on the topic of "unbearable white maleness" of the field of educational technology.[3] I mention how all kinds of lists of digital pioneers, innovators, etc., are overwhelmingly white and male. Keynote speakers at our conferences are often largely white and male. The bibliographies in our writing are often largely white and male (Sara Ahmed also discusses the politics of citation and what inclusions/exclusions in our citations imply).[4] If it is so challenging to find non-white, non-male scholars and leaders in this field, then it is a problem of our field of vision. Is it an issue with the field itself (that in general has barriers towards females and minorities) or in our vision (that we don't see them, that we *selectively* see the white and male ones)? Probably both. Privilege is complex — and many of the minorities we would end up including will be people of some form of privilege in terms of education, class, etc. But what I have found in attempting to be more and more inclusive is that inclusion, while valuable for its own sake, also has practical utility in how it can *enrich* and *transform* a conversation. If we (dominant and subaltern and intersectional) *listen* to the diverse voices we include, if we *include* them by empowering them to include themselves and speak

---

2   Audre Lorde, "The Master's Tools Will Never Dismantle the Master's House," in *Sister Outsider: Essays and Speeches,* 110–14 (Berkeley: Crossing Press, 1984; reprint 2007).

3   Maha Bali, "Unbearable Whiteness, Elusive Exclusivity, and the count-on-one-hand test," *Reflecting Allowed* (blog), 27 March 2015.

4   Sara Ahmed, "Making Feminist Points," *feministkilljoys* (blog), 11 September 2013.

on their own terms, we can potentially subvert the hegemonic whiteness of the digital.

We began our article "Bonds of Difference: Illusions of Inclusion" with a Nepalese proverb: "*A bull that went blind during the monsoon forgets that the world is not always green.*"[5] This old (and somewhat odd) saying provides an analogy for the blind spots around those of us who may be parts of dominant groups and have (moved into) privileged status but, rather ironically, forget what we left behind or how the world is changing. In the first part of the article, subtitled "illusion of inclusion," we credited MOOCs for bringing to light the vast inequities in international higher education when it went virtual and viral, then discussing the lack of inclusiveness in the design and use of ostensibly "global" platforms, tools, pedagogies, and modes of collaboration across national and cultural boundaries. In spite of good intentions (and sometimes blatant pretensions of altruism and respect), platforms like the once wildly popular xMOOCs only exposed and intensified fake universality of design and practices in transnational higher education. Inclusion, we suggested, cannot be achieved by imposing or assuming local values as universal, representing others as tokens, refusing to look beyond those who are already in, denying the hegemony of power, or using stories of those who have bought in to suggest inclusion of everyone from everyone.

Thanks largely to the advent of MOOCs, more scholars around the world are engaged in conversations about cross-border higher education today than ever before. As teachers who are interested in the prospects and pitfalls of emerging academic technologies and pedagogies for learning and teaching across national, social, and cultural contexts, we have been sharing our experiences in different venues. While the hype about the private higher education industry's push for massive open online courses as the future of cross-border education rages on, we find ourselves much more interested in smaller-scale conversations about teaching and learning in all their confusing complexities

---

5   Bali and Sharma, "Bonds of Difference."

in different contexts. Essentially, we were brought together primarily by our different backgrounds, experiences, and perspectives; it is within the interest in difference that we share ideas, interests, and concerns.

Bonds of difference

So, how did the two of us come to share the critical view of MOOCs and what they brought to light about transnational higher education and educational discourse? Maha is a faculty developer and teacher educator at the American University in Cairo who got her PhD from Sheffield, UK; Shyam, is an assistant professor of writing now in New York, a man who hailed originally from the hinterlands of western Nepal (via the routes of education and professional careers in east India, Kathmandu in Nepal, and Kentucky in the US). Because we value (and indeed benefit from) our different identities, ideas, experiences, and perspectives based on our respective backgrounds, we come together in that valuation of difference. However, we are also aware that we are connected by our shared appreciation of difference as it is defined in Western or Westernized academic communities that we are part of.

We started our conversation through a common interest in MOOCs. Our collaboration seems worth noting as a powerful testimony to the idea that networks build communities these days. We had been reading each other's work for about a year, due to our critical, skeptical, "outsider" perspective on MOOCs. The spark for our professional dialogs came when Shyam noticed a twitter conversation Maha was having using the #FutureEd hashtag, which spilled over to emails and responding to blog posts that we and others in our networks wrote. We critique MOOCs from a global South perspective, providing constructive feedback because we are both interested in the possibilities of what could be called a "humane" pedagogy where educators exchange ideas and students participate in learning across contexts, rather than a one-way traffic of information in the name of education. During those email exchanges, we felt that what we were discussing privately needed to be shared with the world, and we started the Google Doc that became this article. [As a slightly humorous side note, we co-wrote this paragraph synchronously while chatting on the margin, and when the paragraph was complete — with the two of us completing each other's sentences — Shyam wrote "Wow" and Maha said "keep it keep it keep it" when he tried to delete it. Several paragraphs in this article have our voices interwoven such that we cannot differentiate where one's voice starts and the other's ends.] The shared dissatisfaction about the vast blind spots created by supposedly shared spaces of MOOCs led us to start a new project called EdConteXts (which I elaborate on below), inviting fellow educators from around the world to share their stories and ideas about teaching in their own contexts.[6] We hoped to highlight to educators the significance of context in terms of geopolitical and material conditions, limitations and opportunities, perspectives and experiences.

My ultimate interest is to contribute constructively to the conversation of emerging academic technologies and pedagogies on the front of cross-border higher education. I am both cognizant of the limitations of technology and passionate about exploring its potential, as "digital agnostics." So, I start by sharing some of the major concerns that temper our enthusiasm about emerging academic technologies for cross-border education, followed by positive potentials of the same developments, giving concrete examples from our personal experiences as academics from the global south who are participating in the emerging spaces for learning and teaching.

In writing the two-part article with Shyam for *Hybrid Pedagogy*, from which I draw some substance here, we were inspired by the journal's interest in pedagogical alterity.[7] The editors invited "a cacophony of voices" in their call for papers. While the variety of voices that the call seemed to envision did not cover

---

6   *EdConteXts,* http://edcontexts.org/.

7   Jesse Stommel, "CFP: Pedagogical Alterity: Stories of Race, Gender, Disability, Sexuality," *Hybrid Pedagogy,* 7 January 2014.

the marginality of academics in and from non-Western contexts/backgrounds, at least our Western-educated selves could identify with the key issues and perspectives in the invited conversation. Like the editors, who draw on the idea that "Difference is not our deficit; it's our operating system" from Fiona Barnett and Cathy Davidson of HASTAC, we were inspired by the idea of looking at difference positively, as a resource and not a problem.[8] And yet, while parts of us wanted to positively respond to the call to challenge the tendency to view difference as a deficiency, we were also keenly aware of the potential risks that the call may embody insofar as it assumes the desirability of difference, originality, reinvention, and such other ideals/objectives as universal rather than local. We shared experiences of how the very attempt at inclusion can inadvertently lead to exclusion. As a simple example, using the metaphor of the "operating system" in order to describe difference as a universal default can exclude many in our communities who will not draw the same inference from the vehicle and/or the tenor of the metaphor. We found it problematic to refer to all humans as having the "same" operating system based on difference (in itself a denial of "difference," if we ever saw one). We are aware that the idea of "difference" itself has different meanings and values for members of different contexts and communities. In certain contexts in our social and professional lives, diversity, divergence, and dissent mean different things for us compared to our colleagues with whom we "theoretically" found common bonds in the celebration of difference.

Different notions of difference

Educators are increasingly embracing the idea of diversity around the world. Many even go further and value dissidence as means of constructive and productive exploration of ideas and rethinking of educational practices. This may make it seem

---

8   Cathy Davidson, "Difference Is Our Operating System," HASTAC (blog), 3 August 2011.

as if educators around the world are teleologically moving toward the idea of promoting difference, reinventing education, challenging conventions and so on. But the challenge arises when the attempts to pursue the above objectives are based on assumptions, contexts, and perspectives assumed universal by those at the global "centers" and hard for others beyond those centers to relate to. So, in spite of all the good intentions, the harder one tries to challenge the current exclusionary systems in favor of accommodating diverse agents and issues, the more entrenched one can be in one's own "local" context, worldview, and frame of reference. For example, a teacher of philosophy who is based or educated in the Euro-American culture may try to promote critical and independent thinking, originality and rejection of convention, novelty and creativity; but a young man in, say, Nepal, may find these "concepts" more fascinating than practically useful. Participating in the philosophy course may still provide him some cultural capital based on learning the ideas; but unbeknownst to the philosophy teacher, the young man may jeopardize his standing in and prospects in his local society, education, and workplace. An Egyptian woman, on the other hand, may be familiar with the notion of criticism on the street, but have no educational experiences of critiquing the authority of the teacher or the text. It may take years, not just a semester or two, for her to be able to behave critically in an educational setting. She may feel a nagging discomfort, a loss of her innocence as she is encouraged to question hidden agendas. Her initial attempts at critical thinking may create social problems as she starts to rebel indiscriminately against other authorities in her life and becomes perceived as "rude."

There are also contexts/times when critical thinking (as understood in the North American context as leaning towards skepticism) as a prerequisite to citizenship is not necessarily the most highly valued approach: during times of political uncertainty and conflict, people might be in need of a more constructive, empathetic approach to citizenship.[9] In these cases, the teacher who is trying to teach the importance of critical and creative thinking might need to learn that these ideas will need to be translated very differently in different contexts. Needless to say, no local value systems can be valid and meaningful universally.

Difference tends to induce discomfort, which individuals and communities try to overcome or avoid in some way. One of the responses to difference — as when scientists deny the influence of context/culture, politics, economics, and material conditions on the shape and direction of their inquiry — is to "deny" it altogether, to say that there is no difference but a universality of subjects, methods, perspectives, and understanding. A second problematic response to difference — as when those who study society and culture "reify" cultural differences — is to seek and find distinctions, creating silos of sociocultural values, norms, and practices. This approach makes people look for difference, and find it, as when they try to understand cultures and societies by "contrasting" them wholesale, instead of paying attention to how people and societies are increasingly malleable and complex. It puts people and societies in containers defined by distinctions. Thus, the celebration of difference goes in the opposite direction of denying it and tends to overshadow complex overlaps between differences and similarities among societies and cultures.

A third response to difference is to try to recognize differences as a normal and default condition of human life and society. At first, this sounds like an absolutely true description of reality, a practical middle ground between the two extremes above. But on a closer look, this view can also easily go too far. This view "universalizes" difference on the basis of certain local conceptualizations and valuations of difference. Those who assume the universal value and meaning of difference don't realize that it means different things in different contexts to different people. The universal valuation of difference starts by assuming one's

---

9   Maha Bali, "Critical Citizenship for Critical Times," *Al-Fanar Media,* 19 August 2013.

own ground as home, one's own terms as the fulcrum around which everything else has to turn, whereas difference is intersectional and dynamic. Different in what context, different from *what*, different from *whom*, different *in what way*?

Thus, we postulate that we cannot find common bonds if we forget the paradox of trying to find similarity in difference. If differences are to be valued, they may need to be understood in their own terms, the confusions that they create being tolerated, the complexities that they give rise to be appreciated. That is, for instance, when we say that we can and should all question conventions, be critical and creative, strive for originality, and so on, certain assumptions and conventions still undergird these ideals and ideas. The attempts to create "bonds" through shared spaces, agreed-upon ideas, common denominators, collaboratively derived perspectives may ultimately fail when the foundation of the entire attempt is one party's familiar territory, when participants of a discussion are from many and vastly different contexts/backgrounds, and when the perspectives are only common via mimicry of those at the center by those in the peripheries. Thus, we urge our readers to be aware that whenever we try to "find" bonds by embracing differences, we might be impoverishing our ideas, weakening the very bases of our bonds.

Fake universality and illusions of inclusion

What are some of the ways we experience exclusion as young non-Western academics? There are the small ways, like the joking cultural reference on Twitter to having a beer or tattoos (excludes Maha as a practicing Muslim). There are the well-intentioned claims that show lack of awareness of global injustice, like a claim that university should not be about getting jobs. In theory, this may sound right almost universally, but claims of "learn before you earn" are slogans that mean nothing for someone who needs to earn so they can live. Some people cannot afford to learn first and earn later. This tendency takes more serious forms when canons of knowledge are assumed to have inherent value and referred to repeatedly. For example, Martha

Nussbaum attempts (in her 1997 book *Cultivating Humanity*) to reform liberal arts education by making it more inclusive of other cultures (including opening up space for knowledges of marginalized peoples) while she continues to refer to liberal arts education from the perspective of the ancient Greeks.[10] She insists that the study of philosophy is inherently valuable for a liberal arts education, despite the fact that the majority of academic philosophers in the US are white and male, who are constantly referring to canons that are largely white and male, becoming largely unwelcoming to difference. And of course, she accepts liberal arts as the ideal approach to "reinvent," rather than one approach of many that could have been explored. This is not unexpected given that she is an American philosopher. What is strange is that she does not recognize that her recommendation of inclusiveness was paradoxically not used in her book even as she advocates it.

Then there were the mind-bending claims that MOOCS would make "quality education" accessible for millions of students around the world who "have not had such access so far." In the absence of considerations about the relevance of content, linguistic, technical, and intellectual accessibility in the delivery, and appropriateness and effectiveness of pedagogy, the idea of "quality education" just becomes absurd. A teacher may use bland and non-context-restricted questions and learning objectives in the course, but such an attempt is also bound to find a common ground by denying, reifying, or universalizing contextual differences. For many disciplines, subjects, and issues, it may be ultimately impossible to educate anyone in the entire world by using a one-size-fits-all course and from the convenience of one's laptop.

We have observed that quite often, in the attempt to make their teaching more inclusive, teachers in the global North include superficial or incidental references to other cultures. What our colleague Dorothy Kim called "referential tokenism" — or a

---

10 Martha C. Nussbaum, *Cultivating Humanity: A Classical Defence of Reform in Liberal Education* (Cambridge: Harvard University Press, 1998).

mention of diversity to claim inclusion — cannot take the place of true inclusion in teaching/learning or professional networking. Attempts at inclusion can only be authentic and meaningful when we make the content, process, and outcome of education more egalitarian, open, and inclusive.

Teachers who want to create meaningful learning environments for participants from vastly different backgrounds must construct and design their courses with an awareness of the fractals of contextual, cultural, and material differences from the ground up — not by treating those differences as an afterthought and by using superficial gestures. As long as the teaching and learning experiences are only envisioned within the dominant worldview, incorporating patchwork elements about or from diverse others will only serve to distort or reify the differences rather than allowing the stakeholders from different backgrounds to truly participate in learning and sharing ideas. Without rethinking the assumed universals underlying the course design, content, and pedagogy, the canon of established Western knowledge will live on, embellished here and there by some exotic accessories.

Of course, we all have our own unconscious, habituated ways of thinking about the "world out there." And of course, what I say above does not diminish my appreciation and regards for teachers/scholars who are trying to be inclusive and respectful toward diverse groups of people from around the world. But I do want to emphasize that the moment teachers try to cross their local contexts and invite participants from other contexts, they should also start becoming aware about how their local worldviews and understanding are bound to be incomplete and insufficient. I want to urge teachers to acknowledge that their ideas and teaching methods may not be very meaningful in many different contexts around the world. If educators from dominant contexts were to cultivate and foster such awareness sufficiently, that would add value and incentive for educators from different places or with knowledge about different places to join cross-border educational initiatives, to share their knowledge, to make greater and more positive impacts.

The coming together of educators across borders clearly promises tremendous possibilities for the advancement of education within and across borders. It is the vision of such positive opportunities and possibilities that inspire me to join the conversations, critiquing constructively where we see opportunities for improvement, appreciating what we find beneficial from our local and global perspectives.

As we discuss in the second part of the article, inshallah, subtitled "participation as inclusion," inclusive communities and collaborations can only happen when we stop assuming that we can do so by simply creating the space and inviting everyone to it. We must ask what principles of learning and sharing the spaces are based on. Whose contexts and values undergird the space and whose voices are being heard/unheard, misunderstood/understood, privileged/marginalized, or stereotyped/glamorized.

Bonds of difference: Participation as inclusion

> We [the minorities] and you [the dominant] do not talk the same language. When we talk to you we use your language: the language of your experience and of your theories. We try to use it to communicate our world of experience. But since your language and your theories are inadequate in expressing our experiences, we only succeed in communicating our experience of exclusion. We cannot talk to you in our language because you do not understand it.[11]

What Lugones and Spelman describe above can be illustrated by considering the Arabic word *mazloum* (مظلوم), which has no direct English translation. It means "the person against whom injustice has been done," but the connotation is so much deeper than that. It is as strong as the word "oppressed" but actually

---

11 María C. Lugones and Elizabeth V. Spelman, "Have We Got a Theory for You! Feminist Theory, Cultural Imperialism and the Demand for the 'Woman's Voice,'" *Women's Studies International Forum* 6, no. 6 (1983): 573–81, at 575.

oppression is a different word in Arabic, *idtihad* (اضطهاد). When educators try to make complex experiences "legible" to diverse communities of learners and colleagues, those attempts can be problematic in ways that belie the sincerity and commitment of the educators. In our attempts to be legible, "the relative simplicity and platonic orderliness of the [simplified, legible] vision [which] represents rationality" that we use may be tantamount to "authoritarian power," a power that "demolish[es] the old reality if necessary."[12] Top-down attempts to educate the world, enlighten the ignorant, liberate the deprived — whether or not such terms are used or accepted by educators whose voices are inevitably heard through the mechanisms of power and privilege — can make our "rational Utopia fail horribly."[13] Any grand vision to develop educational/pedagogical models that fit every society, promote learning and knowledge-making globally, bring educators together that are based on certain local understandings and worldviews can be, in the words of Rao again, "generally dangerous, and a formula for failure, [in] that it does not operate by a thoughtful consideration of local/global trade-offs, but through the imposition of a singular view as 'best for all' in a pseudo-scientific sense."[14] In fact, even as we critique such grand visions, we become keenly aware that the very basis of our critique may be singular, limiting, and exclusive of other critical perspectives. Thus, we urge that any educational initiatives that strive to engage people and ideas from across borders and contexts incorporate people and perspectives from as many contexts as possible in the very construction, development, and promotion of those initiatives.

Full inclusion may be an impossible goal, not just across sociocultural and geopolitical borders but also within those borders. However, educators can and should strive for genuine

---

12 Venkatesh Rao, "A Big Little Idea Called Legibility," *ribbonfarm* (blog), 26 July 2010. Rao cites James C. Scott, *Seeing like a State: How Certain Schemes to Improve the Human Condition Have Failed* (New Haven: Yale University Press, 1999).

13 Ibid.

14 Ibid.

attempts toward inclusion by not assuming the local to be universal, by inviting colleagues and other learners to participate on their own terms, and by developing a high sense of tolerance and openness about difference. Howard describes the importance of culturally relevant pedagogy, because "teachers must be able to construct pedagogical practices that have relevance and meaning to students' social and cultural realities."[15]

Prospects for productive participation

How can educators strive for inclusion? For example, MOOCs, however un-inclusive their practice has become since the idea was taken over by corporate interest, were originally developed and experimented as a "connectivist" pedagogical model, which continues to thrive in parallel with the more famous/funded MOOCs on platforms like Coursera and EdX. Adapting xMOOCs to other languages (like the Arabic Edraak) is a very small and limited step.[16] I have since collaborated on multiple open online educational experiments meant to embody participation as inclusion, which we discuss next.

Inclusion by participation

To promote equity, inclusion, and participation of educators from across geopolitical and cultural borders, two (admittedly imperfect) initiatives stand out: EdContexts.org and Virtually-Connecting.org.

Edcontexts.org is "an informal network and community of teachers, scholars, students, researchers, and others interested in promoting conversations about education in and across con-

---

15 Tyrone Caldwell Howard, "Culturally Relevant Pedagogy: Ingredients for Critical Teacher Reflection," *Theory Into Practice* 42, no. 3 (2003): 195–202, at 195.

16 Maha Bali, "Another Step Forward for Arab MOOCs," *Al-Fanar Media,* 23 February 2014.

texts around the world."[17] Several educators from the global South started this website/blog to counter the largely Western discourse on education, to provide nuanced, localized perspectives from people who are living in different areas of the world. We launched it soon after the Bonds of Difference articles were published (the idea of the website and the articles had emerged in parallel). We invited a diverse team of facilitators, and actively solicit posts by educators from the global South, or some educators from the global North who are highly sensitive to contextual conversations in education. Most of our posts are not written by facilitators; for some time, we posted f5f ("favorite five finds") by linking to the work of others that we appreciated and wished to promote. We recognize that while we have provided this avenue for a different perspective on education, our website still uses English language since all of us (facilitators) do not speak the same language as each other or as our writers. And of course we use WordPress on Reclaim Hosting and Twitter and Facebook for promotion (all of them US-centric technologies).

VirtuallyConnecting.org was co-founded by Rebecca J. Hogue and me "to enliven virtual participation in academic conferences, widening access to a fuller conference experience for those who cannot be physically present at conferences. Using emerging technologies, we connect onsite conference presenters with virtual participants in small groups. This allows virtual conference attendees to meet and talk with conference presenters, something not usually possible."[18]

Conferences are all about networking and building social capital — conferences that livestream sessions for virtual participants do not achieve this as they only allow minimal interaction via text boxes among virtual participants or on Twitter. While this is better than nothing, Virtually Connecting goes beyond that and gives virtual participants an opportunity to have a live conversation with onsite presenters/participants/keynote

---

17 *EdConteXts*, http://edcontexts.org/.
18 "About Us," *Virtually Connecting*, http://virtuallyconnecting.org/about/.

speakers. By doing so, we expand and extend conference conversations — and in turn enrich the onsite conversation by the perspectives coming in from outside.

Compare the situation of a full-time academic and someone who is an adjunct or graduate student or international scholar. The amount of money a full-time academic receives that could pay for three or four conferences a year might be the same as the money needed for an international scholar to attend *one* conference a year, and is likely an amount an adjunct does not get at all, and that graduate students do not regularly have access to. Moreover, many people attend conferences but do not get opportunities to have in-depth conversations with others onsite, particularly keynote speakers. As virtually connecting has evolved, conversations with keynotes and onsite participants have grown to have a stronger participant voice. Rather than virtual participants simply asking onsite speakers questions, we are having multi-way conversations, discussing complex questions, and striving for deeper understanding and even solutions. Keynote speakers have thanked us for helping them expand their thinking or extend the conversation beyond the keynote speech itself.

Even though Virtually Connecting sessions are open to anyone to participate (up to the 10 person limit of a Google hangout on Air), we recognize that the experience remains exclusive in the following ways:

1. Conversations are conducted in English;
2. Synchronous livestreaming video requires a minimum infrastructure that supports it, suitable time zones, and digital literacy and willingness for the person to speak live and recorded. Having a team spanning different time zones expands the possibilities;
3. Despite a growing team of virtually connecting buddies, some people might still be uncomfortable requesting to join one of these events if they do not know any of us personally; we have been largely promoting ourselves on Twitter which not everyone uses;

4. We are using Google hangouts, with all the ethical issues that come with using Google products. Other options for synchronous communication are available but none offer the free, livestream & record (with immediate upload to YouTube) options.

One could argue that Virtually Connecting is on others' terms because most conferences and onsite speakers are Western and the majority of the team (including the co-founder Rebecca) are Western. However, we call it on "our terms" because the initial pilot of #et4buddy was meant to meet a need of mine, and the technical design and approach were all made to fit a developing country context (e.g., Google hangouts poses much fewer technical problems for my internet setup than any other synchronous tool). While Virtually Connecting serves the needs of scholars in the Western world, too, it targets mainly people with limited access to conferences and who are thus less privileged in an academic context — it not only gives us *access to listen*; it gives us *voice* and *presence*.

Inclusion on others' terms: #TvsZ, ConnectedLearning.tv & #DigiWriMo

By inclusion on others' terms, we mean that we joined already-existing initiatives, and while participating in them, attempted to make them more inclusive. By participating as a facilitator of #TvsZ (a Twitter game initially designed by Pete Rorabaugh and Jesse Stommel), the game became more conscious of cultural nuances, timezone differences, bandwidth limitations and potential for linguistic and cultural enrichment of the experience when a large number of participants from Egypt joined.[19] When Shyam and I co-facilitated ConnectedLearning.tv in July 2015, we brought in an internationally and ethnically diverse set of guests to discuss topics such as equity in ed tech, trans-

---

19 Maha Bali, "Embodying Openness as Inclusive Digital Praxis," *Hybrid Pedagogy,* 22 January 2015.

media, educators across contexts, and emerging trends in open scholarship.[20] When I was invited to facilitate Digital Writing Month (originally created as part of *Hybrid Pedagogy*), my first thoughts were to include others to co-facilitate and to invite a diverse group of guest contributors.[21] However, again, we realized that all of our guest contributions were in the English language (and when Maha wrote about this, one participant from Singapore taught us some Chinese).

And yet

Such collaborative, open, online experiences have that potential of including and spreading the power of participants' voices. Educators need to remember that the attempts to work, learn, and teach across vastly different contexts around the world are no small feats, and therefore, the challenges remain sticky and they deserve continued attention and dedicated attempts to address them. In any of the examples above, language and digital fluency as well as technical access can lead to exclusions. In any situation where people supposedly are given "equal voice," vocal or eloquent or influential minority voices can have the power to silence others. People are different on many levels, and those differences should not be generalized, idealized, or viewed in monolithic ways.

Educators can delight in the fact that there are promising developments from the perspective of sensible pedagogy and effective teaching and learning. At the same time, they should also realize that there are still tremendous needs for further thinking in the particular case of engaging learners and educators from different national, cultural, and geopolitical backgrounds.

As I, Maha, tweeted during the Digital Learning Research Network conference:

---

20 "Learning and Leading in a Connected World with Educator Innovator, National Writing Project and EdConteXts," *Connected Learning,* http://connectedlearning.tv/learning-and-leading-connected-world.

21 Maha Bali, "Digital Writing Month: Striving for Inclusion in Open Online Learning," *The Chronicle of Higher Education* (blog), 28 October 2015.

> We shouldn't stop striving for inclusion but we need to recognize our limitations & shifting complexity of ever reaching it[22]

Also:

> Minority voices will almost never be truly heard in a room where dominant perspectives exist. Too much struggle[23]

Looking back at each of the examples above, we have been working with the master's tools, the technologies, as well as the approaches, developed in a Western world, using the English language, but trying to make way for a non-Western perspective to influence how it is used.

Cultivating awareness, empathy, and openness

To echo an old saying, drastic changes demand drastic adaptations. As educators, we are able to share our ideas literally across the world, with thousands of learners and colleagues, and with a great deal of added affordances that emerging technologies provide to educators and learners. However, the same developments have also exponentially increased the need to be aware and tolerant about differences, to be willing to accept failure and even misunderstanding, to cultivate empathy in the face of complexity and confusion. The same developments that have opened up unprecedented opportunities for cross-border education and scholarly discourse have also served to expose, quite frankly, embarrassing realities about the status of cross-border education. Most strikingly, otherwise serious and sensitive educators from dominant societies and academies log on to supposedly "open" spaces online, set up curricular and pedagogical mechanisms on their own terms, then all but forget the vastly different contexts of the majority of participants whom they claim they are benefiting. But on the heels of such exposures of

---

22 Maha Bali (@Bali_Maha), Twitter post, 17 October 2015, 6:47 a.m.
23 Maha Bali (@Bali_Maha), Twitter post, 18 October 2015, 2:32 p.m.

parochialism, insensitivity, and lack of awareness have come a number of new developments, as we discussed above.

The fact that educators can now reach out to thousands also means that they need to slow down, to invite participants from different contexts for genuine participation, to listen and learn from others, to enrich their own understanding. It is also equally necessary to not simply criticize, find faults, and pass judgments when new opportunities bring about new challenges and blind spots. Criticism seems necessary, but that shouldn't be an end; it should be a means. What are the ways in which we can make critique more useful and productive? Thus, we urge that educators across borders offer different perspectives as a necessary, constructive addition and enrichment for the ongoing conversations about cross-border education — and often the lack thereof.

In the *Hybrid Pedagogy* call for proposals we were drawn to the quotation from Paulo Freire: "the great humanistic and historical task of the oppressed [is] to liberate themselves and their oppressors as well." We certainly don't see the many well-intentioned and intellectually inspired colleagues from Europe, North America, and elsewhere as "oppressors" of any kind.[24] We are in fact sympathetic to even those who buy into and promote open education as a means to "save" the world out there from its own ignorance and backwardness; for instance, when we come across courses whose design and execution signals no consideration for how participants from vastly different contexts around the world may partake of the course/community, we simply view that as an opportunity for pointing out the weakness in the pedagogy and curriculum. We write with the understanding that there is a positive need for constructive dialogues in the world of cross-border higher education more than ever before.

There is no guarantee that goodwill of educators in one place will translate into goodwill across contexts, or that goodwill when implemented will result in universal social good. And

---

24 Paulo Freire, *Pedagogy of the Oppressed,* trans. Myra Bergman Ramos (New York: Bloomsbury, 2000), 44.

we are not suggesting we give up on offering education across borders. I am suggesting that such an education cannot be assumed to represent or meet the needs of diverse others unless it involves those diverse others on deeper levels. Even if those diverse others are still a privileged subset of what and whom they represent (as academics often are), we cannot assume that we know. We should always assume there is more to know, and that others might know it better.

As Sidorkin states in "Toward a Pedagogy of Relation," "polyphonic truth is a much more workable concept than any other form of knowledge. Relations thus are not describable by one person. Instead, a group of people can describe relations, and then one person can describe their description."[25] We want to go further than having one person offer their description to the world, like an anthropologist, and instead give each person the space and voice to describe without an intermediary, as in autoethnography (see this "untext," collaborative autoethnographic accounts of the #rhiz014 MOOC).[26]

Working with and through each other should not be seen as a liability, a hassle. It is a process that can transform us. We remember Bakhtin here: "I am conscious of myself and become myself only while revealing myself for another, through another, and with the help of another... I cannot manage without another, I cannot become myself without another."[27] Indeed, we wrote this document through an exploration of our similarities and our differences, through learning about ourselves as we reveal ourselves to each other and to the potential reader. Our experience tells us that we approach all knowledge in this

25 Alexander M. Sidorkin, "Toward a Pedagogy of Relation," *Rhode Island College Faculty Publications,* Paper 17, 4.
26 Keith Hamon, Rebecca J. Hogue, Sarah Honeychurch, Scott Johnson, Apostolos Koutropoulos, Simon Ensor, Sandra Sinfield, and Maha Bali, "Writing the Unreadable Untext: A Collaborative Autoethnography of #RHIZO14," *Hybrid Pedagogy,* 4 June 2015.
27 Mikhail Bakhtin, *The Dialogic Imagination: Four Essays,* ed. Michael Holquist, trans. Caryl Emerson and Michael Holquist (Austin: University of Texas, 1981), 287.

way, recognizing that our own knowledge is *always necessarily* "partial," as Ellsworth suggests: partial as in biased, partial as in incomplete.[28] Jamaica Kincaid reminds us of the impossibility of a (colonized) people giving an accurate account of their own events — the important thing to remember here is that Western/white accounts are also inherently incomplete, inaccurate and biased, and that the only way to get a clearer picture is to continually make room for more lenses, recognizing the power dynamics and intersectionality in doing so.[29]

With Edcontexts.org we envisioned tapping into the experiences and expertise of scholars from different contexts in order to create a shared platform for growing new ideas, forging new relations, and cultivating awareness and empathy. We intended to make knowledge-making and knowledge-sharing in and across many and different contexts a truly open enterprise, open in its many senses — as ongoing, allowing access, exposed to the outside, making the inside exposed, unfolding, and accepting of anyone. We are grateful to all fellow educators from around the world who have contributed to EdContexts.org and we hope to continue promoting (inshallah) the voices of educators who may not feel comfortable, be heard, or taken seriously in transnational platforms that are dominated mostly by the same dominant groups of people, Western white men, regardless of good intentions.

---

28 See Elizabeth Ellsworth, "Why Doesn't This Feel Empowering? Working through the Repressive Myths of Critical Pedagogy," *Harvard Educational Review* 59, no. 3 (1989): 297–324.

29 Jamaica Kincaid, *A Small Place* (New York: Farrar, Straus and Giroux, 1988).

Bibliography

"About Us." *Virtually Connecting*. http://virtuallyconnecting. org/about/.

Ahmed, Sara. "Making Feminist Points." *feministkilljoys* (blog). 11 September 2013. https://feministkilljoys.com/2013/09/11/ making-feminist-points/.

Bakhtin, Mikhail. *The Dialogic Imagination: Four Essays*. Edited by Michael Holquist. Translated by Caryl Emerson and Michael Holquist. Austin: University of Texas, 1981.

Bali, Maha. "Critical Citizenship for Critical Times." *Al-Fanar Media*. 19 August 2013. http://www.al-fanarmedia. org/2013/08/critical-citizenship-for-critical-times/.

———. "Another Step Forward for Arab MOOCs." *Al-Fanar Media*. 23 February 2014. http://www.al-fanarmedia. org/2014/02/another-step-forward-for-arab-moocs/.

——— and Shyam Sharma. "Bonds of Difference: Illusions of Inclusion." *Digital Pedagogy*. 4 April 2014. http://www. digitalpedagogylab.com/hybridped/bonds-difference-illusions-inclusion/.

———. "Embodying Openness as Inclusive Digital Praxis." *Hybrid Pedagogy*. 22 January 2015. http://www. digitalpedagogylab.com/hybridped/embodying-openness/.

———. "Unbearable Whiteness, Elusive Exclusivity, and the Count-on-one-hand Test." *Reflecting Allowed* (blog). 27 March 2015. http://blog.mahabali.me/blog/educational-technology-2/unbearable-whiteness-elusive-inclusivity-the-count-on-one-hand-test/.

——— (@Bali_Maha). Twitter post. 17 October 2015, 6:47 a.m. https://twitter.com/Bali_Maha/ status/655334280302895104?ref_src=twsrc%5Etfw.

——— (@Bali_Maha). Twitter post. 18 October 2015, 2:32 p.m. https://twitter.com/Bali_Maha/ status/655813825535590400?ref_src=twsrc%5Etfw.

———. "Digital Writing Month: Striving for Inclusion in Open Online Learning." *The Chronicle of Higher Education*

(blog). 28 October 2015. http://chronicle.com/blogs/
profhacker/digiwrimo15/61207.

Davidson, Cathy. "Difference Is Our Operating System."
HASTAC (blog). 3 August 2011. https://www.hastac.org/
blogs/cathy-davidson/2011/08/03/difference-our-operating-
system-fiona-barnett.

EdConteXts. http://edcontexts.org/.

Ellsworth, Elizabeth. "Why Doesn't This Feel Empowering?
Working through the Repressive Myths of Critical
Pedagogy." *Harvard Educational Review* 59, no. 3 (1989):
297–324. DOI: 10.17763/haer.59.3.058342114k266250.

Freire, Paulo. *Pedagogy of the Oppressed.* Introduction by
Donaldo Macedo. Translated by Myra Bergman Ramos.
New York: Bloomsbury, 2000.

Hamon, Keith, Rebecca J. Hogue, Sarah Honeychurch, Scott
Johnson, Apostolos Koutropoulos, Simon Ensor, Sandra
Sinfield, and Maha Bali. "Writing the Unreadable Untext:
A Collaborative Autoethnography of #RHIZO14." *Hybrid
Pedagogy.* 4 June 2015. http://www.digitalpedagogylab.com/
hybridped/writing-the-unreadable-untext/.

Howard, Tyrone Caldwell. "Culturally Relevant Pedagogy:
Ingredients for Critical Teacher Reflection." *Theory
Into Practice* 42, no. 3 (2003): 195–202. DOI: 10.1207/
s15430421tip4203_5.

Kincaid, Jamaica. *A Small Place.* New York: Farrar, Straus and
Giroux, 1988.

"Learning and Leading in a Connected World with Educator
Innovator, National Writing Project and EdConteXts."
Connected Learning. http://connectedlearning.tv/learning-
and-leading-connected-world.

Lorde, Audre. "The Master's Tools Will Never Dismantle the
Master's House." In *Sister Outsider: Essays and Speeches,*
110–14. Berkeley: Crossing Press, 1984; reprint 2007.

Lugones, María C., and Elizabeth V. Spelman. "Have We Got
a Theory for You! Feminist Theory, Cultural Imperialism
and the Demand for the 'Woman's Voice.'" *Women's*

Studies International Forum 6, no. 6 (1983): 573–81. DOI:
10.1016/0277-5395(83)90019-5.

Nussbaum, Martha C. *Cultivating Humanity: A Classical
Defence of Reform in Liberal Education.* Cambridge:
Harvard University Press, 1998.

Rao, Venkatesh. "A Big Little Idea Called Legibility."
ribbonfarm (blog). 26 July 2010. http://www.ribbonfarm.
com/2010/07/26/a-big-little-idea-called-legibility/.

Scott, James C. *Seeing like a State: How Certain Schemes to
Improve the Human Condition Have Failed.* New Haven:
Yale University Press, 1999.

Sidorkin, Alexander M. "Toward a Pedagogy of Relation."
Rhode Island College Faculty Publications. Paper 17. http://
digitalcommons.ric.edu/facultypublications/17.

Stommel, Jesse. "CFP: Pedagogical Alterity: Stories of Race,
Gender, Disability, Sexuality." *Hybrid Pedagogy.* 7 January
2014. http://www.digitalpedagogylab.com/hybridped/
cfp-pedagogical-alterity-stories-race-gender-disability-
sexuality/.

18

The Politics of Visibility

Eunsong Kim

*Trending is visibility granted by the algorithms of a
closed, private corporation — De-exceptionalize it!*

*"Twitter determines its trending topics through the site's Trends
algorithm, which Twitter has not released to the public."*[1]

"I'm not afraid, Free Trade!"
— Don Mee Choi[2]

Every time CNN points to "Trending" in order to discuss breaking news, we should laugh. That is, laugh at: CNN, journalists,
experts, the simulacra. Smirking at the notion that privatized,
opaque institutions of selective coverage are working with
other privatized, opaque institutions of selective timelines to
define what's public, what's universal, what's important. Intellectual, digital, and digitized labor is important, and too often
dismissed.[3] This interrogation of visibility coverage and trending isn't a critique of online activism and discourse. Rather, I

---

1   Ross Wilson, "Trending on Twitter: A Look at Algorithms Behind Trending
    Topics," *Ignite Social Media,* 3 December 2012.
2   Don Mee Choi, "Failfail," *Tripwire* 9 (2015): 31–37, at 34.
3   Elizabeth Losh, "Hashtag Feminism and Twitter Activism in India," *Social
    Epistemology Review and Reply Collective* 3, no. 13 (2014): 10–22.

want to examine the political pontification around "trending" and "tagging" — the politics of Twitter algorithms, the bird's eye commentary of "surprise" and "dismay" at certain trends, and ultimately, alternative discourses.

The crisis of visibility

#Solidarityisforwhitewomen was a hashtag created by Mikki Kendall in August of 2013.[4] The tag situated a structural critique of global north feminism and the oppressive dynamics underlying white feminist understandings of "empowerment" and "progress." The tag was also a direct critique against white feminists who defended "male feminist" Hugo Schwyzer. The hashtag trended worldwide, prompted a fury of articles by supporters, surprised journalists and critical commentators.[5]

Criticism of tags like #solidarityisforwhitewomen situated the trending of such conversations as an exceptional and at the same time misdirected, mis-use of digital energy by racialized and gendered users. However, hashtags like #happybirthdaytaylorswift and #happybirthdaydemilevato have yet to received "critical" scholarship, essays, exposes, and op-eds. This is because their themes are viewed as fitting into the dynamic and expectations of what is expectedly visible, what a trend might constitute, and what digital cyborgs are *supposed* to be interested in.

#solidarityisforwhitewomen curated a multitude of responses — I focus on Michelle Goldberg's article in *The Nation* as 1.[6] *The Nation* prides itself on being a platform for progressive news 2. Goldberg's article generated a variety of responses by authors that both agreed and disagreed with her. Michelle Goldberg's "Feminism's Toxic Twitter Wars" focused on criticizing the digital labor of black feminists online, particularly Mikki Kendall

---

4   See Kendall's blog: *hoodfeminism,* https://hoodfeminism.com/.

5   "Twitter Sparks a Serious Discussion About Race and Feminism," NPR, 23 August 2013.

6   Michelle Goldberg, "Feminism's Toxic Twitter Wars," *The Nation,* 29 January 2014.

who created #solidarityisforwhitewomen.[7] The crux of the argument focused on how surprised mainstream feminists were by critiques of their objectives, agendas, methodologies. Goldberg cites Kendall as being the originator of #solidarityisforwhitewomen and as one of the main critical voices of the tag #femfuture. The tag #femfuture came under scrutiny for its erasure of nonwhite voices, questionable Twitter ethics, flattening of difference, and concerns surrounding accessibility.[8] What does it mean when a conference on feminist futures, revolution, and digital technologies at the Barnard Center for Research and Women, organized by Courtney Martin and Vanessa Valenti, does not consider accessibility to be of primary concern to its attendees?[9] Goldberg defended that, "#Femfuture was earnest and studiously politically correct" and that organizers were "floored" to learn that feminists online were critiquing their efforts. How dare feminists disagree with #femfuture, and then go on to trend a movement of their own? Before addressing the claims of the critiques, Goldberg states that such conversations create a "toxic" environment for feminism online.

Goldberg's analysis conveniently ignores how #solidarityisforwhitewomen and criticism of #femfuture were formed to critique white supremacy in feminism (a historical, ongoing, structural claim) — instead in the article Goldberg accuses black feminists of using their online "egos" to play up racial politics, thereby derailing the unification against patriarchy project (as if dismantling white supremacy are side games deployed by black feminists to mess the revolution up!!!). This was and is Goldberg's central and ongoing thesis: criticism of mainstream white feminism are side conversations hurting the "actual" work that feminism needs to do…

---

7   See the #solidarityisforwhitewomen feed at https://twitter.com/search?vertical=default&q=%23solidarityisforwhitewomen&src=typd.

8   Dorothy Kim and Eunsong Kim, "The #TwitterEthics Manifesto," *Model View Culture*, 7 April 2014.

9   Courtney E. Martin and Vanessa Valenti, *#FemFuture: Online Revolution*, vol. 8 (Barnard Center for Research on Women, 2012).

And what might actual feminist work look like? While Goldberg does not address this in her article, writer Jessica Grose notes that #femfuture's overall funding and conference goals were not interrupted.[10] Visible online criticism of mainstream feminist agendas did not prevent Valenti and other organizers from securing necessary grants and proceeding with the conference as planned. Visible and direct critique — though the central concern in *The Nation*'s article — apparently had no bearing on the overall outcome of #femfuture. An analysis of why visible online critique did not affect the funding outcomes of #femfuture should be a central concern for digital humanities scholars — and key to conversations surrounding digital visibility.

The management/displacement of structural critiques

Visibility seems to be the central concern for Goldberg — not the critique of white supremacy within feminist discourse — but the visibility of critique. Goldberg's attachment to the exceptional nature of such visible tags is that they are exceptional, singular. In her view, critiques of white supremacy within feminism are exceptional concerns so they should receive limited visibility. However, irrespective of material outcome (i.e., funding) they have received exceptional attention — and this must be rectified.

While visibility might be a central focus for white feminist's like Goldberg, visibility is not always the central objective for trans, women of color, or black feminisms. Goldberg's concern for visibility might be better interpreted as a PR concern — her critique centrally focused on discussions of certain tags and the "toxic" and "massive" presence of black feminists. The underlying argument seemed to be: Feminism is tenuous as it is — the visibility it receives is hackneyed and deployed haphazardly. Why is the focus on the wrongdoings of white women? Concluding that truly, #solidarityisforwhitewomen.

---

10   Jessica Grose, "Is 'Toxic' Online Culture Paralyzing Feminism?" *XXFactor: What Women Really Think* (blog), *Slate*, 31 January 2014.

Almost a year since "Feminism's Toxic Twitter Wars" was published, Goldberg has taken a slightly more critical approach to the question of visibility.[11] However, questions of how visibility has not been a fixed condition, have historically and currently been taken up by black and woc feminists. At the "We Cannot Live Without Our Lives: A Conversation on Anti-blackness, Trans Resistance and Prison Abolition" forum at University of California, San Diego, activist and artist Reina Gossett problematized the function of visibility for black trans women.[12] Gossett articulated that particularly for black communities, "Visibility is a pillar of criminalization, not a tenant of liberation." In conjunction, Grace Hong has argued in *The Ruptures of American Capital,* that "for women of color feminist practice, visibility is a rupture, an impossible articulation." Hong writes that while some have articulated invisibility as unnatural, "so too, is visibility is unnatural; it is also a kind of violence. [...] [V]isibility is not inclusion, but surveillance."[13]

Visibility — while perhaps essential in the grab for legitimized forms of violence and power (state power, representational power, corporate power) — remains one condition of the expressions of structure. What's visible is crucial because it's a representational element of structure. But as Gossett and Hong have pointed out, to be exposed and figured in the surface has its own limits. Critics like Goldberg fixate on "what has become visible" to protect representations linked to the privileges of the status quo, rather than tend to the ongoing damage of structural violence.

11 Michelle Goldberg, "Feminist writers are so besieged by online abuse that some have begun to retire," *Opinions; Washington Post,* 20 February 2015.
12 "We Cannot Live Without Our Lives: A Conversation on Anti-blackness, Trans Resistance and Prison Abolition," forum, University of California, San Diego, 4 November 2014.
13 Grace Kyungwon Hong, *The Ruptures of American Capital: Women of Color Feminism and the Culture of Immigrant Labor* (Minneapolis: University of Minnesota Press, 2006), xxviii.

Polar opposite examples: The management/displacement of structural critiques

We see this refusal to engage the structural across a breadth of trending topics. #Gamergate is/was a movement entirely unlike #solidarityisforwhitewomen. In fact, they are of polar opposite camps, brought forth by entirely differing subject positions. However, I wish to posit that both hashtags highlight structural concerns (brought forth by entirely differing subject positions) to be managed ultimately as an issue of public relations.

The briefest background possible:

The #gamergate hashtag was formed, more or less, in conjunction with multiple conversations that were happening on Reddit and other media channels, and trended more or less because of actor Adam Baldwin's tweets.[14] The overarching claims of gamergate were criticisms of female game developer Zoe Quinn and feminist critic Anita Sarkeesian, and statements of support for critiques of Quinn's games and Sarkeesian's commentary on gaming culture. The tag demanded a maintenance of our status quo and our patriarchal norms, and directly opposed any potentially feminist commentary. The heart of the issue from the beginning was: who gets to criticize, who does not.

Unlike #solidarityisforwhitewomen, #gamergate was branded as a men's rights, anti-feminist movement. But similar to #solidarityisforwhitewomen and #femfuture, #gamergate is the manifestation of structural violence, of the ongoing, well-documented culture of misogyny, rape, and patriarchy in tech and gaming cultures. If the conversation stemming from #solidarityisforwhitewomen is a structural critique of white supremacy of US feminism and global north feminist movements, the response to #gamergate was also structural critiques of the rampant misogyny of gaming culture. Both issues have been highlighted by

---

14 Caitlin Dewey, "The only guide to Gamergate you will ever need to read," *The Intersect; The Washington Post,* 14 October 2014. See also Adi Robertson, "What's happening in Gamergate?" *The Verge,* 6 October 2014.

visibility, but both issues have been managed as visibility issues and not as material, structural concerns.

The logics of PR were similar in the writings around #gamergate.

1. Obviously #solidarityisforwhitewomen and #gamergate are different, they were started by different people and trended for polar opposite reasons but both tags highlighted a serious structural issue: A. White supremacy in feminism, B. Rampant misogyny in tech and game culture. Rather than grappling with the damage of structural violence, commentators on all sides, across platforms jumped for damage control. Better for business maybe[15]?

2. The logic of "surprise and dismay" applies to both hashtags but in different ways — white feminists are sad that Twitter visibility belongs not to them (at least in that moment), while leading proponents have to be on the offense about their defense of game tech culture.[16]

3. Both parties (the commentators, those that disapprove or are embarrassed by the trend) have an issue with the visibility of the problem, not with the fact that there are structural issues.

4. Gaming executives expressed dismay that the tag was "tarnishing our reputation as gamers," while academics bemoaned: "We have been working for years to make games a legitimate tool for education and for study, and we were making progress... And then came GamerGate... now, when I go

---

15 Gamergate asks companies to boycott *Gawker* and *The Verge* for their reporting that gaming culture is misogynistic — #gamergate stated that such accusations were akin to bullying. In response to this, Adobe calls for "Anti-Bullying" in SUPPORT of the rights of Gamergate, and pulls ads from *Gawker*. Mercedes Benz, Intel pulled from *Gamasutra* "in response to an article by journalist Leigh Alexander that criticized mainstream gaming culture." See Leigh Alexander, "'Gamers' don't have to be your audience. 'Gamers' are over," *Gamasutra*, 28 August 2014. For coverage of this see Adi Robertson, "Adobe's symbolic pro-Gamergate gesture frustrates victims," *The Verge*, 21 October 2014.

16 See Colin Campbell, "Conservative group issues video lambasting gaming's feminist critics," *Polygon*, 16 September 2014 and T.C. Sottek, "Stop Supporting Gamergate," *The Verge*, 8 October 2014.

to talk about games to industry groups or fellow academics, GamerGate always comes up as an example of how terrible and immature people who play games are… It will take years and years to repair the damage…"[17] In both instance the issue seems to be in "damage control" "image" and "PR" rather than care for the damage, care for the issue, structural adjustments.[18]

5. And the logic of Twitter visibility in the case of #gamergate, extends across platforms. Twitter visibility is powerful — it travels. Due to #gamergate wikipedia editors have debated on how to "civilize" the editing process in regards to "gendered" narratives, or to potentially ban or curtail feminist editorship — this is a way to center (once again) anti-feminist narratives and methods.[19] It is the ultimate "We want to be centered (again)" declaration cloaked by calls for civility, objectivity, the same ol' tune too many of us know so well.

The algorithms of visibility

Panic driven by visibility is predictable. Zeynep Tufecki has argued that too often digital humanities or scholarship around the digital, "[R]arely goes beyond exploring big data as a hot, new topic and an exciting new tool, and rarely consider issues of power."[20] The "analyst" in focusing in on the function of technology, completely fails to discuss the structures and dynamics of power materialized every step of the way by this tech, the data and its users. And so it holds that while trends are provoking of such dismay and surprise so worthy of journalistic inquiry and coverage, there is little to no critical analysis of "trending" itself.

---

17 I would like to thank Shanley Kane, editor at *Model View Culture* for finding this quote.

18 Mark Bernstein, "Unanswered," blog post, 15 January 2015.

19 Phillipe Beaudette, "Civility, Wikipedia, and the conversation on Gamergate," *Wikimedia* (blog), 27 January 2015.

20 Zeynep Tufekci, "What Happens to #Ferguson Affects Ferguson: Net Neutrality, Algorithmic Filtering and Ferguson," *The Message,* 14 August 2014.

#Solidarityisforwhitewomen and #gamergate garnered attention for trending. So what are the algorithms of visibility? If trends are so formidable, so important, let's look at the phenomena that's being looked at, by critically defining "Trending":

1. We don't know why something trends. The algorithm is a locked secret, a "black box" (to the point where MIT professors have built algorithms attempting to predict trending tags. Fun fact: the same team have built algorithms predicting bitcoin prices: these are their explicit interests and concerns).[21] The fine print: Trending is visibility granted by a closed, private corporation and their proprietary algorithms. As Tufecki says, "Algorithms have consequences."[22]

2. The visible trending box is supposed to vary according to personal preference. There are algorithms for localized trends, "neutral" US trends, global trends, and other a la carte options. The Fineprint: The algorithms can and should be adjusted according to personal preference — we want our reach to be individualized.

3. The little bit of information the private developers have released is that a "trend" is based on a very specific definition of "now" and "new," that us users do not have access to this precise definition. The fine print: Something cannot trend for too long, this isn't their definition of a "now" and "new." This is why #Ferguson failed to trend after a few days even though it was one of the most widely used hashtags — trending for a few days excluded it from the possibility of trending.[23]

4. Concerns about why certain hashtags don't trend (i.e., #occupywallstreet, #wikileaks, or the various other #occupy's) will lead Twitter developers to tell you that perhaps something is not as popular as you think it is.[24] The fine print: Trending is

21 Adam Conner-Simons, "MIT computer scientists can predict the price of Bitcoin," MIT News, 21 October 2014.

22 Tufekci, "What Happens to #Ferguson Affects Ferguson."

23 Ibid.

24 "FAQs about trends on Twitter," Twitter, https://support.twitter.com/articles/101125.

what they believe is popular, a paradoxical assertion: private formulas declaring what is most public and "new."

Trending is visibility granted by the algorithms of a closed, private corporation — De-exceptionalize it!

Through this lens, trending is merely visibility granted by the algorithms of a closed, private corporation:

Trends — what is happening, new, now according to Twitter the corporation's ever-changing algorithms

Trends — a moment in which the "public" space of the internet becomes concentrated

Trends — a concentration of trolls

Trends — concentrated energy

Trends — visibility granted by a closed, private corporation

Trends — the commons managed by a closed, private corporation

Trends — manifestation of algorithms

Trends — expected

Trends — unexpected

To further illustrate how much we don't understand why something trends, I used Topsy to provide me with analytics on the usage of #FreePalestine, and compared it to the various trending tags on March 1, 2015 and October 19, 2015. This experiment was prompted by users throughout last summer and this year, observing how #KillAllMuslims trended recently but #FreePalestine has been unable to trend. In addition, how #BlackLivesMatter has been consistently and fiercely utilized but has only trended during select moments.

As of January 30th to March 1st, #FreePalestine has been utilized over 84,333 times.

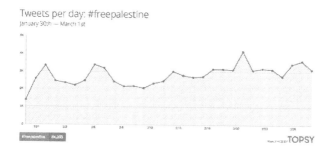

Yet March 1st's United States Trends were:

So how might #WeWantTheCup fare next to #FreePalestine?

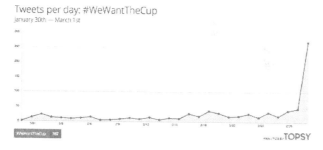

That's 767, total. Not 767,000 shortened.

Or #ExplainAMovieByItsTitle:

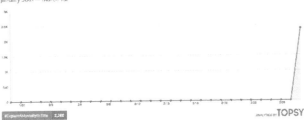

or #KillAllMuslims?

How might all of this fare next to the usage of #Ferguson?

#Ferguson has been used over 2,000 times on March 1st alone... clearly surpassing the usage of the current US trends of#WeWantTheCup and contending with the usage of #ExplainAMovieByItsTitle.

And on October 19th, 2015?

United States Trends · Change

#BoycottStarWarsVII
#MyResumeDoesntMention
Say You Will
#NowPlayingCONFIDENT
John Henson
#TroyeOnVevo
Barry Zito
#mondaymotivation
Young Jesus
Rush Card

Tweets per day: #MyResumeDoesntMention
September 19th — October 19th

5,634 tweets compared to:

Tweets per day: #blacklivesmatter
September 19th — October 19th

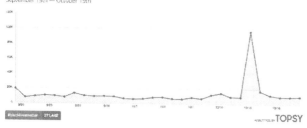

Almost 10k on the 19th and:

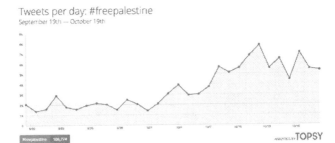

Tweets per day: #freepalestine
September 19th — October 19th

Over 5k — Twitter disavows the continuing conversation?

I have catalogued the usage of #BlackLivesMatter almost every day in 2015, and this pattern is not the anomaly but the standard. Countless, one-time "tags" with half the usages will "trend" — while #BlackLivesMatters will not trend. I encourage everyone to look up usage analytics, but I think you get the picture. Basically, #BlackLivesMatter should be trending everyday.

Of course this Topsy graph is limited to Jan 30–March 1 and from September 19–October 19, so all of these hashtags could've been utilized differently the months and years before. And of course, Twitter developers have already graciously explained that trending is their particular definition of "new" and "now" and so previous tags, or tags used continuously (such as Justin Bieber tags, which have been implicitly banned from trending) will not trend.[25]

I bring this up to interrogate the current framing of visibility via trending, and to point to how the "journalistic scholarship" around visibility and trending is misinformed, and misframed. The exceptional attention given to hashtag discourse by critics, news platforms, and journalists — to what they perceive to be evidence of visibility — takes the focus away from the spaces created by gendered and racialized users, and rewrites it as a singular confrontation racialized/gendered users are having with white audiences within a white space. This rewriting posi-

---

25 Simon Dumenco, "The Real Reason Twitter Radically Reworked Its Trending Topics Algorithm," *AdvertisingAge*, 21 May 2010.

tions trending tags to be isolated explosions. It does not labor through the possibility of communal, ongoing engagement and sustainment, for better or for worse. Though this is clearly their fixation, this fixation should not prevent us from recentering the persistent and ongoing labors involving disobedience, disturbance, and cyborg mutations: alternative discourses.

Rather than treating trending as an exceptional event of well-directed, or misdirected energy (that can be channeled for other, and better purposes!) — I am suggesting that it might be more fruitful to frame current trend algorithms as expressions of de-exceptional events, and to inquire into the idea of the "commons" — a space marked for public debate and protest — instead, so that we may support the tools (information transparency, anti-doxxing, privacy tools to start) users need to claim this ground.

To exceptionalize trending takes it out of the realm of the commons. A trend should not be of public and broadcast interest because it is exceptionally racialized, exceptionally gendered — but because it seems to seamlessly fit into and inside sets of opaque algorithms. #Gamergate is a tag that fits into the discourse of public trending. The tag is rooted in consumer activism, misogyny, and violent calls to preserve patriarchy. However, criticism of #gamergate, similar to criticisms of #femfuture, dramatically highlighted and shifted the hashtags' primary narrative. High five to that. The trending of #solidarityisforwhitewomen #Ferguson #BlackLivesMatter #mediablackout, and the usage of tags like #FreePalestine demonstrate that radical conversations are not exploding or momentary in the Twitter commons: they are ongoing and variegated. In the commons, there are leaps of politicized conversations in what was supposed to be an apolitical, private sphere. Trending highlights what they want us to see, what they're allowing us to see. But the commons is bigger than their grasp, and linked in uncontrollable ways.

"Ancestral, not marginal"

Especially in the case of Goldberg's take on #femfuture and #solidarityisforwhitewomen (and similarly to Wikipedia's Community Advocacy Director Phillippe Beaudette's call for civility) the tackling of visibility posits white liberalism as the only possible ally — while they belittle and besmirch online activism and take no part in such efforts — their instructions, suggestions, and criticisms must be taken seriously and will be aggressively applied (their suggestions: don't be too visible and you must be "civil" in critiquing misogynist gamers). Trusting and utilizing the secret formulas of Twitter and liberal logic, the journalists dissect who is, and is not on our side — why this is a problem, and ultimately, how to appear "better" the next time "we" are visible. This is their primary complaint — we didn't appear pretty enough, nothing was polished, we focused on the wrong issues, we were too angry, we seem childish — and everyone saw.

In this visibility-centric frame — and because utilizing only the result of the trend, the "analyst" misses the genealogy of online conversations, how plentiful they are — as Twitter selected trends or not. Rather than situating certain events as marginal concerns, or the marginal infringing into the dominant — as Goldberg, Wikipedia's Board and others have done — I am more interested in framing the events/conversations as ancestral, rather than marginal (this is a poetic framework I'm borrowing from the poet Lucas de Lima). Rather than fringes, minor and eclipsing — to think of such conversations as digitalized fragments of political ancestry; as ongoing, replenished, connected to and beyond its current framework.

Imagine: sustained communities that congregate through their desire for anti-racists, anti-patriarchal, anti-colonial discourse.[26] Imagine: the continuity of online communities irrespective of normative blessings!

---

26 Sarah Kendzior, "Blame it on the internet," *Opinion; Al Jazeera,* 4 February 2014.

Hacking a commons

Indeed, Twitter is not designed as public even as it fundamentally derives from public input and data, and parades as common grounds. Luis Martín-Cabrera offers that, "Karl Marx saw the appropriation of the commons as one of the elements of 'primitive accumulation,' a 'ground zero' of 'surplus value.'"[27] That is, the shifting of the public, and the taking of the commons is the basis of privatization — a process that exists to exploit the majority for a minor few. However, you don't have to be a Marxist to follow the argument that: privatization is antagonistic to notions of the commons, and explicitly closing the definition of "the public" *to* the public is manipulative and deceitful. Twitter developers might reply that if their algorithms weren't proprietary, then their business model wouldn't be protected. But if your business is about facilitating conversations between people and breaking the news, highlighting the new — could you at least define how you're using the term "new"? They make asking for accountability feel like a business secret, a privilege — and it's not. Trending highlights only what they want us to see, what they're allowing us to see. Fixating on trending/visibility is the secondary layer of their gaze.

But there are some tools being utilized to circumvent Twitter's original design, to disobey aggressively, to claim Twitter as a true "commons" — a space marked for public debate and protest. This commons is bigger than their grasp, linked in uncontrollable ways, and subject to reclamation, to subversion… to hacking. The trending of #solidarityisforwhitewomen was a hack of opaque, proprietary algorithms. Rather than what is being granted visibility, I am interested in these, and related efforts of *affirmative, intimate sabotage:*[28]

---

27 Luis Martín-Cabrera, "The Potentiality of the Commons: A Materialist Critique of Cognitive Capitalism from the Cyberbracer@s to the Ley Sinde," *Hispanic Review* 80, no. 4 (2012): 583–605, at 585.

28 Gayatri Spivak, interviewed by Nazish Brohi, in "Herald exclusive: In conversation with Gayatri Spivak," *Dawn,* 23 December 2014.

- communities that congregate around tags, regardless of their trends
- users that notice when important topic (such as #Ferguson) are not trending, and use alternative tags (such as #mediablackout)
- users who pre-emptively create "when in jail" accounts (planning the heist!!!)
- anti-doxing collectives
- and so much more

This is the ongoing praxis that actively questions the stakes of power in our privatized digital publics. Intimate sabotage: In becoming close to the opaque, being fed by it yet reproducing otherwise. Refusing replication: becoming its expert and traitor. Learning everything about it — disenchanted of its awe, seamlessly wandering inside of it, finding novel ways of attack.

Intimate sabotage of Twitter is a provocation: that any sense of the public in this privatized milieu will have to be reimagined. "The commons" should not be a corporate feel-good initiative, but a provocation that any sense of the public in this privatized milieu will have to be reimagined. In discussing the potentiality of the commons, Martin-Cabrera extends, "[T]he potentiality of the commons can only be actualized when we actively disobey and when we actively 'connect and fight.'" The baseline for a commons as fighting ground? The idea of unmanageable as the commons: the deployment of intimate sabotage as communal building; the algorithms hacked, broken, reworked for purposes beyond the immediate reach of private formulas, private means.

Tufekci argues that we continue to live with ineffective models/tools against our oppression; she suggests that "We need to update our nightmares."[29] The nightmare before may have been — some of us are invisible — before such concerns can be remedied — the update indicates that invisibility can no longer be exchanged evenly for represented visibility. Other updates

---

29 Zeynep Tufekci, "Is the Internet Good or Bad? Yes," *Matter*, 12 February 2014.

may be: our visibility is our surveillance (Hong), our visibility must be contested (Gossett), our visibility is not very particular — and we must find the tools to sabotage it.[30]

Post-script

Since writing this article there are two developments that I've wanted to discuss: 1. The closing of many and most real time analytic companies and 2. Twitter's ongoing stock concerns. I've been thinking about how the very analytic tools that were helpful for me in interrogating the notion of trending have been acquired and made obsolete within a year. For example, Topsy has been "absorbed" into Apple's search functions — or so it is advertised.[31] Going to Topsy.com will now lead you to apple.com where Siri functions are displayed. Topsy was a "real time" social media analytic tool — it was the tool I — and many others — heavily relied on to compare Twitter data — this is not a service that Apple will provide for its users in the foreseeable future. The acquisition of Topsy by Apple is the end of another strain of free, social media analysis for users.

Since learning of Topsy's closing I have been researching into the infrastructure of our digital tools. It turns out that major search engines such as: Google and Bing — search engines previously not known for their "real-time" analytic reports, have since set out to incorporate real time results into their search algorithms. Or in the case of Apple, they are acquiring such companies, and then claiming for their tools to be part of their platforms.

Before there was an interest in "acquiring" or incorporating real time results, real-time analytics had been its own product by companies such as: Oneriot, crowdeye, Topsy — however

---

30 Shanely Kane, "'Internet Famous': Visibility as Violence on Social Media," *Model View Culture,* 30 June 2014.

31 See Matt McGee, "Topsy Social Analytics: Twitter Analytics for the Masses (& Free, too)," *Search Engine Land,* 31 January 2011 and Daisuke Wakabayashi, "Apple Taps Into Twitter, Buying Social Analytics Firm Topsy," *The Wall Street Journal,* 2 December 2013.

these have all closed or been absorbed. Companies such as Collecta (and keyhole.co) have moved into "ad-based" centered firms. Which leads me to my secondary understanding of social media analytic tools (which apparently are being made more obsolete by the day): the marketing for these tools was and is to serve a "branding" purpose. The idea is that there is a real-time analysis of a word, trend, phrase — and this has a marketplace value. Wouldn't branding services like to know what is happening in the moment, in order to capitalize on the issue/thought/ moment? Because there is a "marketplace" for analytics for "business and branding" purposes, sites and apps like Crowdfire may continue to exist. Twitter has taken this approach as well. Twitter's own analytics tool is about the individual brand on Twitter. Analytics on Twitter focus on individual, personal analytics — for personal branding growth we are to assume.

The closing of analytic search firms is perhaps part of the larger narrative of the tech industry — unabashedly neoliberal on all fronts, obsolescent as a rule, and unapologetically on a path to appropriate/conquer. However, the *tools* that we — users, researchers, educators, writers — need in order to think about the web, and web-based interactions, cannot and should not abide by the obsolescent narrative. So what to do?

The algorithms for all of the social media and search sites most popularly used are proprietary, and we very much need the tools that are actively being shut down.

Topsy helped me analyze how Twitter's "trend" is what they have decided *should* trend. With Topsy gone and with Twitter adding its own analytic tools (they seem to want to show you how many tweets are being generated per each tag, who knows what it'll be next month) — it's implied that we — users, educators, researchers — need not double check, need not research further.

Without analytic tools guiding user experience — all methodologies and readings become positivist, fraught, full of more questions. We desperately need public, non-proprietary search functions, and analytic tools for our search endeavors.

Second and relatedly, is Twitter's stock price and the mining of user experience as "future" development projects.[32] Twitter has had a consistent flow of stock problems since it began to be publically traded. In fact, there's been a 65 percent drop in share price from the last year,[33] from an IPO of $26 on November 7, 2013, to around $17 March of 2016.[34] Everything from: growth expectations to growth realities, Twitter — to its dismay — has not been as *profitable*[35] as other companies.

In an attempt to "innovate" its platform, the latest CEO of Twitter Jack Dorsey tweeted in January of 2016 of the possibility of changing its 140 character policy. The decision for this, he stated, is based on user habits. Users are uploading photographs of texts — shouldn't they be able to display the full text in the platform, and be able to search for these texts? Dorsey explained in the tweet,

---

32 Sam Thielman, "Twitter shares hit new low on rumored shift to 10,000-character tweets," *The Guardian*, 5 January 2016.

33 Arjun Kharpal, "Twitter turns 10: What's next?" *CNBC*, 21 March 2016.

34 Philip van Doorn, "It's Twitter's birthday, and its executives are getting huge stock-based gifts," *Opinion; MarketWatch*, 22 March 2016.

35 But why should any of these companies be profitable? Peter Sunde, founder of Pirate Bay questions the relationship between capitalism and the web. He states, "Look at all the biggest companies in the world, they are all based on the internet. Look at what they are selling: nothing. Facebook has no product. Airbnb, the biggest hotel chain in the world, has no hotels. Uber, the biggest taxi company in the world, has no taxis whatsoever…These are insane amounts of money for nothing. That is why the internet and capitalism are so in love with each other." Sunde argues that, "We are trying to recreate this capitalistic society we have on top of the internet. So the internet has been mostly fuel on the capitalistic fire, by kind of pretending to be something which will connect the whole world, but actually having a capitalistic agenda." Twitter not being profitable, or being not *as* profitable could be way for us to think about why the expectation for capital growth of web-based companies is modeled after some kind of capitalist utopian endless growth formula — and how this expectation effects or might effect our understanding of user, tool — or any potentiality of the "commons." See Peter Sunde, interviewed by Joost Mollen, in "Pirate Bay Founder: 'I Have Given Up,'" *Motherboard*, 11 December 2015.

> We've spent a lot of time observing what people are doing on Twitter, and we see them taking screenshots of text and tweeting it. Instead, what if that text… was actually text? Text that could be searched. Text that could be highlighted.[36]

Dorsey's statement is a curious gesture — the "observing" of user behavior and creating "tools" around them — it is a gesture that I hope we can examine closely in the future. On one hand we have the dissolution of user tools, the dismantling of their functions into the platform (as either completely dissolved, or said-to-be-recapitulated, such as Topsy), and on the other hand, we have a mining of user habits for tools in hopes that these new developments will result in a better-functioning and more profitable future company. A completely vertical, top-down relationship with little to no room for inquiry or contest.

In this scenario I am reminded of what mathematician Joel Nishimura has described as a: Technology Dividend. Nishimura argues that "new" technology is made up of research and development that comes inextricably out of public practice and research — and therefore, a "tech dividend" that "functions exactly as an unconditional basic income" could and should be implemented.[37] I am interested in Nishimura's proposal in that it brainstorms a way all kinds of public, visible and illegible engagement might be acknowledged and compensated for partaking in research and development. A "Tech Dividend" moves away from current neoliberal, opaque, top-down economic models and provide us new ways to think about the commons, the center, and the politics of visibility.

---

36 Jack Dorsey (@jack), Twitter post, 5 January 2016, 5:07 p.m.
37 This is from an unpublished article draft provided by Joel Nishimura.

Bibliography

#solidarityisforwhitewomen. Twitter feed. https://twitter.com/
search?vertical=default&q=%23solidarityisforwhitewomen
&src=typd.

Alexander, Leigh. "'Gamers' don't have to be your audience.
'Gamers' are over." *Gamasutra*. 28 August 2014. http://www.
gamasutra.com/view/news/224400/Gamers_dont_have_to_
be_your_audience_Gamers_are_over.php.

Beaudette, Phillipe. "Civility, Wikipedia, and the Conversation
on Gamergate." *Wikimedia* (blog). 27 January 2015.
http://blog.wikimedia.org/2015/01/27/civility-wikipedia-
gamergate/comment-page-3/?more-posts=wmb_posts-2.

Bernstein, Mark. "Unanswered." Blog post. 15 January 2015.
http://www.markbernstein.org/Jan15/Unanswered.html.

Campbell, Colin. "Conservative group issues video lambasting
gaming's feminist critics." *Polygon*. 16 September 2014.
http://www.polygon.com/2014/9/16/6283467/conservative-
group-issues-video-lambasting-gamings-feminist-critics.

Choi, Don Mee. "Failfail." *Tripwire* 9 (2015): 31–37.

Conner-Simons, Adam. "MIT computer scientists can predict
the price of Bitcoin." *MIT News*. 21 October 2014. http://
news.mit.edu/2014/mit-computer-scientists-can-predict-
price-bitcoin.

Dewey, Caitlin. "The only guide to Gamergate you will ever
need to read." *The Intersect; The Washington Post*. 14
October 2014. https://www.washingtonpost.com/news/the-
intersect/wp/2014/10/14/the-only-guide-to-gamergate-you-
will-ever-need-to-read/.

van Doorn, Philip. "It's Twitter's birthday, and its executives
are getting huge stock based gifts." *Opinion; MarketWatch*.
22 March 2016. http://www.marketwatch.com/story/
its-twitters-birthday-and-its-executives-are-getting-huge-
stock-based-gifts-2016-03-21.

Dorsey, Jack (@jack). Twitter post. 5 January 2016, 5:07 p.m.
https://twitter.com/jack/status/684496529621557248.

Dumenco, Simon. "The Real Reason Twitter Radically Reworked Its Trending Topics Algorithm." *AdvertisingAge.* 21 May 2010. http://adage.com/article/trending-topics/twitter-radically-reworked-trending-topics-algorithm/144023/.

"FAQs about trends on Twitter." *Twitter.* https://support.twitter.com/articles/101125.

Goldberg, Michelle. "Feminism's Toxic Twitter Wars." *The Nation.* 29 January 2014. https://www.thenation.com/article/feminisms-toxic-twitter-wars/.

———. "Feminist writers are so besieged by online abuse that some have begun to retire." *Opinions; Washington Post.* 20 February 2015. https://www.washingtonpost.com/opinions/online-feminists-increasingly-ask-are-the-psychic-costs-too-much-to-bear/2015/02/19/3dc4ca6c-b7dd-11e4-a200-c008a01a6692_story.html.

Grose, Jessica. "Is 'Toxic' Online Culture Paralyzing Feminism?" *XXFactor: What Women Really Think* (blog). *Slate.* 31 January 2014. http://www.slate.com/blogs/xx_factor/2014/01/31/_toxic_twitter_wars_is_online_culture_paralyzing_feminism.html.

Hong, Grace Kyungwon. *The Ruptures of American Capital: Women of Color Feminism and the Culture of Immigrant Labor.* Minneapolis: University of Minnesota Press, 2006.

Kane, Shanely. "'Internet Famous': Visibility as Violence on Social Media." *Model View Culture.* 30 June 2014. https://modelviewculture.com/pieces/internet-famous-visibility-as-violence-on-social-media.

Kendall, Mikki. *hoodfeminism* (blog). https://hoodfeminism.com/.

Kendzior, Sarah. "Blame it on the internet." *Opinion; Al Jazeera.* 4 February 2014. http://www.aljazeera.com/indepth/opinion/2014/02/blame-it-internet-20142453122572101.html.

Kharpal, Arjun. "Twitter turns 10: What's next?" CNBC. 21 March 2016. http://www.cnbc.com/2016/03/21/twitter-turns-10-what-next.html.

Kim, Dorothy, and Eunsong Kim. "The #TwitterEthics
    Manifesto." *Model View Culture.* 7 April 2014. https://
    modelviewculture.com/pieces/the-twitterethics-manifesto.

Losh, Elizabeth. "Hashtag Feminism and Twitter Activism in
    India." *Social Epistemology Review and Reply Collectiv*e 3, no.
    13 (2014): 10–22.

Martín-Cabrera, Luis. "The Potentiality of the Commons:
    A Materialist Critique of Cognitive Capitalism from the
    Cyberbracer@s to the Ley Sinde." *Hispanic Review* 80, no. 4
    (2012): 583–605. http://www.jstor.org/stable/23275310.

Martin, Courtney E., and Vanessa Valenti. *#FemFuture: Online
    Revolution,* vol. 8. Barnard Center for Research on Women,
    2012. http://bcrw.barnard.edu/publications/femfuture-
    online-revolution/.

McGee, Matt. "Topsy Social Analytics: Twitter Analytics for the
    Masses (& Free, too)." *Search Engine Land.* 31 January 2011.
    http://searchengineland.com/Topsy-social-analytics-twitter-
    analytics-for-masses-62868.

Robertson, Adi. "Adobe's symbolic pro-Gamergate gesture
    frustrates victims." *The Verge.* 21 October 2014. http://www.
    theverge.com/2014/10/21/7030613/adobe-asks-gawker-to-
    remove-logo-after-gamergate-complaints.

———. "What's happening in Gamergate?" *The Verge.* 6
    October 2014. http://www.theverge.com/2014/10/6/6901013/
    whats-happening-in-gamergate/in/6917590.

Sottek, T. C. "Stop Supporting Gamergate." *The Verge.* 8
    October 2014. http://www.theverge.com/2014/10/8/6919179/
    stop-supporting-gamergate/in/6917590.

Spivak, Gayatri. Interviewed by Nazish Brohi. "Herald
    exclusive: In conversation with Gayatri Spivak." *Dawn.* 23
    December 2014. http://www.dawn.com/news/1152482.

Sunde, Peter. Interviewed by Joost Mollen. "Pirate Bay
    Founder: 'I Have Given Up.'" *Motherboard.* 11 December
    2015. http://motherboard.vice.com/read/pirate-bay-
    founder-peter-sunde-i-have-given-up?utm_source=mbfb.

Thielman, Sam. "Twitter shares hit new low on rumored shift
    to 10,000-character tweets." *The Guardian.* 5 January 2016.

https://www.theguardian.com/technology/2016/jan/05/
twitter-shares-plummet-rumor-10000-character-tweets-
jack-dorsey.

Tufekci, Zeynep. "Is the Internet Good or Bad? Yes." *Matter.* 12
February 2014. https://medium.com/matter/is-the-internet-
good-or-bad-yes-76d9913c6011#.otaovdiae.

———. "What Happens to #Ferguson Affects Ferguson:
Net Neutrality, Algorithmic Filtering and Ferguson." *The
Message.* 14 August 2014. https://medium.com/message/
ferguson-is-also-a-net-neutrality-issue-6d2f3db51eb0#.
n3cwi6dfx.

"Twitter Sparks a Serious Discussion About Race and
Feminism." *NPR.* 23 August 2013. http://www.npr.org/
sections/codeswitch/2013/08/22/214525023/twitter-sparks-a-
serious-discussion-about-race-and-feminism.

Wakabayashi, Daisuke. "Apple Taps Into Twitter, Buying Social
Analytics Firm Topsy." *The Wall Street Journal.* 2 December
2013. http://www.wsj.com/articles/SB10001424052702304854
804579234450633315742.

"We Cannot Live Without Our Lives: A Conversation on
Anti-blackness, Trans Resistance and Prison Abolition."
Forum, University of California, San Diego, 4 November
2014. http://www.reinagossett.com/events/live-without-
lives-conversation-anti-blackness-trans-resistance-prison-
abolition-cece-mcdonald-janetta-johnson-eric-stanley/.

Wilson, Ross. "Trending on Twitter: A Look at Algorithms
Behind Trending Topics." *Ignite Social Media.* 3 December
2012. http://www.ignitesocialmedia.com/twitter-marketing/
trending-on-twitter-a-look-at-algorithms-behind-trending-
topics/.

19

Academic Influence:
The Sea Change

Bonnie Stewart

Sometimes things shift when you're not looking. One morning, I woke up and discovered I was in style. Or, at least, what I *do* was in style: digital and networked scholarship had suddenly been discovered by higher ed media. For this fifteen minutes of fame, practically every *Chronicle of Higher Education* link on my Twitter feed was about some aspect of online identity or networked scholarship.[1] The LSE blog and *Inside Higher Ed,* too.

I peered about, waiting for the punch line. I am accustomed, when I get up in front of fellow educators and academics and say "I study scholarship and... Twitter," to getting the reception of a failed stand-up comic. *"Really? Twitter?"* people communicate with their eyebrows. I am becoming a great student of arched eyebrows.

---

1   Brian Croxall, "How to Overcome what Scares Us about our Online Identities," *Chronicle of Higher Education,* 21 April 2014, http://chronicle.com/article/How-to-Overcome-What-Scares-Us/145967/; Megan O'Neil, "Confronting the Myth of the 'Digital Native,'" *Chronicle of Higher Education,* 21 April 2014, http://chronicle.com/article/Confronting-the-Myth-of-the/145949/; Seth Zweifler "For Professors, Online Presence Brings Promise (and Peril)," *Chronicle of Higher Education,* 21 April 2014, http://chronicle.com/article/For-ProfessorsOnline/145961/.

Yet recently, casual readers of mainstream academic publications — and their eyebrows — would be hard-pressed not to come away with the impression that academic identities in social media are actually Something to Care About, as a profession.

The sense of critical mass is energizing to me. The work of research that is not legible to others always feels, rhetorically, like lifting stones uphill: constantly establishing premises rather than moving on to the deep exploration of that one particular thing. No matter how important a conversation may be, people cannot engage it if there are no shared premises.

The more the conversation about networks and identities and academia grows and pervades people's consciousness, the less of that Sisyphean phase of the lifting I need to do. Still, I recognize the backlash already burbling. People's eyebrows generally do not *like* to be beaten about the head with the idea they should care about something just because suddenly it's the Flavor of the Month. Nor should they. I feel you, eyebrows of the world. But networked scholarship's surge of visibility doesn't mean you have to use Twitter. Or any other social networking platforms. Nor do you need to get personal online if you don't want to. But your concepts of academic identity and academic reputation do need to expand: Twitter and social media and digital platforms are now a part of scholarship, as modes of communication and of scholarly practice. So if I tell you I'm exploring the part they now play in academic influence… try not to arch so hard you hurt yourself.

Because this is *not* a Flavor of the Month, folks. This is a cultural shift, one part of the sea change in contemporary higher education. The once institutionally centered prestige economy of academia has a shadow sibling, now; an alternate yet intersecting prestige economy of its own that does not adhere to the terms most scholars agreed to play by. Thus, the premises surrounding what counts as academic influence need to be renegotiated, waggling eyebrows or not.

The math of influence

Academic influence has never been a simple measure, even long before scholars began building names for themselves on extra-institutional platforms. Influence is a complex, messy, slightly socially discomfiting, catch-all equation for how people determine the reputation and credibility and essentially the status of a scholar. There are two ways influence tends to get assessed, in scholarship: there's the teensy little group of people who actually *understand* what your work really means… and then there's everybody else, from different fields, who piece together the picture from external signals: what journals you publish in, what school you went to, your citation count, your h-index, your last grant. It's credibility math, gatekeeping math. It's founded in names and organizations people recognize and trust, with a running caveat of Your Mileage May Vary.

And now, in the mix, there's Twitter. And blogs. And measures like alt-metrics. How can something that the general population is convinced about, such as what people had for lunch be a factor in changing what counts as academic influence? By generating a parallel prestige economy that the academy can neither subsume nor ignore, that's how.

Beyond gatekeeping: Networked influence signals

Networked scholarship, at its core, is the social, public, relational practice of engaging online as a scholarly identity, about issues of scholarly as well as personal interest. Going online and talking to people you may not know about areas of shared scholarly interest opens up your reach and reputation for what you do, as your name becomes associated with the conversations you contribute to. These contributions, in turn, open up your capacity to build communities of practice around those areas of interest, enriching your own knowledge and networks in those fields. And over time, sustained engagement opens up the possibility that when people in the academy — the people reviewing your panel or on your next granting committee or looking for a keynote on a given topic — hear your name, it will be one of those they already recognize and trust. Maybe. There's a *lot* of Your Mileage that May Vary here. Think of a Venn diagram — one circle is how scholars traditionally share their work, the other is what people had for lunch — and in the middle there are scholarly ideas that circulate according to the open, self-published, non-gate-kept logics of social media. This emergent prestige economy values many of the ideals that scholarship purports to hold dear — new knowledge, open debate, public dissemination — but the terms and processes on which it is premised are the antithesis of the carefully credentialed, hierarchical operations of the academy.

And yet that doesn't mean they're a free-for-all; there are patterns and commonalities in how scholars use Twitter, particularly to build influence. The oft-touted "social media increases your dissemination and citations!" factor *is* important in shaping scholars' practices, but in my research,[2] most active networked scholars reported the citation bump more as a side effect of networked participation than a reason in itself. Community and connection and space to address marginalities on many fronts factored more powerfully in participants' accounts of their reasons for networked practice, particularly for those who used Twitter for more than broadcast purposes. And when I asked networked scholars to assess each other's influence based on the signals they read from Twitter profiles and participation, this is what I found:

1. **The conversation is what counts.** A concept of "The Conversation" — meaning discussions of import both in their particular fields and across higher ed — circulates widely amongst active Twitter scholars. All participants in the study were engaged in curating and contributing resources to a broader conversation in their field or area of interest.

---

2   Bonnie Stewart, "Open to Influence: What Counts as Academic Influence in Scholarly Networked Twitter Participation," *Learning, Media, and Technology* 40, no. 3 (2015): 287–309.

It was capacity for contribution to this larger conversation that counted most in participants' assessments of others' influence.

2. **Assessments of influence in networks are individually centered, rather than institutionally centered.** This may be an interim or transitional feature of networked scholarly influence, while the platforms and their place in scholarship emerge and mature, but while the signals on which actively networked scholars base their judgements are still quick proxies for quality, they are proxies interpreted against individual understandings of "The Conversation," rather than generic and hierarchical ideals of scholarly role.

3. **Metrics matter, but not that much.** Participants in the study tended to assess size of account — over 10,000 followers, in particular — as a general signal of influence, but perception of capacity for contribution was far more important to scholars' assessments of who they would follow, and why. Number of tweets –understood to indicate longevity and thus likelihood of ongoing contribution — mattered more in participants' estimation of an account's influence and value than number of followers.

4. **Commonalities are key.** The perception of a scholar's credibility and capacity for contribution is created and amplified by common interests, disciplines, and share ties and peers. Participants were most likely to assess accounts as credible and likely to make a contribution if they were followed by users the participant already knew and respected. Professional and personal commonalities were also central to perceptions of others' capacity to contribute, but less visible in assessments of credibility.

5. **Institutional signals and affiliations aren't that important.** Except in the case of one profile with an Oxford university affiliation, institutional status signals were not accorded significant value in assessments of networked influence. Though all participants were institutionally affiliated and well aware of the prestige of academic ranks, journal titles, and institutional brands, these were not interpreted as intersecting

meaningfully with capacity to contribute to the networked conversation. In fact, profiles that emphasized institutional status were understood by a number of participants as signaling their lack of interest in participatory engagement.

6. **Automated signals indicate low influence.** Automated daily tweets and link aggregators such as paper.li were seen as indicators of low engagement and low networked influence, in part because these services are seen as violating implicit social contracts of active, personal curation and direct citation within academic Twitter.

What these findings suggest, all together, is that scholars assess the networked profiles and behaviors of peers through a logic of influence that is — at least as yet — neither codified nor especially numeric. Instead, while academic Twitter's concept of influence recognizes status, standing, and scale, it appears to focus primarily on contribution and capacity to build and disseminate knowledge. This suggests that so long as senior scholars and administrators and tenure committees *think* Twitter is what people had for lunch, there's a gap in our understanding of how new ideas are actually spreading through academe, especially in fields that are changing rapidly.

Enter capitalism

At the same time, networked participation and its non-institutional logics also bring more fraught elements overtly into play in the influence equation. Let's not pretend that academic institutions are not capitalist institutions. They are, and increasingly so: capital equations of scarcity and commodity are very much a part of the institutionalized and gate-kept versions of academic influence signals that have gained traction over recent generations. But the individual scholar in these equations, except in superstar instances, plays an institutional role rather than operates as an economic entity unto him or herself. In networks, individual identity operates more like a brand, particularly as the scale of attention on an individual grows.

This allows junior scholars and adjuncts and grad students and otherwise institutionally marginalized identities to build voices and audiences even with minimal institutional status or sanction. It allows people to join the conversation about what's happening in their field or in higher ed in general; to make contributions for which channels do not exist at the local level. Networked platforms act as hosts for public resistance to the irreconcilable contradictions of contemporary academia, as well as society more broadly. But networked platforms are still corporate platforms, and cannot be seen as neutral identity playgrounds. Rather, like institutions, they are complex sites of entrenched power relations. They operate on logics of media and attention rather than academic hierarchy, thus creating alternative channels for the emergence of voices that may not find amplification in institutional prestige economies. But in both spheres, participation and contribution must still be legible to dominant conversations and interests in order to be taken up and validated by peers. Thus, while the tenor of academic Twitter may differ from the formal outputs of academia — and conversations certainly emerge in social networks more quickly than the publication cycle allows — in both spaces, cultivation of identity and influence is constrained by what already counts to those who are established in that space.

Power in networks

Being visible in networks *can* create access to visibility and voice in broadcast media, which sometimes lends perceived credibility to the way a scholar's work is taken up… or at least amplifies his or her name recognition. The power relations of scale are complex, though: the racism and sexism and hetero-sexism and able-ism and Anglo-centrism of our contemporary world are in many ways replicated in the way voices get heard, online,[3] and the backlash for women and people of color who

---

3   Jaime Nesbitt-Golden, "Why I'm Masquerading as a White, Bearded Hipster Guy on Twitter (Despite being a Black Woman)," *XOJane,* 4 April 2014.

dare to speak can be vicious. The constant identity positioning and lack of transparency and understanding about how visibility works can also make the world of academic Twitter into mean streets.

The biggest factor in building influence in networks — one that should assuage some of the arched eyebrows — is that it tends to take, like all scholarship, a great deal of time and work. Twitter is not a magical path to fame, or to celebrity academic status. In fact, on its own, it's created few superstars: the traditional, institutional halls of power and high status still do far more to thrust scholars into influential circles of attention and public regard. Noam Chomsky's speaking fees are not especially under threat from Twitter upstarts, and Twitter and blogging alone do not often result in New York Times gigs. But they are, now, indubitably a part of that picture, in ever-expanding circles.

I see the networked version of academic influence as what Audrey Watters calls "a cyborg tactic"[4]: the illegitimate offspring of complex totalizing equations, and yet potentially subversive to them. This potential lies, as Haraway would put it, in the fact that illegitimate offspring are often "exceedingly unfaithful to their origins."[5] According to Costa, digital scholarship is often perceived within the academy as a trajectory of deviance; from this perspective, networked scholars can be seen as cyborg scholars, deviants within the academy whose networked participation exceeds institutional boundaries.[6] But academia is no longer an ivory tower, if it ever was; it recognizes visibility and publicity — in the right venues — with its own accolades. Thus, cyborg engagement beyond the boundaries of institutional scholarship can enable the development of footholds that span the two worlds. Case in point: I blog sporadically about my research, as

4   Audrey Watters, "Beneath the Cobblestones...A Domain of One's Own," *Hack Education,* 25 April 2014.

5   Donna Haraway, "A Cyborg Manifesto: Science, Technology, and Socialist-feminism in the Late Twentieth Century," in *Simians, Cyborgs and Women: The Reinvention of Nature,* 149–82 (New York: Routledge, 1991).

6   Christina Costa, "Outcasts on the Inside: Academics Reinventing Themselves Online," *International Journal of Lifelong Education* (2014): 1–17.

well as publishing in more conventional peer-reviewed venues. Awhile back, *The Atlantic* published an article based in large part on a year-old speculative blog post I'd shared on Twitter,[7] which had been seen by a writer for the magazine. His piece came out on a Monday. By Friday, I'd been offered two invited talks at leading Canadian universities as well as an appearance on public radio and a consulting gig; all opportunities that otherwise might not have emerged within the increasingly circumscribed professional horizons of contemporary higher ed.

This cyborg version of influence, then, comes with both potential advantages and with risks. In a higher ed landscape familiar with the stories of Steven Salaita and Saida Grundy, wherein scholars can find casual, conversational tweets taken up by institutional and public audiences as if they were professional communications, the cultivation of a networked identity remains less safe than that of a conventional academic identity, except where networked influence may open doors that would otherwise remain shut. As a development in how scholars understand each other's signals of credibility and reputation, then, networked influence is neither good nor bad, and certainly not neutral. But it *exists,* and it is only growing, and if we are to steer the ship of higher ed towards a future where neither the logics of gatekeeping and tradition nor business and media dominate it entirely, network influence is important for all of us to try to understand.

And to those who would raise their eyebrows at this assertion, I say: sometimes, folks, things shift when you're not looking.

---

7   Robinson Meyer, "The Decay of Twitter," *The Atlantic,* 2 November 2015.

Bibliography

Costa, Christina. "Outcasts on the Inside: Academics Reinventing Themselves Online." *International Journal of Lifelong Education* (2014): 1–17. DOI: 10.1080/02601370.2014.985752.

Croxall, Brian. "How to Overcome what Scares Us about our Online Identities." *Chronicle of Higher Education.* 21 April 2014. http://chronicle.com/article/How-to-Overcome-What-Scares-Us/145967/.

Haraway, Donna. "A Cyborg Manifesto: Science, Technology, and Socialist Feminism in the Late Twentieth Century." In *Simians, Cyborgs and Women: The Reinvention of Nature,* 149–82. New York: Routledge, 1991.

Meyer, Robinson "The Decay of Twitter." *The Atlantic.* 2 November 2015. nhttp://www.theatlantic.com/technology/archive/2015/11/conversation-smoosh-twitter-decay/412867/.

Nesbitt-Golden, Jaime. "Why I'm Masquerading as a White, Bearded Hipster Guy on Twitter (Despite being a Black Woman)." *XOJane.* 4 April 2014. http://www.xojane.com/issues/why-im-masquerading-as-a-bearded-white-hipster-guy-on-twitter.

O'Neil, Megan. "Confronting the Myth of the 'Digital Native.'" *The Chronicle of Higher Education.* 21 April 2014. http://chronicle.com/article/Confronting-the-Myth-of-the/145949/.

Stewart, Bonnie. "Open to Influence: What Counts as Academic Influence in Scholarly Networked Twitter Participation." *Learning, Media, and Technology* 40, no. 3 (2015): 287–309. DOI: 10.1080/17439884.2015.1015547.

Watters, Audrey. "Beneath the Cobblestones… A Domain of One's Own." *Hack Education.* 25 April 2014. http://hackeducation.com/2014/04/25/domain-of-ones-own-incubator-emory/.

Zweifler, Seth. "For Professors, Online Presence Brings Promise (and Peril)." *Chronicle of Higher Education.* 21 April 2014. http://chronicle.com/article/For-ProfessorsOnline/145961.

20

Playing as Making

Edmond Y. Chang

Johanna Drucker in "Humanistic Theory and Digital Scholarship" articulates as a key transformation (and bone of contention) in the variegated and interdisciplinary terrains of the digital humanities, saying,

> I am trying to call for a next phase of digital humanities [that] synthesize method and theory into ways of doing as thinking. [...] The challenge is to shift humanistic study from attention to the *effects* of technology [...] to a humanistically informed theory of the *making* of technology.[1]

But what does it mean to *do,* to *make*? And what sorts of *doing* and *making* are privileged over others? In other words, what counts in this shift?

For some, digital humanities making comes down to code, programming, and working in the back end. Stephen Ramsay has famously provoked, "Do you have to know how to code? I'm a tenured professor of digital humanities and I say 'yes.' So if you come to my program, you're going to have to learn to do that

---

1 Johanna Drucker, "Humanistic Theory and Digital Scholarship." In *Debates in the Digital Humanities,* ed. Matthew K. Gold, 85–95 (Minneapolis: University of Minnesota Press, 2012), 87.

Fig. 1. Opening screen. *Lim.* merritt kopas, 2012. Game still.

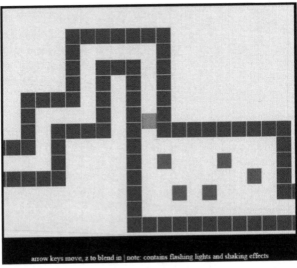

Fig. 2. Entering the first room. *Lim.* merritt kopas, 2012. Game still.

eventually."[2] Like Drucker, Ramsay argues that digital humanities "involves moving from reading and critiquing to building and making...Media studies, game studies, critical code studies, and various other disciplines have brought wonderful new things to humanistic study, but I will say (at my peril) that none of these represent as radical a shift as the move from reading to making."[3] It is this foregrounding of *doing* and *making* that I want to take up, think about, and tinker with, not necessarily to rehash old debates or to pick (or pit) sides. Rather, I hope to articulate alternative modes and forms of doing that engage the *modus operandi* of making without depending on specialized or exclusionary barriers of entry. As Ramsay qualifies, "Personally, I think Digital Humanities is about building things. I'm willing to entertain highly expansive definitions of what it means to build something."[4] In other words, I hope to take advantage of and take for granted that doing, making, and building can and must include a range of practices, processes, and materialities, many of which are accessible, every day, even vernacular. Specifically, I want to argue that playing a digital game is critical doing, that playing *is* making, and to embrace playing as making.

For example, I open a browser and enter a game's URL. The window starts #000000 black and a moment later, the game begins. merritt kopas's *Lim* opens with a pink background, the suggestion of walls made of dark gray squares, and the game's title spelled out in burnt orange squares (see Figure 1). White text below offers terse directions and cautions. Using the arrow keys, I discover that my avatar is one of the orangey squares, which I slide around the 2D world and which is indeed bounded by walls, the walls of a maze. I glide around the walls, making a circuit of the initial "room." I cannot move the other squares no matter how hard I press on an arrow key. An opening in the gray walls on the lower right-hand corner of the screen offers

2   Stephen Ramsay, "Who's In and Who's Out?" blog post, 8 January 2011.
3   Stephen Ramsay, "On Building," blog post, 11 January 2011.
4   Ramsay, "Who's In and Who's Out?"

the only direction and exit. At this point, as I experiment, pressing the Z key does nothing.

Even in the first few moments of *Lim,* the player has explored, discovered, and learned a great deal. Though they are not privy to the lines of code, the programming churning under the *mise-en-scène* of the game's interface and aesthetics, the player is deducing, aggregating, experimenting with a kind of algorithmic sense, a procedural literacy. Movement, keystrokes, boundary, collision, direction, timing, sound, silence, color, darkness, and so forth are made intelligible through the player's interaction with the game, with the code of the game. The player learns what is allowed by the game, and perhaps more importantly, what is not allowed by the game. It is this enactment, this negotiation, this relationship between player and game, action and algorithm that I want to define as a creative and constitutive act. It creates, it makes.

Alexander Galloway says, "If photographs are images, and films are moving images, then *video games are actions.*"[5] He continues, "Without action, games remain only in the pages of an abstract rule book. Without the active participation of players and machines, video games exist only as static computer code. Video games come into being when the machine is powered up and the software is executed; they exist when enacted."[6] This coming-into-being through the cybernetic loop of player input and game feedback reveals that digital games are never disembodied, immaterial experiences.

In *Lim,* I maneuver my little square through the exit from the initial area, leading it through a narrow but crooked path. As soon as I leave the first room, a gray block appears blocking any retreat like a door or gate shutting behind me. My avatar starts to flash, change color, rapidly blinking red, bright green, blue, sky blue, pink, magenta, and purple. The narrow "hall" opens into a second room populated by a scattering of orange-brown

---

5   Alexander Galloway, *Gaming: Essays on Algorithmic Culture* (Minneapolis: University of Minnesota Press, 2006), 2.

6   Ibid.

squares. As soon as my avatar enters the second area, its color stops changing and becomes burnt orange just like the other squares. Again, there is only one obvious exit. Once I leave the room, the exit is shut, and my avatar returns to its gleeful riot of colors. But the flashing colors stops as soon as I approach the third room. This time, my little square turns a dark blue. Three orangey squares live in the third room, and as soon as I enter, forcibly attack me, running into me like an offensive tackle. Every time I am hit, the screen shakes and there is a brash sound like a burst of static. Getting by the orange squares is a challenge; the pummel into my avatar knocking it in different directions. However, by holding down the Z key, my little square turns the same color as the other and the battering stops. While "blending" in, the game zooms into the avatar and movement is painstakingly slow. *Lim*'s renders the difficulty and toll of "passing," of fitting in, of trying to be just like everyone else. By the fourth and subsequent rooms, the ability to blend is no guarantee of safe passage. Even while passing, the other blocks respond violently as if seeing through the performance and disguise.

The premise and mechanics of *Lim* are simple but elegant, and they straightforwardly demonstrate that games are simultaneously a "designed experience"[7] and an "action-based medium."[8] The algorithm, the code of a game is executed, turned from information to machine states, from one electromagnetic form to another, from data to light, sound, traces on a hard disk, to vibrations in a game controller. Games are an "active medium [...] whose very materiality moves and restructures itself — pixels turning on and off, bits shifting in hardware registers, disks spinning up and spinning down."[9] But algorithm and code also get transformed into raised heart rates, cramped fingers, sweaty palms, full bladders, strained eyes, curse words, competitive spirits, and piqued curiosities. As Galloway says, "One *plays* a

---

7   Kurt Squire, "From Content to Context: Video Games as Designed Experiences," *Educational Researcher* 35, no. 8 (2006): 19–29, at 24.
8   Galloway, *Gaming: Essays on Algorithmic Culture*, 3.
9   Ibid.

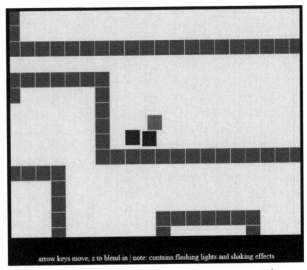

Fig. 3. Being attacked by unfriendly squares. *Lim.* merritt kopas, 2012. Game still.

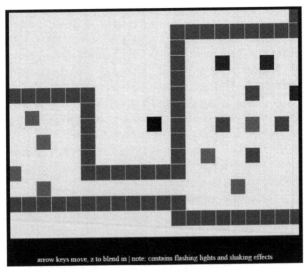

Fig. 4. Glitch: escaping the maze. *Lim.* merritt kopas, 2012. Game still.

game. And the software runs. The operator and the machine play the video game together, step by step, move by move. Here "the 'work' is not as solid or integral as in other media" but it still has some substance, some effect within and without the technologies of the game.[10]

Playing a game is ultimately about learning the rules, the affordances and limitations of the platform, interface, and program, about understanding not the code itself but sensing and manipulating the contours, the structures of the code. Playing reveals what the code is doing (or not doing) and understanding the totality of how each line, function, routine, and sets of programming work together. A piece of code never becomes fully legible and intelligible until it has run, failed, tested, revised, and run again. Though not unique to video games, in a deep sense, playing is never not playtesting; and playtesting is always the bringing into being the game, particularly for algorithmic objects. Running the code turns program to phosphorescent pixel, code to haptic controller, platform to player performance, machine potential to virtual reality. As Ian Bogost argues, "[a]rtifacts like [a video game] suggest that procedural literacy can be cultured not only through authorship, such as learning to program, but also through the consumption or enactment of procedural artifacts themselves. In other words, we can become procedurally literate through play."[11] Drucker extends this notion of procedural literacy by recognizing that code and algorithms are also "performative materiality," which "shifts the emphasis from acknowledgement of and attention to material conditions and structures towards analysis of the production of a text, program, or other interpretive event. After all, no matter how detailed a description of material substrates or systems we have, their use is performative whether this is a reading by an individual, the processing of code, the transmission of signals through a system, the viewing of a film, performance of a play, or a musi-

---

10 Ibid., 2.
11 Ian Bogost, "Procedural Literacy: Problem Solving with Programming, Systems, and Play," *Telemedium* (Winter/Spring 2005): 32–36, at 34–35.

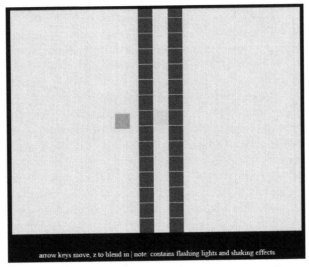

arrow keys move, z to blend in | note contains flashing lights and shaking effects

Fig. 5. End of the game. *Lim.* merritt kopas, 2012. Game still.

cal work and so on."[12] Here I would add playing a digital game. Much in the same way a reader must learn the interface of a book (something that has become naturalized), interfacing with a game requires navigating, parsing, and interacting with multiple layers of mediation and information, feedback loops of input, output, image, sound, text, and trial. Under the rubric of performative materiality, the "idea of a user-consumer is replaced by a maker-producer, a performer, whose performance changes the game."[13]

Returning to *Lim,* there is a curious and wonderful moment when the bullying blocks hit so hard and fast that they inadvertently knock my little square outside of the maze walls as if my 2D avatar somehow turned sideways, became a line, and slid through the thin, pink gap between blocks of gray. The world

12 Johanna Drucker, "Performative Materiality and Theoretical Approaches to Interface," *Digital Humanities Quarterly* 7, no. 1 (2013), par. 8.

13 Ibid., par. 36.

outside of the walls is wide-open and safe from all attack. In fact, I can maneuver up to the wall of the maze, close enough that some of the blocks sense my avatar and move to intercept, but they are trapped and unable to reach me. It is smooth sailing outside of the maze, though the pink expanse is still bounded. My avatar cannot fall off the map as it were. While free of the dangers and confines of the maze, my little square is also free to flash its multi-colored self. Eventually, if I follow the shape of the maze, I can reach the proscribed end of the game — another multi-hued square waiting at the end of the path. As I near the kindred square, even though the wall, the gameplay stops, both cubes blinking at one another, there are a few sounds like the thumping of a heartbeat, and the game ends and goes black.

There is no clearer moment of a game's performative materiality then when the game fails, glitches, freezes, or otherwise breaks the illusion of seamless and invisible code, computation, and other machine processes. Glitches reveal the formality and fragility of the rules of the game and of the algorithm. Although merritt kopas has been decidedly cagey about whether or not the above glitch is intentional or not (if that is even possible), the ability to escape the maze and the "proper" or "correct" way to play and win the game marks alternative even radical opportunities in game play and design. They allow resistance and refusal of the "basic conclusion that to play in a digital sandbox one had to follow the rules of computation."[14] This glitch in *Lim*, as I have written elsewhere, is a queer moment, a queer mechanic that challenges the normativity of code and digital technologies. Drucker says, "Algorithms are instructions for processes, for performances, whose outcomes may usually be predictable, but of course, are as open to error and random uncertainties in their execution as they are too uncertain outcomes in their use."[15] The glitch defies planning, coherence, homogeneity, and perfect control, and most importantly, a glitch cannot come

---

14 Drucker, "Humanistic Theory and Digital Scholarship," 88.
15 Drucker, "Performative Materiality and Theoretical Approaches to Interface," par. 11.

into being without the running of the program and the playing of the game. By playing through *Lim,* by working through its obvious, unobvious, and accidental rules, I have gained an understanding of the mechanics and mechanisms of the game without necessarily parsing the code itself. This understanding, this experience and knowledge-making can then become the ways and methods of playing, breaking, and (eventually) making other games. Without immediate access to the lines of code itself, I can still recognize, interpret, even feel, like an algorithmic kinesthetic sense, the IF/THEN/ELSE statements, the links between key and command, and the subtle (and not so subtle) shifts between subroutines for exploration, exposition, combat, cut scenes, success, failure, reward, even death. Kopas says,

> Games can serve as sites for us to gesture towards queer utopias, to imagine alternative ways of being and living. For that to happen, we have to interrogate and rethink the work of playing. Mutating, breaking, and twisting games are valuable actions insofar as they help make visible our assumptions about play.[16]

All in all, it is this working, breaking, twisting, and imagining that underscores the myriad of ways that playing is making, a different kind of creating. Games are a constellation of textual, narrative, digital, mechanical, and cultural practices. Ask any player of digital games whether or not they are makers or builders and the response will be loud and affirmative. Playing and gaming makes all sorts of things beyond and in response to scholarly critique and theorization: high scores, fan fiction, save files, game art, machinima, strategy guides, bug reports, muscle memory, maps, notes, mods, reviews, communities, museums, schools, and not to forget heat, dust, radiation, noise, and tons and tons of e-waste. That said, I return to Ramsay's invitation to expand and explore definitions of making and building. In this

---

16 Naomi Clark and Merritt Kopas, "Queering Human-Game Relations: Exploring Queer Mechanics and Play," *First Person Scholar,* 18 February 2015.

case, the playing of games is one form of procedural literacy and performative materiality, which has

> often been relegated to the domain of computer programming. [...] But the value of procedural literacy goes far beyond the realm of programming alone; indeed any activity that encourages active experimentation with basic building blocks in new combinations contributes to procedural literacy.[17]

Or as Tara McPherson argues,

> In extending our critical methodologies, we must have at least a passing familiarity with code languages, operating systems, algorithmic thinking, and systems design. We need database literacies, algorithmic literacies, computational literacies, interface literacies.[18]

By playing digital games, I think we develop these "passing familiarities" but simply have not put names and numbers to what we take as "just a game." Like my glitched square in *Lim*, it is possible to understand the inner workings of a system by looking at it from vantage points outside of it. Understanding code, understanding algorithms, understanding the ways digital technologies work need not be reduced to a Hamletian dilemma: to code or not to code. Rather, playing with code, experimenting with technologies can offer entrances and experiences that rely on an inclusive and everyday form of doing: playing a game.

---

17 Bogost, "Procedural Literacy," 36.
18 Tara McPherson, "Why Are the Digital Humanities So White?" in *Debates in the Digital Humanities,* ed. Matthew K. Gold, 139–60 (Minneapolis: University of Minnesota Press, 2012), 154.

Bibliography

Bogost, Ian. "Procedural Literacy: Problem Solving with Programming, Systems, and Play." *Telemedium* (Winter/Spring 2005): 32–36.

Clark, Naomi, and Merritt Kopas. "Queering Human-Game Relations: Exploring Queer Mechanics and Play." *First Person Scholar.* 18 February 2015. http://www.firstpersonscholar.com/queering-human-game-relations/.

Drucker, Johanna. "Humanistic Theory and Digital Scholarship." In *Debates in the Digital Humanities,* edited by Matthew K. Gold, 85–95. Minneapolis: University of Minnesota Press, 2012.

———. "Performative Materiality and Theoretical Approaches to Interface." *Digital Humanities Quarterly* 7, no. 1 (2013). http://digitalhumanities.org:8081/dhq/vol/7/1/000143/000143.html.

Galloway, Alexander. *Gaming: Essays on Algorithmic Culture.* Minneapolis: University of Minnesota Press, 2006.

kopas, merritt. *Lim.* July 2016. https://a-dire-fawn.itch.io/lim.

McPherson, Tara. "Why Are the Digital Humanities So White?" In *Debates in the Digital Humanities,* edited by Matthew K. Gold, 139–60. Minneapolis: University of Minnesota Press, 2012.

Ramsay, Stephen. "Who's In and Who's Out?" Blog post. 8 January 2011. http://stephenramsay.us/text/2011/01/08/whos-in-and-whos-out/.

———. "On Building." Blog post. 11 January 2011. http://stephenramsay.us/text/2011/01/11/on-building/.

Squire, Kurt. "From Content to Context: Video Games as Designed Experiences." *Educational Researcher* 35, no. 8 (2006): 19–29. DOI: 10.3102/0013189X035008019.

21

Humanizing the Interface

Kat Lecky[1]

Oppression is inherently spatial. Governments use biopolitical mechanisms such as urban zoning and prisons to keep undesirable populations fixed in place; institutions use office location to distinguish permanent from contingent faculty; houses of worship physically separate believers from infidels. These structures all classify exclusion as a topography dividing "us" from "them." Resistance is also spatial: Rooms of one's own and brave new worlds constitute alternate landscapes that restage the relation of the person to society. These oppositional spaces protect us from the onslaught of the myriad forms of social coding that define us as objects rather than selves.

The Graphical User Interface (or GUI for short) is a high-stakes battlefield in this struggle between the oppressed and the powers-that-be. The general term for the proliferating technologies that communicate with their human users through images rather than text, the GUI expands the concept of digital literacy to include those excluded from traditional forms of technological aptitude. Smartphones and tablets depend upon rebuses

---

1 This piece was originally published on *Hybrid Pedagogy*. It was reviewed by Jesse Stommel and Adam Heidebrink. See Kat Lecky, "Humanizing the Interface," *Hybrid Pedagogy*, 27 March 2014, http://www.digitalpedagogylab. com/hybridped/humanizing-interface/.

(image-words) rather than lexicons to generate meaning. Hegemony and alterity converge on their topography, which offers people the same playing field regardless of literacy level. The GUI opens a space within which dwell ever-increasing numbers of individuals from diverse classes, educational backgrounds, races, nationalities, genders, and political, sexual, or religious orientations. It forges a common identity grounded in nothing other than its shared love of this user-friendly technology. This hybrid technology opens the same world up to the excluded and powerful alike. And this world is pocket-sized: virtually anyone can own it, carry it, use it. As a pedagogical device, the GUI erodes distinctions between those privy to elite education and those without access to basic learning.

My own work studies the authority of the user in Renaissance England, which saw the birth of royally commissioned national atlases collected by the upper classes. The same era spawned a thriving industry in pocket maps that were produced cheaply and sold widely to ordinary people with no part to play in official governance. I read portable cartography as a premodern avatar of the GUI. These serviceable artifacts were marked only by bare necessities like simple icons for shires, rivers, and markets to allow even illiterate travellers to use them. Designed for heavy use by a broad audience, these maps also engendered a powerful mode of resistant pedagogy for those excluded from traditional conduits to power. The predecessors of today's graphic interface, pocket maps taught their possessors to be agents rather than receptors: to read the signs, choose their paths, and learn about their country. Renaissance pocket maps undercut sovereign visions of England by allowing them to see their country on their own terms. Modern GUIs afford their users an open, adaptable landscape upon which they may chart their own forms of resistance, their own sense of individual sovereignty.

Take Minecraft, for example.[2] It is one of seemingly countless programs that one may choose to install on the GUI, and it begins with a bare landscape and rudimentary building blocks.

---

2   *Minecraft,* https://minecraft.net/.

My four-year-old, who is only beginning to learn how to read (and cannot write without direct guidance) is already a virtuoso at writing his own world within this game. Minecraft has a basic frame and structure, but no hard limit: its terrain stretches potentially to infinitude, reconfiguring and adapting to its users' commands. And it does not distinguish the elite from the uneducated, but rather brings the marginally literate and illiterate onto the same common ground as the programmers who enjoy this game with a highly sophisticated understanding of its code.

However, digital humanists have recently theorized the GUI as a mechanism of hegemonic repression that forces its users into pre-existing epistemes. They challenge scientific discourses that grant creative agency to the user while defining technology as a tool whose affordances facilitate the user's comprehension of and action in their environment. Instead, these digital humanists interpret the interface as the newest layer of cultural oppression weighing down a human consciousness already burdened by the overarching nexus of power that subjugates the individual to society. Some critics have even stripped the users of their personhood: Johanna Drucker asserts, "the very term 'user' has to be jettisoned — since it implies an autonomy and agency independent of the circumstances of cognition — in favor of the 'subject' familiar from critical theory."[3] Unfortunately, the familiar subject of theory is the one subjected / subjugated to the social structure, who collapses into preexisting spatial classifications. These theoretical moves often dehumanize both this image-driven interface and its users.

This dehumanization of the user stems from the current drive to theorize digital humanities. For instance, Alan Liu explains that the field is the primary beneficiary of the post-May 1968 rise of cultural theory that defined humanity as "congenitally structural, epistemic, class based, identity-group based

---

3   Johanna Drucker, "Humanities Approaches to Interface Theory," *Culture Machine* 12 (2011): 1–20, at 8.

(gendered, racial, ethnic), and so on."[4] Although this critical move reveals the investments of the digital humanities in interrogating social inequities, it also reduces the individual to a mere element of the massive intersecting networks of cultural formation that Liu describes as "the total machine of historical, material, and social determinism that is both the condition and the dilemma of modernity."[5] It defines digital humanities as a method that views people as products rather than producers. The same move that makes DH speak more intimately to the interests of humanities scholars paradoxically strips this field of its humanity.

The hegemony of the digital world becomes simply another panopticon containing a human mass living in quiet desperation. Mark Sample's post mourns the death of the fugitive in a media culture so panoptic that it is impossible to evade. Instead, Sample explains, "in a world of digital, synchronized communication we have what amounts to infinite tracking, deep searching, and persistent indexing. Of everyone."[6] From this perspective, the proliferation of information technologies that gave birth to the realm of digital humanities simultaneously killed the independent spirit of the people who created that space.

Perhaps on some level Sample is mourning the movement of this once-marginal field into the center of academe. Until recently, digital humanists were outlaws of the system; now, they shape it. But this very movement of digital humanities from the periphery to the center signals the basic and incessant cultural effect of resistance. Antonio Gramsci argued that hegemony is always shaped by those who oppose it: as those excluded by a society voice their malcontent, they shift the conduits of power in their favor until they in turn pull the strings — at least for a moment, since by the time it is recognized as hegemony it is

---

4 Alan Liu, "Where is Cultural Criticism in the Digital Humanities?" in *Debates in the Digital Humanities,* ed. Matthew K. Gold, 490–510 (Minneapolis: University of Minnesota Press, 2012), 493.

5 Ibid., 500.

6 Mark Sample, "The Century of the Fugitive and the Secret of the Detainee," *samplereality* (blog), 22 April 2013.

always already slipping toward the interests of other rising sub-altern groups.[7] The fugitive of hegemony is fundamentally so-cial: she captures our collective imagination when she refuses to slink into oblivion. She is the outlaw central to juridical rein-scription, the criminal who is queen. We love fugitives not for their anonymity, but for their power to change society by carv-ing their own place into it.

For this reason, the move of the digital humanities from the periphery to the center is not tragic but promising, because the field is gaining the power to remap the very ways we think about the human. Digital humanists should not clothe them-selves in the tired theoretical habits that subsume the individual under repressive social constellations. The new hegemony of digital humanities in academe exposes a broader readiness to shed current conceptions of the humanities and explore new realms of thought about the ordinary person's agency vis-à-vis the pedagogical and the political. Why should digital humanists conform to existing discourses, when the field of DH is fecund precisely because it does not fit those outmoded categories?

As digital humanists move into the center of the debate about what a humanities education is and does, they have the freedom to redefine the hegemonic theories binding it to the past. GUIS are the perfect proving ground for this retheorization: they au-thorize the self (the fugitive par excellence of the dreary world of unrelenting compartmentalized labor) to escape her subjection to totalizing systems via a rhizomatic engagement with a new topography of the common that encourages free play. Michel de Certeau complicates producer-driven ideas about how in-formation shapes its recipients by highlighting how ordinary consumers repurpose media to serve their own interests.[8] The GUI exemplifies this model of eccentric consumption. Although the interface constrains its user to a certain set of actions, that

---

7   See, for instance, Antonio Gramsci, *Selections from the Prison Notebooks*, ed. and trans. Quentin Hoare and Geoffrey Nowell Smith (London: Law-rence and Wishart, 1971).

8   Michel de Certeau, *The Practice of Everyday Life*, trans. Steven Randall (Berkeley: University of California Press, 1984).

set is so expansive that it affords remarkable freedom within a very large frame. Since the icons shaping the rhetoric are gestural rather than comprehensive, they invite meaning rather than invoking it. The GUI invites us to take pleasure in exploring its world, clicking on links that strike our fancy and wandering through strange landscapes. In the process, our enjoyment translates into a cognitive process that blocks our complete assimilation into any superstructure. The "we" produced by the interface is not monolithic, since the individualistic nature of various users' relations to the GUI cannot be reduced to that of a singular "subject."

Renaissance pocket maps show us that the fugitive and the outlaw are not the last bastion against the postmodern annihilation of the self in service to the subject. They reveal that there have always been many ways of looking at the world, and the perspective of the outlaw has often vied with (and even become) that of the monarch. Giant maps were panoptic, but in everyday life they were crowded out by the tiny maps that proliferated in the nation's retail bookstalls. Similarly, the ideological place of the modern subject is shaped by systems that put Orwellian models to shame; but myriad individuals create pockets of space within that place that weaken its very infrastructure. Yes, "they" can track "us," but that very phrase creates a common ground for countless fugitive users united by little else than our shared resistance to oppression. Our dismissal of those who would contain us marginalizes repressive power systems: we decentralize the center that would subjugate us. And this "us" is potent because it celebrates individual differences rather than flattening all users into a single identifiable populace. The shared world of the GUI creates strange bedfellows: it is, after all, the place where the politics of far left and extreme right converge in their shared revulsion of surveillance and censorship. As Liu points out, the field of DH has the unique potential to forge new lines of communication between the academic study of the humanities and the digital media upon which a shifting and riven public inscribes its real-world concerns.

Resisting the dehumanization of the interface thus entails the paradoxical return of the concept of the user: the autonomous, absentminded, bodied entity who learns haptically by touching the icons that beckon to him, who forges his own curricula extemporaneously between bouts of Candy Crush, who learns as much or more about the world when he digresses as when he stays on his intended track. In the classroom, the quotidian nature of ordinary resistance might translate into a rejection of the professorial mistrust that often attends technology, paired with an open invitation for students to click the hyperlinks, Google terminology, and search for visualization aids without external prompts. If the user is central to the pedagogical experience of the interface, then a student should not be subject to someone else's usage maps. They will likely Snapchat in class; but they will also undoubtedly find ways through the subject matter that will teach them the value of developing their unique authority within (and without) that topic.

For many of us, the ideal endpoint of a humanities education is to train our students to step into society as active and engaged citizens of the world. This goal is fundamentally pragmatic, of course, since the good life in this case necessitates the transition from uncritical subjecthood to civic engagement, to action in the world. Humanist pedagogy thus works best when it aims not for transcendence but utility, and for lived experience rather than abstraction. The field of digital humanities is the terrain upon which we may best embrace this prosaic, everyday approach to humanism. As students become users rather than passive receptacles, they become fugitives of the hegemony in their rejection of others' dictates (including those concerning the "proper" pedagogical use of technology). They shed the strictures of subjection to step playfully into the fullness of humanity. For students as well as scholars, the GUI maps this path into the pleasure of the unknown. The interface teaches us all to leave behind the baggage of what we should be and do and say, and simply embrace our individual roles in the creation of brave new worlds.

Bibliography

de Certeau, Michel. *The Practice of Everyday Life*. Translated by Steven Randall. Berkeley: University of California Press, 1984.

Drucker, Johanna. "Humanities Approaches to Interface Theory." *Culture Machine* 12 (2011): 1–20. https://www.culturemachine.net/index.php/cm/article/viewArticle/434.

Gramsci, Antonio. *Selections from the Prison Notebooks*. Edited and translated by Quentin Hoare and Geoffrey Nowell Smith. London: Lawrence and Wishart, 1971.

Lecky, Kat. "Humanizing the Interface." *Hybrid Pedagogy*. 27 March 2014. http://www.digitalpedagogylab.com/hybridped/humanizing-interface/.

Liu, Alan. "Where is Cultural Criticism in the Digital Humanities?" In *Debates in the Digital Humanities,* edited by Matthew K. Gold, 490–510. Minneapolis: University of Minnesota Press, 2012. http://dhdebates.gc.cuny.edu/debates/text/20.

Minecraft. https://minecraft.net/.

Sample, Mark. "The Century of the Fugitive and the Secret of the Detainee." *samplereality* (blog). 22 April 2013. http://www.samplereality.com/2013/04/22/the-century-of-the-fugitive-and-the-secret-of-the-detainee/.

22

Bend Until It Breaks

Robin Wharton[1]

"*[T]he problem is to begin with a conception of power relations that grants that resistance is always possible but not always successful.*"
— David Sholle[2]

Why resist?

The capability inherent in digital humanities for resistance is part of what makes digital humanities "humanistic" — rather than, say, techno-utopian or neoliberal — it's what connects the digital humanities to the humanities. Alan Liu and Stephen Ramsay have both argued for the necessity of theorizing "resistance" and its place in the work of digital humanists. Ramsay gets to the heart of what "resistance" might look like in this context when, in an eloquent defense of the humanities in general, he describes the humanities as a discursive space in which we answer the pressing question, "How do we become individuals who move through the world with awareness, empathy, and thoughtful-

---

1   Originally published as Robin Wharton. "Bend Until It Breaks: Digital Humanities and Resistance," *Hybrid Pedagogy*, 19 February 2014.
2   David Sholle, "Resistance: Pinning Down a Wandering Concept in Social Studies Discourse," *Journal of Urban and Cultural Studies* 1, no. 1 (1990): 87–105, at 98.

ness, and who know how to act upon those dispositions?"[3] What if, "we can resist" were at least a partial answer to Ramsay's question? If this is what resistance can do for us, for the project of being or becoming human, then I think we can see pretty clearly why it matters, why digital humanists should be investing time and resources in activities of resistance. As Liu observes, however, "[h]ow the digital humanities advances, channels, or resists today's great postindustrial, neoliberal, corporate, and global flows of information-cum-capital is […] a question rarely heard in the digital humanities associations, conferences, journals, and projects with which I am familiar."[4]

What is resistance?

In thinking about resistance and the role it has to play in humanistic inquiry, I am using the definition David Sholle offered more than twenty years ago: "At its root, as an activity, resistance is a defensive contestation, an act of refusal. If we strip away the inessential from the inflections of resistance that have been described thus far, it can be seen that they are all, at root, defensive activities in that they work to limit the capacity of power to define the parameters of action."[5] As Sholle describes it, resistance in this sense involves first understanding how power relations — social, economic, political — work to perpetuate things like inequality or poverty or racism or "evil," in part, through the fragmentation of "all points of view and all values [so] as to render them without meaning beyond their value as commodities."[6] Resistance then requires a "refusal to participate in a strategic contest that power dictates," i.e., "a rule-breaking activity" aimed at "encourag[ing] or discourag[ing] other activi-

---

3   Stephen Ramsay, "Why I'm In It," blog post, 12 September 2013.

4   Alan Liu, "Where Is Cultural Criticism in the Digital Humanities?" in *Debates in the Digital Humanities,* ed. Matthew K. Gold (Minneapolis: University of Minnesota Press, 2012), http://dhdebates.gc.cuny.edu/debates/text/20.

5   Sholle, "Resistance," 99.

6   Ibid.

ties aimed at altering power effects."[7] To the extent humanities scholarship is itself an attempt at resistance, "[s]imply finding resistance already operating is not enough." Rather, scholars "must be interested in pointing out where and when particular struggles point to potential alliances that lead to further transformative action."[8]

Although in Sholle's formulation successful resistance might be rare, opportunities for resistance abound. Success is uncertain because power itself is uncertain. Power is not hegemonic, consolidated, or even consciously coordinated. Rather it is shifting, changeable, contingent, and diffuse. Success depends in part upon carefully thinking through how power relations may evolve in response to a particular strategy of resistance, including understanding how and when new power relations may emerge to contain or neutralize it. In this way, we can identify strategies that not only work to change how power is distributed within the system, but also — and this is the real trick if it can be accomplished — disrupt or destabilize the "means of bringing power relations into being"[9] in the first instance.

Why resistance and DH?

Given a working definition of resistance that highlights the necessity of rule-breaking, of interfering with and perhaps even restructuring power relationships, Jesse Stommel's declaration, "The digital humanities is about breaking stuff,"[10] can be seen for what it is. More than a provocation, it is a call to action. In particular, I would like to explore how the tendency towards, even perhaps the necessity of "breaking stuff" in the digital humanities plays out in the regulatory context. The idiomatic expressions we use to describe criminal activity in English are telling here. We describe criminals as those who "bend" or "break" the

---

7   Ibid.
8   Ibid., 97.
9   Ibid., 99.
10 Jesse Stommel, "The Digital Humanities Is About Breaking Stuff," *Hybrid Pedagogy,* 2 September 2013.

law. Obviously, breaking the law in this sense, while it may be resistance in one definition, isn't effective resistance of the sort Sholle describes and for which Liu and Ramsay are advocating. That's because criminal activity is, returning to Sholle again, a form of resistance that is already "managed" and "limited"[11] within the law's discursive space for the most part (though important exceptions to this general observation do exist, of course, including civil disobedience of the sort we saw during the US civil rights movement and have seen in Nigeria,[12] Egypt, Turkey, and, even more recently, the rallies and "die-ins" that have organized to protest racialized police violence against people of color in the US).[13]

When is DH resistance?

What does it mean, though, to engage in professional practices whose end is, at least in part, to "bend," "deform," or even "break" the law? What happens when digital humanists — or any humanists, for that matter — rather than treading lightly,[14] instead run roughshod through the carefully cultivated regulatory landscape in which formal, aesthetic distinctions between art and scholarship, between creating and critiquing, and between pedagogy and artistry must be maintained and reproduced in order for everything to work?

I have written elsewhere about how the discursive representation within legal decision-making of literary concepts such as "authorship," "scholarship," "utility," and "ornament" have influenced the evolution of copyright law.[15] These legally significant

---

11 Sholle, "Resistance," 97–100.
12 Teju Cole, "The White-Savior Industrial Complex," *The Atlantic,* 21 March 2012.
13 David M. Perry, "#FergusonSyllabus," *The Chronicle of Higher Education,* 25 November 2014.
14 Collen Flaherty, "Timid About Fair Use?," *Inside Higher Ed,* 30 January 2014.
15 Robin Wharton, "Digital Humanities, Copyright Law, and the Literary," *Digital Humanities Quarterly 7,* no. 1 (2013).

concepts are often borrowed almost wholesale from the discourse of literary studies or humanistic scholarship more generally. Thus, for example, the "authorship" copyright rewards does not include "sweat of the brow" collection and rote organization of information — however much labor that might involve, and however useful or interesting the resulting artifact — but only original, creative "invention."[16] In the digital humanities, ethics of collaboration, such as those expressed in the "Collaborators' Bill of Rights," arguably require us to push back against this institutionalized elevation of invention over the other forms of essential labor involved in the creative process: "All kinds of work on a project are equally deserving of credit (though the amount of work and expression of credit may differ). And all collaborators should be empowered to take credit for their work."[17]

The law in turn has influenced and shaped the work we do as scholars and the forms our scholarship can take. The fair use analysis, for instance, often relies on a presumption that the items enumerated in the fair use preamble — "criticism, comment, news reporting, teaching [...], scholarship, or research"[18] — will not look like creative artifacts, that literary scholarship will not resemble, except in the most superficial way, the literary objects with which it engages.[19] To the extent humanistic inquiry redefines things like "authorship" and "scholarship," and also begins to shift its practices in ways the law has not already anticipated, these transformations become part of the objective reality the law must come to terms with and regulate in subsequent deci-

---

16 See, e.g., *Feist Publications, Inc. v. Rural Telephone Service Company*, 499 US 340 (1991), in which the Supreme Court held: "The revisions [to the 1976 Copyright Act, United States Code Title 17] explain with painstaking clarity that copyright requires originality, §102(a); that facts are never original, §102(b); that the copyright in a compilation does not extend to the facts it contains, §103(b); and that a compilation is copyrightable only to the extent that it features an original selection, coordination, or arrangement, §101."

17 Tanya Clement et al., "Collaborators' Bill of Rights," in *Off the Tracks: Laying New Lines for Digital Humanities Scholars* (Media Commons Press, 2011).

18 United States Code, Title 17, Section 107.

19 Wharton, "Digital Humanities."

sions. By understanding the relationship between the law and the objects and activities towards which it is directed, and the complex discursive exchange among law and literary studies, we begin to reveal pressure points where opportunities for resistance, as I've described it here, arise.

How does DH break the law?

In his essay "The Law Wishes to Have a Formal Existence," collected in *There's No Such Thing as Free Speech: And It's a Good Thing, Too,* Stanley Fish observes,

> [A] legal system whose conclusions clashed with our moral intuitions at every point so that the categories legally valid and morally right never (or almost never) coincided would immediately be suspect; but a legal system whose judgments perfectly meshed with our moral intuitions would thereby be rendered superfluous."[20]

While there may be much to disagree with in Fish's analysis of the law's rhetorical operation, that analysis — and particularly this point — nevertheless manages to isolate a potentially useful discursive tension produced when the law casts human activity into narrative in order to accomplish its aim. Certain kinds of scholarly activity within the digital humanities exploit that tension to resist the process of disciplinary regulation through legal narrativization.

> All discourses (not only Marx's, but also Smith's or Ricardo's or the discourse of neoclassical economists) can be read as comprising different orders in which signifiers are articulated into discourse in order to produce different meanings; that is, different discourses are different constitutions of signs

---

20 Stanley Fish, "The Law Wishes to Have a Formal Existence," in *There's No Such Thing as Free Speech: And It's a Good Thing, Too,* 141–80 (Oxford: Oxford University Press, 1994), at 141–42.

rather than different interpretations of an empirically given object of analysis.

In order to make sense of the set of meanings produced in a particular discourse, we must take into account the relations of confrontation between this particular discourse and other discourses. These relations of confrontation are themselves the partial result of the operation of different politico-theoretical priorities in discourse. Thus, for example, the meanings that Marx ascribed to his key concepts (e.g., value, economy, ideology) do not have fixed, empirical referents. These meanings are construed as they are because Marx, facing the meanings produced by classical political economy, confronted this school of thought-its politico-theoretical priorities-by setting out to produce a different set of meanings exhibiting a different set of priorities.[21]

Much more recently, Stephen Best has asked us to consider,

> How does the "form" in the commodity form generate social phenomena in ways that are neither mechanical (historical causality) nor fully contingent (analogy)? When I ask, what is the generative power of form? I am not asking the conventional question, what does form cause? […] It is more a question of what form produces, what form generates.[22]

As Jack Ameriglio and Antonio Callari argue, because of how discourses function, we can always expect slippage between an "empirically given object of analysis" and each of the various discursive forms we use to talk about that object. Further, again because meaning within discursive systems depends more upon the relationship among signifiers, rather than the relationship between any given signified and its signifier(s), some discursive

---

21 Jack Amariglio and Antonio Callari, "Marxian Value Theory and the Problem of the Subject: The Role of Commodity Fetishism," *Rethinking Marxism* 2, no. 3 (1989): 31–60, at 42.

22 Stephen Best, *The Fugitive's Properties: Law and the Poetics of Possession* (Chicago: University of Chicago Press, 2004), 20.

systems will actually tolerate quite a bit of slippage between an object and its discursive representation. Finally, in the "relations of confrontation" between discourses, such as those Fish identifies between legality and morality, lies a generative potential, one that is neither — to draw from Best — "mechanical" nor "fully contingent" in its operation, when the forms of one discourse are drawn into, or in conversation with another.

Because the law — like most discourses probably — is a hybrid, blending legal and non-legal discursive forms, legal narratives simultaneously comprise a variety of narrative registers or modes, including documentary, fiction, and speculative realism. Legal statutes and court cases simultaneously describe the world as it is, how it needs to be in order for laws and legal doctrines to make sense, and how it might be if everything goes according to the regulatory game plan. As I've argued previously,

> [court decisions] can, of course, be read as judicial responses to historically specific social, political, and economic pressures. They may also [...] be read as judicial efforts to create an internally coherent legal epistemology that not only reacts to the world it is intended to regulate but also proactively creates the terms by which that world will subsequently be known and understood.[23]

Forms, both objective and discursive, matter in the law. Legal narrative can tolerate a certain amount of fictional or even speculative realism without losing its authority. When, however, the slippage between the "real" objects of regulation and their discursive representation becomes too great, or, even, when the discursive representation of things in statutes and legal decisions strays too far from how those same things are represented in other discourses — like scholarly essays or the popular press — the law's ability to "define the parameters of action" is tested, and occasionally even disrupted.

---

23 Wharton, "Digital Humanities."

When digital humanities scholars make scholarly artifacts, when any scholars make scholarly artifacts that don't conform to aesthetic expectations baked into the law (or into more localized regulatory apparatus such as campus intellectual property policies or review, promotion, and tenure procedures), they are potentially engaging in a kind of resistance. Such artifacts present opportunities to construct new disciplinary, discursive, and professional alliances and to write new narratives within which such forms can be discursively articulated as "scholarship" or "fair use" or "non-infringing" or, perhaps most tellingly, "transformative."[24] Similarly, when digital humanities projects establish new working relations among collaborators that not only ignore but in some cases actively defy institutionalized hierarchies of labor, we create pockets of jurisdictional uncertainty where regulatory assumptions about whose contributions will count for what and why are called into question.[25] Simply making or building these things, however, is not in and of itself going to lead to the creation of a world in which "individuals [...] move through the world with awareness, empathy, and thoughtfulness, and [...] know how to act upon those dispositions." Rather, we must also be able to cast these objective forms into discursive forms that can act within, can transform the complex regulatory narratives that constrain the field of human activity. And finally, because no one should be expected to two jobs for the price of one, we must be open to new collaborative

---

24 See, for example, Katherine D. Harris, "Explaining Digital Humanities in Promotion Documents," *Journal of Digital Humanities* 1, no. 4 (2012).

25 Julia Flanders, "Time, Labor, and 'Alternate Careers' in Digital Humanities Knowledge Work," in *Debates in the Digital Humanities,* ed. Matthew K. Gold (Minneapolis: University of Minnesota Press, 2012). Flanders does an excellent job of exploring the ways in which "alternative academic" or "alt-ac" work within DH labs and centers contravenes, undermines, and calls into question traditional models of academic labor. She also offers suggestions about how graduate education and academic work environments could be restructured to better accommodate, value, and reward the variety of labor that makes DH scholarship possible.

relationships and new definitions of what counts in academic hiring, tenure, and promotion.[26]

How does resistance become reform (Pt. 1, Pedagogy)?

In "Unpacking My Library," and in his other work on the figure of the collector, Walter Benjamin argues for the existence of subject–object (or maybe object–object) relations that, even though they cannot exist outside of the exchange economy, nevertheless resist the ontological consequences of that economy in highly productive ways.[27] I see similar potential in Mark Sample's description of non-consumptive reading.[28] Non-consumptive reading resists the ontological consequences of the current regulatory system, transgressing distinctions between the products and objects of literary analysis upon which its operation depends. It blurs the distinction between producer and consumer, or artist and critic — categories with substantial legal, social, and economic significance. Because non-consumptive reading causes us to re-examine foundational and often implicit discursive assumptions, it has potential value not only as a scholarly practice in the digital humanities, but also as a critical pedagogical practice in the humanities, and perhaps even other disciplines more broadly.

---

26 See for example, Trevor Muñoz, "Digital humanities in the library isn't a service," blog post 19 August 2012. Muñoz makes a compelling argument that, "Having those who work on digital projects claim identities as researchers rather than as some other kind of academic employees who serve faculty research is important for addressing the issues of power balance within the academy."

27 Walter Benjamin, "Unpacking My Library," in *Illuminations,* ed. Hannah Arendt, trans. Harry Zohn, 59–68 (Berlin: Schocken Books, 1969). See also, Benjamin's essay, "Edward Fuchs: Collector and Historian," *New German Critique* 5 (1975): 27–58, and his commentary in "The Collector," in *The Arcades Project,* ed. Rolf Tiedmann, trans. Howard Eiland and Kevin McLaughlin, 203–11 (Cambridge: Belknap Press, 2003).

28 Mark Sample, "The Poetics of Non-Consumptive Reading," *Sample Reality* (blog), 22 May 2013.

For children can accomplish the renewal of existence in a hundred unfailing ways. Among children, collecting is only one process of renewal; other processes are the painting of objects, the cutting out of figures, the application of decals — the whole range of childlike modes of acquisition, from touching things to giving them names.[29]

I want to advocate for a poetics of non-consumptive reading in the digital humanities. Scholars and students of art, literature, history, and culture ought to transform more of our non-consumptive research into expressive objects. Non-expressive use of texts is a dead-end for the humanities. A computer model surrounded by a wall of explanatory words is not enough. Make the computer model itself an expressive object. Turn your data into a story, into a game, into art.[30]

To put it another way, digital humanities scholarship has caused us to examine more carefully how the discursive forms — including the channels of distribution — produced within humanities disciplines are deployed in regulatory discourses to perpetuate a copyright regime that chills speech and significantly restricts academic freedom, as well as to maintain structural inequality throughout the academy. Classroom praxis in the humanities comprises discursive forms that can be just as rigidly predictable, and as closely entwined with regulatory apparatus, as those we find in traditional print journal articles and scholarly monographs. Too often in post-secondary pedagogy we ask students to iterate discursive forms without asking whether that is the best way to teach them — either the forms or the students. Yes, certainly, it may be the best way to train students to become members of our own professional discourse communities as they are currently configured, but given the precarity of the academic labor market, we should at least be questioning the wisdom of that justification.

---

29 Benjamin, "The Collector," 61.
30 Sample, "The Poetics of Non-Consumptive Reading."

A digital humanities pedagogy of resistance cannot simply take for granted that the whole purpose of K–12 education is to prepare students for college and a job, and also that the social, political, and economic functions of the academy are all unquestionably good. Instead, DH practitioners should participate in a genuine, dialogic conversation[31] about what the purposes of lifelong learning should be and how best to design our pedagogy to fulfill those purposes at every stage in a learner's experience. Rather than presuming serious discussion should be the model for every seminar meeting, we should be much more mindful of how what we (a we that includes instructors and well as students) want to accomplish in a given period, students' learning preferences, and the material under consideration should determine the methods we employ. We should be open to the possibility that field trips, games, physical activity, show and tell, and other "childish" things need not be left behind once students enter college. Examples of pedagogical practices that emphasize and value play, emergence, and collaborative processes over rules, structure, and individual work product include Adeline Koh's "Trading Races,"[32] Pete Rorabaugh's and Jesse Stommel's "Twitter vs. Zombies,"[33] Frederick Cope's and Michelle Kassorla's *Generative Literature Project*[34] and the sprawling, semi-mythical *DS106* that began at the University of Mary Washington.[35] We should be aware that the spaces in which learning takes place may be as variable as the activities that take place within them. Finally, we must be attentive to the affective dimensions of the learning experience. Ensuring learning is pleasant, engaging, and pleasurable — as well as challenging, sometimes difficult,

---

31 Richard D. Kahlenberg, "What Colleges Can Learn From K-12 Education," *The Conversation on Chronicle of Higher Education,* 22 May 2013.

32 Adeline Koh, "Trading Races," http://tradingraces.adelinekoh.org/.

33 Pete Rorabaugh and Jesse Stommel, "TvsZ," http://www.twittervszombies.com/.

34 See the *Generative Literature Project* archives on *Hybrid Pedagogy: A Digital Journal of Learning, Teaching, and Technology* for a history and overview of the project: http://www.hybridpedagogy.com/tag/generative-literature-project/.

35 The *DS106* project site can be accessed at http://ds106.us/.

and transformative — does not necessarily reduce a pedagogue to an entertainer.

Looking beyond college and university classrooms to education reform policy debates taking place in legislative sessions and the popular press,[36] we should be worried by unreflective calls for increased "rigor" and greater "accountability" in K–12 education. Standardized testing, the controversial common core standards,[37] proposed MOOC-ification of remedial education,[38] these "innovations" are all arguably attempts to address students' "underpreparedness" for college and the workplace.[39] Meanwhile, art, music, physical education, and recess are disappearing from the curriculum. In their zeal for "reform," policymakers may be eradicating the very things about K–12 education that might teach us and our students about where practices like curation,[40] building,[41] and creative production[42] fit in humanistic inquiry.

Further, even where we have begun to acknowledge their value, our obsessive emphasis on end results may actually empty out the resistive potential of these pedagogical strategies. For Benjamin, "childlike processes" and collecting — processes that work against or at cross-purposes with the logic of capital — are strategies of material and ontological renewal precisely because they are done for themselves rather than as a means to a consciously articulated and pre-determined end. The end result of

---

36 Audrey Watters, "Hacking at Education: TED, Technology, Entrepreneurship, Uncollege, and the Hole in the Wall," *Hack Education*, 3 March 2013.

37 *Common Core State Standards Initiative*, http://www.corestandards.org/.

38 Tyler Kingkade, "San Jose State University Begins MOOC Partnership As California Schools Pushed To Online Education," *Huffington Post*, 16 January 2013.

39 Eric Kelderman, "Colleges Must Help Prepare Students for Higher Education," *Chronicle of Higher Education*, 30 October 2012.

40 Dominik Lukeš, "Zero pedagogy: A hyperbolic case for curation and creation over education in the age of the MOOC (#moocmooc)," *Researchity — Exploring Open Research and Open Education*, 15 August 2012.

41 Stephen Ramsay, "On Building," blog post, 11 January 2011.

42 Dale Dougherty, "Georgia Tech's Makerspace is a Model for Higher Education," *Make:*, 28 March 2013.

education reform focused on learning as a product that serves individual interests, rather than a process that enriches the community as a whole as a public education system — and, increasingly, a system of higher education — that exacerbate income inequality and the effects of institutionalized racism.[43] Parents and students with the means to pay for it can still pursue an education filled with opportunities for open-ended exploration through play, building, performance, and making. Parents and students without that option are often forced to settle for an education designed to enforce compliance rather than empower citizens.[44] If wealthy, white suburban parents are entitled to "opt-out" of the relentless cycle of high stakes testing in public schools on behalf of their children,[45] then why the parents and teachers of children of color must be disciplined so harshly for their own vital acts of resistance in relation to the very same system?[46] To understand digital humanities pedagogy as resistance is to understand that technology can be used to oppress as well as liberate, and to be wary of institutional conversations that construct an us against them relationship between those who use technology in the classroom and those who do not. As Sholle reminds us, the forms of resistance are various and highly context-dependent. We should build communities of pedagogical practice around the shared goal of student empowerment, not fascination with the latest gadgets and technological solutionism.

As a lawyer and legal scholar, I am absolutely aware of how essential the ability to interpret and reproduce the discursive

---

43 Suzanne Mettler, "More bad news for millenials: College is actually making inequality worse," *Salon*, 14 March 2014. Tressie McMillan Cottom, "Why Free College Is Necessary," *Dissent* (Fall 2015).

44 David Perry, "The Corrosive Cult of Compliance in Our Schools," *Special Needs Digest*, 21 April 2015, http://www.specialneedsdigest.com/2015/04/the-corrosive-cult-of-compliance-in-our.html, and "Ferguson and the Cult of Compliance," *Al Jazeera America*, 15 August 2014.

45 "What galvanized standardized testing's opt-out movement," *PBS News Hour*, 26 May 2015.

46 Emma Brown, "Atlanta judge reduces sentences for three educators in cheating case," *The Washington Post*, 30 April 2015. "Where school boundary-hopping can mean time in jail," *Al Jazeera America*, 21 January 2015.

forms in which power speaks to power can be. Yet, what is truly empowering is understanding such forms are constructed, contingent, open to interpretation, negotiable, and also knowing where they fail and when other forms are better suited to the task at hand. I have seen in a variety of contexts how process and methodology work to establish personal and professional identity in ways that can be liberating and also limiting. We should constantly be re-examining how our own processes and methodologies as teachers, students, scholars, and artists position us in relation to one another and the subjects/objects of study within our classrooms. Rather than simply allowing social, economic, political, legal, and disciplinary regulatory structures to dictate the shape of what we do, we should be more mindful of how what we do helps give rise to and reinforce such structures.

How does resistance become reform (Pt. 2, Scholarship)?

Sholle's definition of resistance has been in circulation in media studies since the early nineties. Resistance, like many of the other things digital humanists do, is something media studies and cultural studies folks have also been doing for thirty years or more. I wonder, though, if part of the reluctance Liu has observed on the part of digital humanists to do cultural criticism is attributable, at least in part, to scholars' general reluctance to engage with entrenched power dynamics and structural inequality within the academy itself, however astutely they may critique their manifestation beyond the ivory tower.

> [I]t is essential those involved in promotion and tenure reform recognize that excellence is a socially constructed notion. As human beings in social systems within universities, we are flawed. Efforts to become a more diverse, inclusive community are intimately tied to the kinds of work our aca-

demic reward systems value, how we evaluate it, and how conscious we can be about the biases we bring to the table.[47]

The redesign of scholarship to allow for participation is an enormous undertaking, not yet much beyond prototypes, none of which have yet proved fully viable except the wiki. And the difference between a book chapter that lays out a well informed and studied discussion of new research and a set of guided activities for the acquisition of that knowledge is the difference between research and pedagogy. They perform different roles.[48]

We must understand the conversations about things like getting rid of the dissertation — or accepting digital projects in lieu of graduate theses or print monographs — as arguments about whether and how to resist. We need to acknowledge how new modes of open access and open source scholarship and publication may involve reconfiguring the means of academic production, and the relationship between author and producer, reader and consumer, text and commodity in ontologically significant ways. Similarly, debates about the relationship between pedagogy and scholarship, and their relative value in the academy, are directly relevant to the project of theorizing resistance in the digital humanities.[49] When we build things, we are also building networks and relationships. When we define "digital humanities" and "scholarship" in ways that exclude "pedagogy," we are articulating relations of power that will govern the working con-

---

47 Kerry Ann O'Meara, "Change the System," *Inside Higher Ed,* 14 January 2014.

48 Johanna Drucker, "Pixel Dust: Illusions of Innovation in Scholarly Publishing," *Los Angeles Review of Books,* 16 January 2014.

49 In "Play, Collaborate, Break, Build, Share: 'Screwing Around' in Digital Pedagogy, The Debate to Define Digital Humanities... Again," *Polymath: An Interdisciplinary Arts and Sciences Journal* 3, no. 3 (2013): 1–24. Katherine Harris, for example, argues that definitions of "digital humanities" as a field that exclude teaching and pedagogy "silence" the only "brand of Digital Humanities" in which most scholars at teaching-intensive institutions engage, given the institutional circumstances in which they perform scholarly labor.

ditions of our colleagues, shaping and perhaps constraining the field of intellectual activity within and perhaps even beyond the academy for years or decades to come.

At the same time, we must do more than pay lip service to scholarly or disciplinary innovation. We must walk the walk as well as talk the talk, hack as well as yack. The formal risks digital humanists take with their scholarship are every bit as important to the project of resistance as are theoretical and institutional advocacy that help to justify such work as scholarship.[50] If the digital humanities are to be an effective path of most resistance in the academy, then the "digital" in digital humanities needs to refer to more than just the methods scholars employ, and the digital forms they produce must do more than simply iterate the aesthetics and conventions of print scholarship. Advisors and dissertation committees must be willing to let graduate students take formal risks with their work, and hiring and tenure and promotion committees must be willing to accept innovative forms not just in addition to,[51] but in lieu of the print monograph.

The goal of resistance in the digital humanities should not, I think, be to replace one kind of thing with another kind of

---

50 See, for example, Adeline Koh's discussion, "More Hack, Less Yack? Modularity, Theory and Habitus in the Digital Humanities," blog post, 21 May 2012, in which she maintains, "[J]ust as UNIX programmers relied, in [Tara] McPherson's argument [in "U.S. Operating Systems at Mid-Century: The Intertwining of Race and UNIX"], on a common-sense modular 'lenticular logic' to avoid speaking about the socio-political origins and conditions that allowed for their 'common sense' to come into being, perhaps the same logic has underwritten our resistance to theory within the digital humanities." Bethany Nowviskie, in "Resistance in the Materials" (http://nowviskie.org/2013/resistance-in-the-materials/), a talk she gave at the Modern Language Association convention in 2013, observed that, "[c]asualized labor begets commodity toolsets, frictionless and uncritical engagement with content, and shallow practices of use." Although the stereotypical call for "more hack, less yack," tries to suggest otherwise, theory and practice within the digital humanities are inextricably intertwined. Together, both Koh and Nowviskie reveal how a lack of critical reflection or "theorizing" may work to maintain exploitative labor practices and unfair distribution of resources, as well as result in inferior digital tools and bad scholarship.

51 Sydni Dunn, "Digital Humanists: If You Want Tenure, Do Double the Work," *Vitae,* 5 January 2014.

thing in academic work, but to open up the field of possibilities. Further, we must be open to critique that points out unintended consequences, and be wary of the "old wine in new bottles" problem in which forms that seem innovative at first glance simply repackage and recirculate the familiar damaged goods of socio-economic stratification, political alienation, contingency, ivory tower isolationism, and exclusion or disenfranchisement of people of color, those who identify as queer, women, and many others who don't fit into a dominant Western, white, male, heteronormative paradigm. To resist, we must refuse to accept as given and even be willing to break the existing rules, and we should also be careful we don't intentionally or unintentionally replace them with something worse.

Bibliography

Amariglio, Jack, and Antonio Callari. "Marxian Value Theory and the Problem of the Subject: The Role of Commodity Fetishism." *Rethinking Marxism* 2, no. 3 (1989): 31–60. DOI: 10.1080/08935698908657868.

"Articles Tagged: Generative Literature Project." *Hybrid Pedagogy: A Digital Journal of Learning, Teaching, and Technology.* http://www.hybridpedagogy.com/tag/generative-literature-project/.

Benjamin, Walter. "Edward Fuchs: Collector and Historian," *New German Critique* 5 (1975): 27–58.

———. "Unpacking My Library." In *Illuminations,* edited by Hannah Arendt, translated by Harry Zohn, 59–68. Berlin: Schocken Books, 1969.

———. "The Collector." In *The Arcades Project,* edited by Rolf Tiedmann, translated by Howard Eiland and Kevin McLaughlin, 203–11. Cambridge: Belknap Press, 2003.

Best, Stephen. *The Fugitive's Properties: Law and the Poetics of Possession.* Chicago: University of Chicago Press, 2004.

Brown, Emma. "Atlanta judge reduces sentences for three educators in cheating case." *The Washington Post.* 30 April 2015. https://www.washingtonpost.com/local/education/atlanta-judge-reduces-sentences-for-three-educators-in-cheating-case/2015/04/30/25ce05f6-ef3f-11e4-8abc-d6aa3bad79dd_story.html.

Clement, Tanya, et al. "Collaborators' Bill of Rights." In *Off the Tracks: Laying New Lines for Digital Humanities Scholars.* Media Commons Press, 2011. http://mcpress.media-commons.org/offthetracks/.

Cole, Teju. "The White-Savior Industrial Complex." *The Atlantic.* 21 March 2012. http://www.theatlantic.com/international/archive/2012/03/the-white-savior-industrial-complex/254843/.

Common Core State Standards Initiative. http://www.corestandards.org/.

Cottom, Tressie McMillan. "Why Free College Is Necessary." *Dissent* (Fall 2015). https://www.dissentmagazine.org/article/tressie-mcmillan-cottom-why-free-college-necessary.

Doughety, Dale. "Georgia Tech's Makerspace is a Model for Higher Education." *Make:*. 28 March 2013. http://makezine.com/2013/03/28/georgia-techs-makerspace-is-a-model-for-higher-education/.

Drucker, Johanna. "Pixel Dust: Illusions of Innovation in Scholarly Publishing." *Los Angeles Review of Books*. 16 January 2014. http://lareviewofbooks.org/essay/pixel-dust-illusions-Innovation-scholarly-publishing.

DS106. http://ds106.us/.

Dunn, Sydni. "Digital Humanists: If You Want Tenure, Do Double the Work." *Vitae.* 5 January 2014. https://chroniclevitae.com/news/249-digital-humanists-if-you-want-tenure-do-double-the-work.

Feist Publications, Inc. v. Rural Telephone Service Company, 499 US 340 (1991).

Fish, Stanley. "The Law Wishes to Have a Formal Existence." In *There's No Such Thingas Free Speech: And It's a Good Thing, Too,* 141–80. Oxford: Oxford University Press, 1994.

Flaherty, Collen. "Timid About Fair Use?" *Inside Higher Ed.* 30 January 2014. https://www.insidehighered.com/news/2014/01/30/experts-say-academics-are-timid-about-fair-use-laws.

Flanders, Julia. "Time, Labor, and 'Alternate Careers' in Digital Humanities Knowledge Work." In *Debates in the Digital Humanities,* edited by Matthew K. Gold. Minneapolis: University of Minnesota Press, 2012. http://dhdebates.gc.cuny.edu/debates/text/26.

Harris, Katherine D. "Play, Collaborate, Break, Build, Share: 'Screwing Around' in Digital Pedagogy, The Debate to Define Digital Humanities… Again." *Polymath: An Interdisciplinary Arts and Sciences Journal* 3, no. 3 (2013): 1–24. https://ojcs.siue.edu/ojs/index.php/polymath/article/view/2853.

Harris, Katherine, D. "Explaining Digital Humanities in Promotion Documents." *Journal of Digital Humanities* 1, no. 4 (2012). http://journalofdigitalhumanities.org/1–4/explaining-digital-humanities-in-promotion-documents-by-katherine-harris/.

Kahlenberg, Richard D. "What Colleges Can Learn From K–12 Education." *The Conversation; Chronicle of Higher Education.* 22 May 2013. http://chronicle.com/blogs/conversation/2013/05/22/what-colleges-can-learn-from-k-12-education/.

Kelderman, Eric. "Colleges Must Help Prepare Students for Higher Education." *Chronicle of Higher Education.* 30 October 2012. http://chronicle.com/article/Colleges-Must-Help-Prepare/135480/.

Kingkade, Tyler. "San Jose State University Begins MOOC Partnership As California Schools Pushed To Online Education." *Huffington Post.* 16 January 2013. http://www.huffingtonpost.com/2013/01/16/san-jose-state-state-univ_n_2488734.html.

Koh, Adeline. "More Hack, Less Yack? Modularity, Theory and Habitus in the Digital Humanities." Blog post. 21 May 2012. http://www.adelinekoh.org/blog/2012/05/21/more-hack-less-yack-modularity-theory-and-habitus-in-the-digital-humanities/.

———. "Trading Races." http://tradingraces.adelinekoh.org/.

Liu, Alan. "Where Is Cultural Criticism in the Digital Humanities?" In *Debates in the Digital Humanities,* edited by Matthew K. Gold. Minneapolis: University of Minnesota Press, 2012. http://dhdebates.gc.cuny.edu/debates/text/20.

Lukeš, Dominik. "Zero pedagogy: A hyperbolic case for curation and creation over education in the age of the MOOC (#moocmooc)." *Researchity — Exploring Open Research and Open Education.* 15 August 2012. http://researchity.net/2012/08/15/zero-pedagogy-a-hyperbolic-case-for-curation-and-creation-over-education/.

Mettler, Suzanne. "More bad news for millenials: College is actually making inequality worse." *Salon.* 14 March 2014.

http://www.salon.com/2014/03/15/more_bad_news_for_
millennials_college_is_actually_making_inequality_worse/.

Muñoz, Trevor. "Digital humanities in the library isn't a
service." Blog post. 19 August 2012. http://trevormunoz.
com/notebook/2012/08/19/doing-dh-in-the-library.html.

Nowviskie, Bethany. "Resistance in the Materials." Lecture,
Modern Language Association Convention, 2013. http://
nowviskie.org/2013/resistance-in-the-materials/.

O'Meara, Kerry Ann. "Change the System." *Inside Higher
Ed.* 14 January 2014. https://www.insidehighered.com/
advice/2014/01/13/essay-calls-reform-tenure-and-
promotion-system.

Perry, David. "The Corrosive Cult of Compliance in Our
Schools." *Special Needs Digest.* 21 April 2015. http://www.
specialneedsdigest.com/2015/04/the-corrosive-cult-of-
compliance-in-our.html.

———. "Ferguson and the Cult of Compliance." *Al Jazeera
America.* 15 August 2014. http://america.aljazeera.com/
opinions/2014/8/ferguson-police-shootracismcompliance.
html.

———. "#FergusonSyllabus." *The Chronicle of Higher
Education.* 25 November 2014. http://chronicle.com/blogs/
conversation/2014/11/25/fergusonsyllabus/.

Ramsay, Stephen. "On Building." Blog post. 11 January 2011.
http://stephenramsay.us/text/2011/01/11/on-building/.

———. "Why I'm In It." Blog post. 12 September 2013. http://
stephenramsay.us/2013/09/12/why_im_in_it/.

Rorabaugh, Pete, and Jesse Stommel. "TvsZ." http://www.
twittervszombies.com/.

Sample, Mark. "The Poetics of Non-Consumptive Reading."
Sample Reality (blog). 22 May 2013. http://www.
samplereality.com/2013/05/22/the-poetics-of-non-
consumptive-reading/.

Sholle, David. "Resistance: Pinning Down a Wandering
Concept in Social Studies Discourse." *Journal of Urban and
Cultural Studies* 1, no. 1 (1990): 87–105.

Stommel, Jesse. "The Digital Humanities Is About Breaking Stuff." *Hybrid Pedagogy*. 2 September 2013. http://www.hybridpedagogy.com/journal/the-digital-humanities-is-about-breaking-stuff/.

United States Code, Title 17, Section 107.

Watters, Audrey. "Hacking at Education: TED, Technology, Entrepreneurship, Uncollege, and the Hole in the Wall." *Hack Education*. 3 March 2013. http://hackeducation.com/2013/03/03/hacking-your-education-stephens-hole-in-the-wall-mitra/.

Wharton, Robin. "Bend Until It Breaks: Digital Humanities and Resistance." *Hybrid Pedagogy*. 19 February 2014. http://hybridpedagogy.org/bend-breaks-digital-humanities-resistance/.

———. "Digital Humanities, Copyright Law, and the Literary." *Digital Humanities Quarterly* 7, no. 1 (2013). http://www.digitalhumanities.org/dhq/vol/7/1/000147/000147.html.

"What galvanized standardized testing's opt-out movement." *PBS News Hour*. 26 May 2015. http://www.pbs.org/newshour/bb/galvanized-standardized-testings-opt-movement/.

"Where school boundary-hopping can mean time in jail." *Al Jazeera America*, 21 January 2015, http://america.aljazeera.com/watch/shows/america-tonight/america-tonight-blog/2014/1/21/where-school-boundaryhoppingcanmeantimeinjail.html.

23

Outsiders, All: Connecting the Pasts and Futures of Digital Humanities and Composition

Christopher R. Friend

This sentence you are now reading was, of course, written on a computer. Myriad technologies influenced the process of writing that sentence, and still more worked to let you read it. Considering how much writing today is done on/with/by/for computers, studying how technology and writing interact should warrant significant academic attention. Such attention would allow us to better understand how modern composition works and help our students better understand how to put it to good use. Writing on/with/by/for computers does indeed garner academic attention, but which discipline rightly claims that kind of writing as its central concern? The answer here is not the Digital Humanities, despite the nominal fidelity. Even though the belletristic arts are under the purview of the humanist tradition, the Digital Humanities are, in practice, far removed from the concerns of composition.[1]

---

1   For the purposes of this discussion, the various names used to identify subtly different approaches to the discipline of writing studies — "rhetoric and composition," "first-year composition," "first-year writing," etc. focus

Instead, the study of how technology influences the creation of text is housed in a subfield of composition known simply as "computers & writing." This subfield has struggled since its inception in the 1980s to gain traction within composition studies, to say nothing of the academy at large. Many computers & writing scholars, with Cynthia Selfe and Gail Hawischer perhaps the most prominent and vociferous, repeatedly encourage composition researchers and teachers to consider the benefits, implications, and dangers inherent in using computers in the writing process and in the writing classroom. Yet they issue their warnings in publications of and for scholars who identify within computers & writing. Conversations about the influence of technology are less common in more mainstream publications designed for composition studies at large.

So why is Digital Humanities not home to scholarship about writing on/with/by/for computers? Today, Digital Humanities studies extant texts — the product, not the process — while the creation of writing is the subject of distinct disciplines with a fraught relationship to DH. Examining how composition and computers & writing took shape reveals the historical precedent that led to the separation of those fields from the Digital Humanities. This overview of composition's pedagogical history will show that rhetoric and composition as a field struggled to define and distinguish its identity before the Digital Humanities went through a similar process, defining its own membership. As Kim and Stommel discuss in their introduction to this volume, the Digital Humanities formed around outcasts, giving a professional home to scholars whose research and teaching stood apart from traditional forms. But as the Digital Humanities has taken shape, built a community, and solidified as a field in its own right, composition studies grew increasingly separate from traditional humanities departments, creating a rift that

---

attention on semantic debates internal to the discipline that, while valid and worth consideration, do not directly relate to the discussion at hand. I will use the term "composition" or "composition studies" as a generalized term intended to encompass all those listed above.

is now counterproductive and difficult to bridge. The separation between the Digital Humanities and composition is both a historical artifact and a temporary state of affairs contrary to the aims of the modern academy. Today's scholarship (and, for that matter, intellectual labor) is increasingly interdisciplinary. Collaboration facilitates insights at the intersections of various fields, and complex phenomena or systems need the perspective of sophisticated analytical frameworks. For instance, biology, chemistry, or psychology alone are each insufficient to explain the inner workings of the brain, but neurophysiology, a combination of the three fields, begins to answer some of today's questions about human behavior. When it comes to the complexity of writing in various situations, composition and the Digital Humanities need to buttress one another by combining forces, strengthening the diversity of DH and the reputation of rhet/comp. The natural point of connection already exists but is not commonly recognized as the needed bridge: the computers & writing field.

Rhetoric & Composition as an outcast field

Despite my claim above that separating DH from rhet/comp is contrary to the modern academy, the academy's nature is itself responsible for rhetoric and composition's less-than-desirable position. Specifically, the needs of the academy situate rhet/comp in the service of other disciplines, with writing classes (particularly at the undergraduate level) frequently working to meet the demands of courses in other fields, rather than as a field of study in its own right.

In what Randall Collins calls "the university revolution," distinct disciplines splintered and re-formed in academia throughout the 18th century.[2] In this educational restructuring, writing

---

2 See Randall Collins, "The Transformation of Philosophy," in *The Rise of the Social Sciences and the Formation of Modernity: Conceptual Change in Context, 1750–1850*, eds. Johan Heilbron, Lars Magnusson, and Björn Wittrock, 141–62 (Dordrecht: Kluwer Academic Publishers, 1998).

did not enjoy equal standing, being seen as a tool used by other more legitimate fields. Maureen Daly Goggin details the creation of writing as a discipline in her thorough historical volume *Authoring A Discipline: Scholarly Journals and the Post-World War II Emergence of Rhetoric and Composition.*[3] According to Goggin, the 19th century solidified the service position of rhet/comp because, while other disciplines worked to *build* students' knowledge, "writing served as a demonstration of knowledge" (emphasis added).[4] By serving as a tool to be used by other disciplines, rhet/comp held merely support status. And because rhet/comp wasn't viewed as its own field of study, a hierarchy developed within English departments. Goggin explains that "at the top, scholarship and research were privileged as the real work; at the bottom, practice and pedagogy resided, divorced materially and politically from the real work at the top."[5] Composition became common in both senses of the word: It was everywhere, and it was non-distinctive. The tool used by all disciplines on campus became the grunt work of English, rather than a project worthy of resources and theory.

So long as writing is viewed as a tool, rather than a subject of study, it cannot take hold as a scholarly pursuit, essentially getting the short end of the academic stick. John Dewey comments on the preference for the intellectual over the practical, which he traces back to the ancient Greek "preference for studies which obviously demarcated the aristocratic class from the lower classes."[6] Because rhet/comp addresses a skill useful in every discipline, it often loses its ability to be differentiated. Its ubiquitous usefulness becomes its own liability, tarnishing its luster by virtue of being commonplace. That I distinguished

---

3   Maureen Daly Goggin, *Authoring A Discipline: Scholarly Journals and the Post-World War II Emergence of Rhetoric and Composition* (Mahwah: Lawrence Erlbaum, 2000).

4   Ibid., 8.

5   Ibid., 28.

6   John Dewey, "Intellectual and Practical Studies," in *John Dewey: The Middle Works, 1899–1924,* vol. 9: 1916, ed. Jo Ann Boydston, introd. Sidney Hook (Carbondale: Southern Illinois University Press, 1980), 275.

reputation has beleaguered rhetoric and composition virtually since its inception. As Peter Vandenberg observes, "writing instructors languished in the lower strata of the college system as disseminators of 'practical' knowledge rather than creators of theory."[7] Rhet/comp scholars settled in a position of being outcasts, and the field held a position of being beneath the need for rigorous academic attention.

As disciplinary boundaries further solidified and fields grew more insular, composition suffered due to its nature as "dynamic, multidimensional, and contingent, making it open to fragmentation."[8] At the highest order, the division between arts and sciences subverted the position of composition. Literary studies takes a reasoned approach to the analysis and explication of texts, its scholars making and supporting arguments *about* art, rather than *as* art. Composition, then, lacking an established methodological foundation, appeared to itself be an art, bereft of rigor, taking the work of other fields as the content of writing. This art/science divide reinforces the notion of composition as tool.

Composition even lost its traditional home within English departments in the first-ever issue of *PMLA*. Theodore W. Hunt wrote to bolster the standing of those departments, lamenting that "no department of college work has so suffered as the English at the hands of novices."[9] However, he aired his concerns at the expense of composition, which earned barely a mention. The exclusion of composition from the MLA was complete in 1903, when the pedagogical section — which often dealt with rhetoric — was removed. By 1910, papers about composition pedagogy "disappeared entirely" from the conferences and literature of English studies.[10] The justification of English depart-

---

7   Peter Vandenberg, "The Politics of Knowledge Dissemination: Academic Journals in Composition," (PhD diss., Texas Christian University, 1993), 55. Cited by Goggin, *Authoring a Discipline*, 15.

8   Goggin, Authoring a Discipline, 12–13.

9   Th.W. Hunt, "The Place of English in the College Curriculum," *PMLA* 1 (1884–85): 118–32, at 119.

10 Goggin, *Authoring a Discipline*, 21.

ments put the scholarship of writing and literature officially at odds, with literature gaining prestige and composition being further relegated to a service role. The history and development of composition has been replete with marginalization.

Digital Humanities as turf war

Despite its origins in rhetoric, English as a field transitioned to revere literary studies as its torch-bearer. As computers entered education — particularly multimedia technologies offering significant increases in storage capacity — distributing whole libraries of texts on single optical discs became popular. Search, analytics, and other linguistic technologies improved, and scholars could analyze trends across an entire corpus. Speed and storage made these new forms of analysis possible. Newer location-based technologies like GPS tagging, hyper-accurate mapping systems, and high-resolution imaging systems broaden the scope and scale of data tagging and mining by adding additional layers of information and additional opportunities for visualization of results. Technological developments changed not what was studied, but how it was studied, and that change created a tension. Traditional literary studies preserve the importance of close reading, text explication, and the cultural commentaries/insights that follow. By contrast, computational analysis of texts uncovers trends in authors' or societies' language use. The source material is the same, but the tools and conclusions are different.

The Digital Humanities have grown out of technological developments offering new ways to examine, dissect, and combine the materials of humanities scholarship. Each new tool requires a nontrivial amount of technical knowledge and skill, so the separation between the traditional and the digital continues to grow. As a result, DH scholars maintain a technical competence that sets them apart from traditional humanities scholars, and the separation grows more noticeable as the tools develop. Matthew Kirschenbaum, himself a strong advocate of Digital Humanities, says that the field "has accumulated a robust professional apparatus that is probably more rooted in English than any other departmental home."[11] These forces tying DH to English departments are stronger — with more historical precedent — than those permitting DH to exist as a separate field in its own right. As much as tools contribute to a scholar's identity, the affordances of DH serve to split literary studies in two. But the origins of DH problematize such a simple separation.

The tension between a traditional connection with English studies on one hand and the modern separation on account of technological tool use on the other echoes a division in higher education that Laurence Veysey says began in the early 19th century.[12] According to Veysey, universities held one of three ideals for the goals of education: research, liberal culture, or utility. Literary studies in the belletristic tradition are well-suited to the liberal-culture ideal, helping to open students to the issues and perspectives that inform and enrich human life.[13] But the technological focus of DH and the analyses abstracted from direct human interpretation of a limited number of documents mean this work is better suited to research-driven endeavors.

Viewing the relations of DH and literary studies in terms of research and liberal-culture ideals may help explain an identity or public-relations problem faced by DH: On one hand, many scholars proclaim the open, accepting, "big tent" notion of DH working for anyone wishing to apply technical tools to humanistic interests. On the other hand, the cutting-edge projects that garner attention for DH require specialized knowledge, powerful tools, and massive datasets, each the specialized purview of large research institutions and outside the reach of more resource-limited players in the academy. David Golumbia notes a concerning dynamic among DH scholars regarding how the field is defined: "one unthreatening, expansive definition when

---

11 Matthew Kirschenbaum, "What is Digital Humanities and What's It Doing in English Departments?" in *Debates in the Digital Humanities*, ed. Matthew K. Gold, 3–11 (Minneapolis: University of Minnesota Press, 2012), 3.

12 See Laurence R. Veysey, *The Emergence of the American University* (Chicago: University of Chicago Press, 1965).

13 Ibid.

outsiders look in, another, exclusionary, imposed by a small but powerful and influential subset of DHers, forcefully advocated behind the scenes."[14] The identity of DH exists as a tug-of-war between English and other programs, between humanities and computer science, and between an open and an exclusive self-image. As a result, labeling a scholar or a project as being part of DH applies an automatic degree of separation, and as a result, marginalization.

Computers & Writing as struggling subfield

Just as English programs encountered disruption with the arrival of digital technologies, so too did composition suffer a split in identity and purpose. Computers & writing[15] came about as some scholars, most notably Cynthia Selfe and Gail Hawisher, began questioning how technology reforms — rather than merely influences — writing and writing instruction. As computer use has become more common for the creation, publication, and consumption of texts, the computers & writing subfield has gained momentum and recognition within composition studies. Indeed, Karl Stolley, in his 2013 Computers & Writing conference keynote, called for the Conference on College Composition and Communication (the major annual writing-studies conference) to become a subconference of Computers & Writing, reversing the current scenario where the "Computer Connection" holds marginal status within the greater CCCC event and the Computers & Writing conference remains relatively unknown.[16]

---

14 David Golumbia, "'Digital Humanities': Two Definitions," *uncomputing* (blog), 20 January 2013.

15 Much like the earlier footnote about "composition studies," the term "computers & writing" warrants discussion, as it is but one of several possible names for the subfield under discussion. "Computers and composition" is another. The former phrase is also the name of the major conference of the subfield, whereas the latter phrase is also the name of the major journal of the subfield. I use the word "writing" here to nominally distinguish between the subfield and the broader parent field.

16 Karl Stolley, "In Search of Troublesome Digital Writing: A Meditation on Difficulty" (Keynote Speech, Computers and Writing Conference, 2013).

Scholarship about computers & writing started with the sort of enthusiasm that is now uncomfortably familiar to readers who witnessed the burst of the "dot-com bubble" in the early 2000s: Many articles proclaimed the benefits and advances possible through the incorporation of technology. Authors served as advocates for incorporating the digital into the classroom, and the focus and methodology of traditional composition research faded to the background. Eventually, as Clay Spinuzzi points out, "the optimism wears off" in the late 1990s.[17] At this point, social concerns of equity and access move to the forefront, and a discussion begins about the benefits and assumptions implicit in computer-aided instruction in composition courses. That discussion continues to this day, and for good reason: Technology continues to advance, infiltrate social norms, and change expectations in the classroom. With each change, renewed critical analysis is required to ensure awareness of the complexities of digital composition instruction.

Throughout its history, computers & writing has existed as this somewhat-marginalized, somewhat-progressive entity that is attached to, yet separate from, composition studies as a whole. In its thirty-year history, computers & writing has seen shifting emphases in teaching and scholarship, yet it maintains a precarious and difficult-to-define relationship with composition studies overall. Such challenging definitions effectively form a connection among the various offshoot fields discussed so far. Recognizing those connections and employing them to develop support and mutual benefit could bring the recognition that each field has struggled to maintain over their brief histories.

---

17 Clay Spinuzzi, "Computers and writing: History, theory, philosophy," *Spinuzzi* (blog), 13 July 2007.

The divergent alignment

> The dangers of being an array of divergent practices, rather than a unified field comes when your work is scattered.[18]

Nystrand, et al. argue that composition really took hold in the 1970s, as a response to the larger social literacy crisis.[19] Many historical views of composition, including that of Goggin heavily cited above, generally consider the late 1880s as the starting point, failing to separate composition the course from composition the discipline. Composition classes have existed for well over a century in one form or another, but it has not been until recently that composition has existed as what Nystrand, et al. call "an interdisciplinary writing research community as well as a pedagogical forum."[20] That community and forum now exist within a larger context of scholarship that examines — and is inscribed by — technology. This technological moment necessitates a reconsideration of academic disciplines.

The distinct developmental histories of the Digital Humanities and computers & writing mean that the two fields are considered together or related only as an exception. Yet the fields are most certainly not at cross-purposes. Scholars who typically position their work in Computers & Writing occasionally — when professionally profitable — assert a position in the Digital Humanities, as well. This positioning leads to some identity politics among those in Computers & Writing. In her contribution to the 2011 "Are You a Digital Humanist?" Town Hall session at the Computers & Writing conference, Cheryl E. Ball succinctly stated that financial incentives lead her to publicly adopt the title: "When I'm talking to the NEH, I'm a digital human-

---

18 Virginia Kuhn, "Are You a Digital Humanist?" (Town Hall Session, Computers and Writing Conference, 2011).

19 Martin Nystrand, Stuart Greene, and Jeffrey Wiemelt, "Where Did Composition Studies Come From? An Intellectual History," *Written Communication* (July 1993): 267–333, at 267–68.

20 Ibid., 314.

ist. When I'm talking to y'all, I'm me."[21] For Ball, the DH title is one of opportunity, rather than identity. The Digital Humanities currently affords caché with hiring and funding committees. As Doug Eyman said in his contribution to that same 2011 Town Hall session, "It's not just this label you put on anybody who can make a website. But that's what's getting money now."[22] But the absence of a clear public understanding of what constitutes the Digital Humanities allows for its adoption by those who do not generally consider themselves members of the field — scholars who might be considered disciplinary imposters.

But who defines the discipline and determines who is a member and who, an imposter? Ball shared her observations of how academics on the job market draw the proverbial line in the sand: "Everyone has read *Remediation* and can quote from it in their job talk, and apparently that's what makes you a digital humanist."[23] Though the oversimplification served to make her point, defining a field by its source texts is common practice among those attempting to segregate the Digital Humanities from computers & writing — a point that Alex Reid took up in his blog post, "Digital Humanities Tactics":

> I don't want to make an argument about what English de-
> partments "should" look like. I don't want to make an argu-
> ment about what DH should look like. It's just a mispercep-
> tion of who is in the room.[24]

As a movement toward remedying the divide, Doug Walls suggests that we use "digital trade routes" to connect DH with rhet-

---

21 Cheryl E. Ball, "Are You a Digital Humanist?" (Town Hall Session, Computers and Writing Conference, 2011).
22 Doug Eyman, "Are You a Digital Humanist?" (Town Hall Session, Computers and Writing Conference, 2011).
23 Ball, "Are You a Digital Humanist?"
24 Alex Reid, "Digital Humanities Tactics," *digital digs: an archaeology of the
future* (blog), 17 June 2011.

oric so they can benefit from one another's work.[25] While I support his idealism and agree that open communication between the fields is necessary, his suggestion does not go far enough to ensure future growth in both fields. We need to loosen the grip of current definitions and embrace the role of these offshoot, marginalized subfields. The Digital Humanities and Computers & Writing need to abandon the pretense of separation or animosity and combine efforts to improve mutual standings within the academy.

Mutually assured disruption

The technology-focused subfield of computers & writing is familiar with ostracism, developing over a series of evictions spanning several decades. After being disowned by English departments when composition programs formed and fighting for recognition as computers gained prominence in the act of composition, Computers & Writing knows how to exist independently. But as the field moves toward greater inclusion into the rhetoric and composition academic space, we need to think about a larger, more universally beneficial goal: re-integrating writing and literature studies. This call for integration is nothing new, as Peter Elbow argued for it in a 1993 issue of *Rhetoric Review*:

> The dominance of reading at all levels of education reinforces the problematic banking metaphor of learning: the assumption that students are vessels to be filled. But when we give equal emphasis to writing, we are more likely to assume the contrasting metaphor: learning is the making of meaning.

---

25 Doug Walls, "In/Between Programs: Forging a Curriculum between Rhetoric and the Digital Humanities," in *Rhetoric and the Digital Humanities*, eds. Jim Ridolfo and William Hart-Davidson, 210–23 (Chicago: University of Chicago Press, 2015), 213–15.

[…] When we stop privileging reading over writing, we stop privileging passivity over activity.[26]

The current calls for inclusion and acceptance between c&w and DH are merely a new iteration of Elbow's argument infused with the digital. Indeed, this discussion may have been encouraged by the gradual success c&w has seen in moving into rhet/comp programs. But simply playing nice is insufficient. c&w and DH have too much to gain from one another; they can no longer afford to work as separate entities. Reading and writing — literature and composition — must be seen as complementary components of good thinking.

Connecting reading and writing as acts seems simple. Connecting literature and composition as teachable content *should* be simple, but their histories diverge too much. Writing studies came into being when traditional English departments jettisoned a concern for pedagogy so they could specialize in literary analysis. Today's departments cannot afford to ignore pedagogy in light of our changing educational landscape. While MOOCs, learning academies, and other edtech solutions work to eliminate the personal support teachers provide their students, the corporations building these resources — rather than the local institutions/communities — get to dictate pedagogy. Without scholarly attention to pedagogy, academics will lose their ability to make informed decisions about how their field is represented in the classroom.

At the same time, the continual focus on the pedagogies of composition have confined writing studies to placement as a second-tier academic interest. By more directly, regularly, and publicly aligning with the Digital Humanities, computers & writing would benefit from name recognition (and funding) that comes with the currently popular field. While it would move from a subfield of rhet/comp to a subfield of DH, by preserving its ties with the rhet/comp community, c&w would be

---

26 Peter Elbow, "The War between Reading and Writing — And How to End It," *Rhetoric Review* 12, no. 1 (1993): 5–24, at 16.

supported by the rhetorical tradition of comp and the academic tradition of the humanities.

While it may sound at first like a corporate merger, this blending would be neither forced, disingenuous, nor imbalanced. The connections between c&w and DH are critical — if not essential — for the continued development of both fields. We have the opportunity to strengthen the reputation of c&w and the teaching of DH, allowing those of us in each subfield to benefit from the strengths of the other. For those in what can now be called traditional DH studies, working with c&w scholars can highlight pedagogical concerns that too often go unaddressed in DH conferences or general discussion. For those in c&w, working with DH scholars can highlight historical connections with traditional scholarship and bring greater awareness from public institutions and funding agencies.

We need to disrupt the Digital Humanities by incorporating computers & writing into the "big tent" categorization. Cheryl E. Ball struggled to incorporate the two in her professional identity, debating whether she fits in as a digital humanist. Her solution is expressed through an integration of disciplines in what she calls an "editorial pedagogy."[27] Those who study c&w should not question whether they are digital humanists. We have a deep interest in humanistic concerns of education, advocacy, and social conditions. Yet we clearly work in the digital realm of technology and media interactions. We cannot and must not avoid that label. The division between c&w and DH is a false separation created by our two fields' respective histories, not their current conditions. c&w scholars have the opportunity — the necessity — to disrupt the DH establishment by taking on the identity as their own, changing and expanding both fields as they do so. DH itself must disrupt c&w by connecting it with the coding, analytic resources, and large project management experience common in DH circles, adding

---

27 Cheryl E. Ball, "Editorial Pedagogy, pt. 1: A Professional Philosophy," *Hybrid Pedagogy*, 5 November 2012.

breadth to the reach of c&w research and collaboration. It is time for DH and c&w to work together to raise hell.

Bibliography

Ball, Cheryl E. "Are You a Digital Humanist?" Town Hall Session, Computers and Writing Conference, 2011.
———. "Editorial Pedagogy, pt. 1: A Professional Philosophy." *Hybrid Pedagogy*. 5 November 2012. http://www. digitalpedagogylab.com/hybridped/editorial-pedagogy-pt-1-a-professional-philosophy/.

Collins, Randall. "The Transformation of Philosophy." In *The Rise of the Social Sciences and the Formation of Modernity: Conceptual Change in Context, 1750–1850,* edited by Johan Heilbron, Lars Magnusson, and Björn Wittrock, 141–62. Dordrecht: Kluwer Academic Publishers, 1998.

Dewey, John. "Intellectual and Practical Studies." In *John Dewey: The Middle Works, 1899–1924,* vol. 9: 1916, edited by Jo Ann Boydston, introduction by Sidney Hook. Carbondale: Southern Illinois University Press, 1980.

Elbow, Peter. "The War between Reading and Writing — And How to End It." *Rhetoric Review* 12, no. 1 (1993): 5–24. DOI: 10.1080/07350199309389024.

Eyman, Doug. "Are You a Digital Humanist?" Town Hall Session, Computers and Writing Conference, 2011.

Goggin, Maureen Daly. *Authoring A Discipline: Scholarly Journals and the Post-World War II Emergence of Rhetoric and Composition.* Mahwah: Lawrence Erlbaum, 2000.

Golumbia, David. "'Digital Humanities': Two Definitions." *uncomputing* (blog). 20 January 2013. http://www. uncomputing.org/?p=203&cpage=1.

Hunt, Th. W. "The Place of English in the College Curriculum." PMLA 1 (1884–85): 118–32.

Kirschenbaum, Matthew. "What is Digital Humanities and What's It Doing in English Departments?" In *Debates in the Digital Humanities,* edited by Matthew K. Gold, 3–11. Minneapolis: University of Minnesota Press, 2012. http://dhdebates.gc.cuny.edu/debates/part/2.

Kuhn, Virginia. "Are You a Digital Humanist?" Town Hall Session, Computers and Writing Conference, 2011.

Nystrand, Martin, Stuart Greene, and Jeffrey Wiemelt, "Where Did Composition Studies Come From? An Intellectual History." *Written Communication* (July 1993): 267–333.

Reid, Alex. "Digital Humanities Tactics." *digital digs: an archaeology of the future* (blog). 17 June 2011. http://alex-reid.net/2011/06/digital-humanities-tactics.html.

Spinuzzi, Clay. "Computers and writing: History, theory, philosophy." *Spinuzzi* (blog). 13 July 2007. http://spinuzzi.blogspot.com/2007/07/computers-and-writing-history-theory.html.

Stolley, Karl. "In Search of Troublesome Digital Writing: A Meditation on Difficulty." Keynote Speech, Computers and Writing Conference, 2013.

Vandenberg, Peter. "The Politics of Knowledge Dissemination: Academic Journals in Composition." PhD diss., Texas Christian University, 1993.

Veysey, Laurence R. *The Emergence of the American University.* Chicago: University of Chicago Press, 1965.

Walls, Doug. "In/Between Programs: Forging a Curriculum between Rhetoric and the Digital Humanities." In *Rhetoric and the Digital Humanities,* edited by Jim Ridolfo and William Hart-Davidson, 210–23. Chicago: University of Chicago Press, 2015.

24

W(h)ither DH? New Tensions, Directions, and Evolutions in the Digital Humanities

Lee Skallerup Bessette

I will readily admit to feeling imposter syndrome while trying to write this. Who am I to stand here and tell you about Digital Humanities? I am, to a certain extent, an outsider looking into DH; I stand at a periphery; I hover around the margins. I was not trained in DH at any of the well known Centers (which include but are not limited to University of Virginia, MITH at the University of Maryland, MATRIX at Michigan State, the University of Nebraska (Lincoln), or The Center for History and New Media at George Mason University, or UCLA), nor am I a part of a growing group of new DH centers, projects, or initiatives popping up all over the country.[1] I've never "built" anything (which

---

1 Links to their websites are, respectively: *The Institute of Advanced Technology in the Humanities,* University of Virginia, http://www.iath.virginia.edu/; *Maryland Institute for Technology in the Humanities,* University of Maryland, http://mith.umd.edu/; *Center for Digital Humanities and Social Sciences,* Michigan State University, http://www2.matrix.msu.edu/; *Center for Digital Research in the Humanities,* University of Nebraska-Lincoln, http://cdrh.unl.edu/; and the *Center for History and New Media,* George Mason University, http://chnm.gmu.edu/.

is a tension within digital humanities I will come back to): there are no databases or digital archives or tools or maps or digital editions or other pieces of born-digital scholarship that bear my name anywhere, in collaboration or otherwise.

Some context: I came to DH "late" and wholly by accident. I am a self-taught digital humanist. I was stuck in a contingent faculty position outside of my area of expertise at an institution with neither the capital nor the motivation to get involved in this growing trend both in research and teaching. This is not to say that my learning happened alone; in fact, it happened because of a vast community of fellow DH enthusiasts who tweeted, blogged, and made available through many different means and mediums their work within the field. And, they at first graciously tolerated my presence online, helped me to learn, mentored me, and supported me.

Because there was never a "Center" for me, both literally *and* figuratively, I have a different perspective to DH. I am particularly attuned to finding a lot of "stuff" on and around and about DH because I had to do it all myself. And because I was coming to the discipline unaware, I didn't know to distinguish between the "big names" at the center and the smaller voices in the margins. Everyone that I encountered in DH was, to use Tom Scheinfeldt's language, so nice, but I encountered them in a space that was itself, at that moment, still a marginal one, and relatively (but not unproblematically) democratizing: Twitter. I was exposed at once to the broad world of DH through a very select group of DH practitioners who were active on Twitter.

My perspective, or perhaps to use a more technical term, "filter bubble," is one that is firmly informed by my embodied self; I provide this personal history for you because I think it is important that you know that this is just one perspective. I embody a position that stems from a very specific academic training and professional experience; I have a PhD in Comparative Literature and my research interests have been on major marginal authors, while I have worked largely in contingent positions at regional state institutions that primarily focus on teaching. My outsider status also allows me to listen in different ways and have people reveal things that they might not otherwise. All of these perspectives inform my view of where DH is going.

Microhistories of DH

I plan to examine how microhistories inform our understanding of DH the various "centers"; and what draws DH participants to these gravitational areas. In her essay, "Literary History as Microhistory," Heather Murray states that microhistory "is rooted in the attempt to incorporate peripheral or marginal events, figures, and communities into the historical picture."[2] It is "history with a human face: and that face is the face of the daily, the ordinary, the subaltern."[3] I believe that DH generally could use a more microhistorical approach given the current obsession with Big Data. In particular, as it relates to the subaltern, the more data we have does not necessarily make everything more visible but instead risks subsuming individual stories and subtle nuances, important for academic insight and critical reflection. DH has the possibility of telling both the "Big Stories" and also the microhistories in innovative and interesting ways.

Microhistory, primarily, looks to "focus on the anomalous rather than the typical."[4] Now, the question should be asked, is the history and evolution of DH the anomalous or is it the typical? And within the history of DH, what are the microhistories that are forgotten or are silenced? These are questions beyond the scope of this article, but it is important to keep questioning these master narratives, to remember that there are few typical stories these days, in either DH or in higher education more generally. This fragmentary nature, this lack of master narrative, in particular one with any kind of happy ending, can be disillusioning and, let's face it, depressing. But I think digital humanists have given us a potential alternative path to follow in order

---

2   Heather Murray, "Literary History as Microhistory," in *Home-Work: Postcolonialism, Pedagogy, and Canadian Literature,* ed. Cynthia Sugars, 405–22 (Ottawa, Ont.: University of Ottawa Press, 2004), 406.

3   Ibid., 411.

4   Ibid., 415.

to make sense of these fragments, of these realities, by using the technology that many blame for this situation in the first place.

W(h)ither DH

I chose my title playfully, emphasizing my literary bent and love of language. If something can mean two or more things, or be easily misunderstood whether read or spoken... But this pun might strike you as being strange to be asking around DH. Wither the academic job market in the humanities — sure. Wither state support for public higher education — definitely. But wither DH? We know where DH is: *It is everywhere!* And it's *new now and funded!* It's at the MLA, the AHA, even the *New York Times*! There's no wondering where DH is, or a pulling back from DH. It appears to be experiencing the opposite of withering; it would appear to be flourishing!

But, do we really know where DH is happening, or how it's happening, or where to find DH? Or, rather, do we know where to put interdisciplinary programs and centers devoted to DH? The question has been raised over and over and over. The most well-known missive is from Matthew G. Kirschenbaum in an essay "What Is Digital Humanities and What's It Doing in English Departments?" from 2010: "Whatever else it might be then, the digital humanities today is about a scholarship (and a pedagogy) that is publicly visible in ways to which we are generally unaccustomed, a scholarship and pedagogy that are bound up with infrastructure in ways that are deeper and more explicit than we are generally accustomed to, a scholarship and pedagogy that are collaborative and depend on networks of people that live an active 24/7 life online. Isn't that something you want in your English department?"[5]

I can think of quite a few people who wouldn't want DH in their departments given this description. 24/7 online? Public?

---

5   Matthew G. Kirschenbaum, "What Is Digital Humanities and What's It Doing in English Departments?" in *Debates in the Digital Humanities,* ed. Matthew K. Gold, 3–11 (Minneapolis: University of Minnesota Press, 2012), 6.

Think about infrastructure and technology? And then DH gets so big and so popular that you begin to get inflammatory (or click-bait) articles like "In the Near Future, Only Very Wealthy Colleges Will Have English Departments: Adapt (not publish) or perish" that appeared in *The New Republic*.[6] Here is the opening paragraph: "Within a few decades, contemporary literature departments (e.g., English) will be largely extinct — they'll be as large and vibrant as Classics departments are today, which is to say, not very active at all. Only wealthy institutions will be able to afford the luxury of faculty devoted to studying written and printed text. Communications, rhetoric/composition, and media studies will take English's place. The change isn't necessarily an evil to be decried but simply reflects how most people now generate and read narratives and text — they do it on digitally based multimedia platforms."[7]

This is what the rise in DH has given us: click-bait. Unfair click-bait that understands DH in a narrow way because they couldn't "find" DH anywhere but English departments, but that nonetheless drives traffic and feeds into technological paranoia and perceived academic decline. This article is completely unfair to Classics, a discipline that is still a) a vibrant field and b) avid practitioners of a form of DH, where we can now 3D print artifacts, scan ancient documents, and read them using technology. They are using technology to invigorate their research and their teaching. But it should not be surprising that we have reached this stage.

So one side doesn't want DH because, ugh, the digital, but the other side of that coin is that *everyone* wants DH: why should English have all the fun? Where should DH "live" within larger colleges and universities? Often there will be an academic discipline where DH is strong and the center or services spring up around that group of faculty and graduate students. One exam-

---

6   James Pulizzi, "In the Near Future, Only Very Wealthy Colleges Will Have English Departments: Adapt (Not Publish) or Perish," *New Republic*, 8 June 2014.

7   Ibid.

ple would be the Center for History and New Media at George Mason.[8] This is a major center located, clearly, within the field of history. But where does this leave professors and students from other disciplines? Now, this isn't to say that CHNM hasn't benefitted the entire field of DH, and they have developed a number of research tools that benefit any researcher, digital or not, such as the research and citation tool Zotero. And they have set best-practices, supported numerous researchers and junior scholars, as well as been the incubator for ideas and approaches to DH and technology. But, massive academic turf wars have been waged over less. Then money, support, and new hires have been dangled in front of faculty members, chairs, and deans, well, of course people are going to fight over it and feel resentment over these decisions.

One solution — or maybe it isn't a solution so much as a natural outgrowth of the growth in DH — has been to house DH centers in libraries. This makes sense: libraries are in a position to help serve diverse needs of multiple disciplines and fields; librarians have experience in issues like digital storage and archiving and as well as with metadata; libraries are where the archives live; and finding things is what librarians do. As Chris Bourg summarizes these strengths and traditions, "That's our job it's the kind of thinking and work that is a distinct strength of librarians." ("The Once and Future Librarian").[9]

The partnership between libraries and DH isn't wholly utopian; librarians and scholars such as Chris Bourg and nina de jesus have written extensively on the so-called "neutrality" of the library. de jesus, in her piece "Locating the Library in Institutional Oppression," positions the library as a tool of liberalism, and thus a part of systems of white supremacy, slavery, genocide, and Orientalism: "When we look into the collections, the actual 'information' contained in libraries and how it is organ-

---

8   *Center for History and New Media,* George Mason University, http://chnm. gmu.edu/.

9   Chris Bourg, "The Once and Future Librarian," *Feral Librarian* (blog), 18 March 2015.

ized, we can see that it (surely by accident) somehow manages to construct a reality wherein whiteness is default, normal, civilized and everything else is Other."[10] Bourg extends this critical look at the library towards the digital tools libraries build — "often gendered and/or racist, frequently ableist, and almost always developed with built-in assumptions about binary gender categories."[11] These are similar critiques to what DH is facing, and de jesus's advice could be heeded by both librarians and digital humanists: "Realizing the emancipatory potential of the library as institution would require breaking and disrupting the system of intellectual property and other aspects of capitalism, especially the publishing industry. It would require disrupting the empire's mechanisms for creating 'knowledge' by being more than a repository for imperial knowledge products. It would require supporting Indigenous resistance to the settle state and working towards dismantling anti-Blackness."[12]

There are examples. The best example I can give of a DH center linked to the library is at the University of Virginia, overseen for a long period by Bethany Nowviskie. She writes about the utopian possibilities of DH centers being located in libraries:

> However, where the two models exist in tandem — that is, where digital research is robustly supported throughout the library as the norm for humanities research itself and where the institution is resourced adequately to support a dedicated, library-based DH center — an enviable opportunity exists…When a library can both support basic digital scholarship needs through distributed services and create a critical mass of staffing and intellectual energy in something like a center (however conceived), it has set the conditions for the

---

10 nina de jesus, "Locating the Library in Institutional Oppression," *In the Library with the Lead Pipe* (blog), 24 September 2014.

11 Chris Bourg, "Never Neutral: Libraries, Technology, and Inclusion," *Feral Librarian* (blog), 28 January 2015.

12 de jesus, "Locating the Library in Institutional Oppression."

advancement of knowledge itself, through the fulfillment of research desires yet unknown, un-expressed.[13]

But Nowviskie puts an important caveat on her vision for library-based DH centers:

> Monolithic approaches to the digital humanities function well precisely nowhere — not even, in fact, at the places where they are first instantiated and from which "the model" emerges for future labs and centers. [...] A monolithic approach, I say, doesn't even work at the monolith, because changing local conditions and the very advancement of scholarship and scholarly methods mean that every center must evolve — evolve, or die. In DH, as elsewhere, the center(s) cannot hold.[14]

Each DH center, wherever it may be located must be willing to shift, to evolve, but also be equipped to nurture and support the research and work scholars and graduate students are seeking to do, to be willing and able to help shepherd that work into the world.

A good example is DH at the University of Kentucky. The institution has a long and robust history of DH work. In 1999, the Collaboratory for Research in Computing for Humanities was formed, and it brought together the work of historians, computational linguists, and anthropologists. It is an impressive list of digital research and projects, and would seem to offer some resources and support to interested scholars. But the webpage hasn't been updated in two years, and it doesn't include other kinds of DH work going on at the university including Critical GIS and Digital Writing, two areas of great strength within the university. There is a lot of great work being done in DH at the University of Kentucky, but there doesn't seem to be a vibrant

---

13   Bethany Nowviskie, "Asking for It," blog post, 8 February 2014.
14   Ibid.

and strong center or community around DH itself, just around disciplines or projects.

And maybe the name itself of the Collaboratory is a give away: it is still focused on humanities computing, an earlier iteration of what is now known as Digital Humanities. Names matter. I ask w(h)ither DH because it was once humanities computing, not that long ago. There are still those who hold on to humanities computing because it more accurately describes the work they do, but also because it is more exclusionary. "Digital" is a more inclusive term, encompassing the wide variety of possibilities afforded to us with the internet, mobile technology, and (yes) computers. Humanities computing, on the other hand, focuses on exactly that: computational applications to humanities disciplines. Of course we still do that in DH, and with Big Data we are probably doing more of that than ever. But it is a more narrow, more exclusionary definition of DH, and one that does not necessarily open up the possibilities of the growing and changing discipline(s) of DH.

Ted Underwood, in a blog post entitled, Digital Humanities Might Never be Evenly Distributed, explores this phenomenon I'm describing here at his own large campus:

> I rapidly discover the size of this campus, and the huge range of digitally-human projects already scattered across it, already moving (quite successfully) in diametrically opposed directions — and it occurs to me, first, that it would take superhuman effort to herd them into the same room, and second, that maybe UIUC doesn't have a digital humanities center because it doesn't *need* one. I'm finding all the resources I need over at GSLIS and NCSA; other kinds of projects are also humming along; maybe we've never developed a single center precisely because our various distributed centers are so strong.[15]

---

15 Ted Underwood, "Digital Humanities Might Never Be Evenly Distributed," *The Stone and the Shell* (blog), 7 September 2015.

He goes on to describe the various shortcomings of this model, including difficulty promoting the activities to the larger community, and working to connect undergraduates in particular to this kind of work.

This brings up another point about names and naming; think back to the quote from The New Republic article: "Communications, rhetoric/composition, and media studies will take English's place."[16] These are disciplines that already exist alongside English, and ones that have an uneasy relationship with digital humanities. There is a lot of debate around whether or not these disciplines and fields "belong" in digital humanities, but also whether or not these fields and disciplines even *want* to be a part of DH. There is a growing push within fields that have been working on "the digital" or technology more broadly for a long time (like Computers and Writing for example) to differentiate themselves and resist the temptation or the pressure to be subsumed under the Big DH Tent. What these three fields share is often the perception and attitude in the past (and often still in the present) that they were "less than" the traditional humanities field, and now that we've all embraced the digital turn, they want the respect and recognition they feel they deserve, without having to become a part of the latest and greatest digital humanities trend, a "trend" they have been a part of and to a certain extent, leading.

We are also seeing a growth in new kinds of programs that might look or sound a little like DH, but that are calling themselves something entirely new and different. One example is the new department in Computational Media at UC Santa Cruz, the first of its kind. There is already a degree in Computational Media at Georgia Tech and at the University of Calgary, but this is the first stand-alone department rather than interdisciplinary program. Ian Bogost, a professor at Georgia Tech observes: "There is sometimes a sense that we've decoupled computing from its cultural and artistic and humanistic context, and

---

16 James Pulizzi, "In the Near Future, Only Very Wealthy Colleges Will Have English Departments."

some of the trouble we might point to in the world we are living in — run by Wall Street and Silicon Valley — is perhaps a result of thinking of everything as just an engineering problem."[17]
Sounds a lot like DH, but maybe not.

We have moved from Humanities Computing to Digital
Humanities, which broadened the field, to Computational Media, which would seem to narrow the field yet again. Or at least
breaks off a piece of what could be considered digital humanities (but maybe isn't) and stakes a new place for itself within
the various fields of study and disciplines and departments.
W(h)ither DH indeed. We find ourselves within DH continually
trying to simultaneously define and resist the Center, but the
Center has always been the dominant model for DH. Is DH the
prism, then, where we refract into a rainbow of colors? Should
we think, then, of DH work and research as a spectrum? But,
depending on what side of the prism you're on, you don't see the
refracted light, but instead see the monolithic white light. Is this
a good analogy for DH right now?

Who does DH?

We've spent a good deal of time talking about the center(s) of
DH, but I want to move on now to examine the question: *Who*
does DH? Adeline Koh posited a pretty provocative position on
this subject when she wrote:

> You are already a digital humanist, whether or not you know
> it. [...] But while digital humanities may seem like an in-
> timidating, exponentially growing field with varying ideas
> of "insiders" and "outsiders," you and your students are all
> already digital humanists, because you all use technology in
> your daily lives. At it's best, the digital humanities is about
> engaging more critically with the intersections between tech-
> nology and how we act, think and learn. Without knowing

---

17 Ian Bogost, "A New Department Marks the Rise of a Discipline: 'Computational Media,'" *Wired Campus* (blog), 13 October 2014.

it, you're probably already using many of the techniques of digital humanists in your life and in your classroom.[18]

On one hand, this statement could be misread to argue for subsuming long-active disciplines, like digital media or rhetoric, into the larger whole of DH. There is indeed something downright *colonial* about the tendency for DH to claim other fields and disciplines. But, that is not what Koh is doing here. Rather, she is resisting the rigid definition of DH that keeps appearing around the question of building, a question I myself I have struggled with. David Golumbia most recently addressed this tension between the "big tent" tendencies of DH versus the narrowly defined and, I might add, *disciplining* definition focused exclusively on building. This tension is almost wholly original within higher education, particularly in the humanities:

> The difference in DH, and the reason definitions of it matter so much, is that from its inception, some very powerful people and institutions have insisted on one definition, even when many others do not accept or endorse that definition, and these persons and institutions have been able to enforce that definition in one critically important sphere that has no parallel in queer theory, deconstruction, or any other recent movement in literary studies: newly available, large-scale, field-defining grant funding. Further, the availability of unprecedented amounts of grant funding to English professors has had a follow-on deformative effect in perhaps an even more critical venue: hiring. These, in turn, have had consequences (though, I think, less obviously dramatic ones) for promotion and tenure standards, although I'll leave those aside for the time being.[19]

---

18 Adeline Koh, "Introducing Digital Humanities Work to Undergraduates: An Overview," *Hybrid Pedagogy,* 14 August 2014.

19 David Golumbia, "Definitions That Matter (Of 'Digital Humanities')," *uncomputing* (blog), 21 March 2013.

Building in DH is a privilege. We should all be thankful for the builders, those who digitized, built the databases, created the interfaces and tools, wrote the algorithms, and allowed for DH to enter the mainstream because the tools and the research became increasingly visible and accessible. However, we should never forget that building is a privilege. It requires infrastructure, support, and a team of skilled programmers, archivists, and others. It requires a large institution and institutional support. It demands capital. And it requires, in a lot, but not all cases, tenure. These are elements that are in short supply in today's higher education landscape.

It should also be pointed out that there is building being done *outside* of the academy, which is also excluded from being considered a legitimate form of DH scholarship. But there is also a tendency to assume that all work and knowledge production seeks the support and approbation of the academy. Returning to nina de jesus's work, she connects a long history of appropriation by institutions as an extension of slavery.[20] One recent example is the multiple attempts by institutions to appropriate the community-based database and map of missing and murdered Indigenous women called Save Wiyabi Mapping Project. Founder Lauren Chief Elk recently tweeted: "You're the ones who need to prey on young women's words & labor for articles and funding."[21] The bitter irony of using this tweet in this essay are not lost on me. But grassroots, digital efforts by different communities are excluded, appropriated, and their contributions often erased.

It's also important to make the point that a DH narrowly defined as "only those who build" within the narrow confines of a center privileges only a certain kind of scholar that comes from a certain kind of graduate program. It also, almost cannibalistically, means that any tool that is developed to automate the work

---

20 de jesus, "Locating the Library in Institutional Oppression."

21 Lauren Chief Elk (@chiefelk), Twitter post, 25 March 2015, 12:23 p.m. See also Dorothy Kim, "Social Media and Academic Surveillance: The Ethics of Digital Bodies," *Model View Culture*, 7 October 2014.

of DH automatically makes that no longer a part of DH. An example Golumbia gives is around HTML or even topic modeling. As he points out,

> I don't deny that there will probably remain new topic modeling tools to build. What I am hoping to point out is that the very usefulness of topic modeling suggests it will become part of the scholar's toolkit, and that if we then arbitrarily deem that success to mean it is no longer part of our research enterprise, we are cutting off our nose to spite our face. Wide adoption and use is success, and interesting results produced with digital tools deserve to be called digital humanities.[22]

The more we demand of those who want to "do" digital humanities, the more we privilege and the more we exclude. Ernest Priego fears that we are seeing the rise in the demand for what he calls "the super humanist" — the humanist who is the top of his or her traditional humanistic field *and* can code, program, do network analysis, etc, etc, etc.[23] Priego has the same concerns I do around the digital divide, the scarcity of resources, and how DH can be used to reinforce traditional institutional hierarchies, between the "haves" and "have-nots." These super-humanists, and the bigger and better tools they build, are but a pipe dream for many graduate students, not to mention contingent faculty. But it also neglects scholars in the developing world, whose institutions are even more starved, while their infrastructure can't support the bandwidth necessary to use the tools created elsewhere, or access even their own archives, held and digitized elsewhere. In response, Global Outlook DH (or GO::DH) created a working group around Minimal Computing, in part in response to the difficulty participants were faced with when THATCamp Caribe was held in Cuba.

---

22 David Golumbia, "Building and (Not) Using Tools in Digital Humanities," *uncomputing* (blog), 5 February 2013.

23 Ernest Priego, "Various Shades of Digital Literacy," HASTAC (blog), 22 October 2012.

The GO::DH group itself "acts to foster collaboration and co-operation across regions and economies; it coordinates research on and in support of the use of technology in these areas across the globe; and it advocates for a global perspective on work in this sector."[24] GO:DH advocates for a global perspective. This has been another point of conflict within DH, particularly in defining who does DH, and what has been, and continues to be funded in DH. Looking at the biggest DH projects, they typically revolve around the already established canon of literature and history. Big Data leaves little room for microhistory; the majority can drown out and silence those smaller voices and stories. The language of DH is, in most cases, English. The priorities of what has been digitized are that which was already catalogued, already visible, already known. And, it is largely old because it is out of copyright. These are not insignificant concerns, and they are concerns that are beginning to be voiced within DH.

When Adeline Koh writes about how we all already do DH, she is also including herself in her recovery work with "Digitizing 'Chinese Englishmen'" and advocacy work with Postcolonial Digital Humanities. She is looking to "decolonize the archive" or counter the master narrative the main, dominant archive and "big data" of digitization often provides and presents.[25] This work used to be more prominent in digital humanities. Amy E. Earhart, in her article "Can Information be Unfettered? Race and the New Digital Humanities Canon," which appeared in the book *Debates in Digital Humanities,* points out that many early "recovery" projects around race have been lost, due in large part because they were passion projects done by individual or small groups of graduate students, and as Earhart notes: "Digital humanists are fond of talking about sustainability as a problem for current and future works, but it is clear that

---

24 "About," *Global Outlook::Digital Humanities,* http://www.globaloutlookdh.org/.

25 Adeline Koh, "Decolonizing the Archive," *Digitizing "Chinese Englishmen": Representations of Race and Empire in the Nineteenth Century,* http://chineseenglishmen.adelinekoh.org/decolonizing-the-archive/.

we already have sustained a good deal of loss within the broadly defined digital canon."[26]

This also raises the question of *language* particularly if DH programs are seen as being traditionally housed in English departments. This is a topic near and dear to my heart, having grown up in Quebec, where questions of language and translation were always driving conversations. How do we "translate" in a literal and figurative way into different languages, and perhaps more importantly, cultural traditions? Looking at the major DH projects from around the world, it is hard to find ones that look to preserve works done in minority languages. Of course there are exceptions within computational linguistics and anthropology, but they are not the mainstream DH projects that get much of the attention. GO::DH has done great work trying to translate DH into a variety of languages, while databases like Mukurtu bring an Indigenous approach to knowledge, remembering, and archiving.[27] This is an instructive example of a people and culture making archives for themselves, rather than letting the traditions of and traditional archive continue to dictate and define them. The Center(s) are being challenged.

The collective Transform DH is interested in all of these issues and more: gender, sexuality, race, and class issues are all important considerations that those who are a part of the Transform DH collective look to examine.[28] It also seeks to make visible the labor that women, LGBTQA, and people of color have done in DH and the digital more broadly, and that has often been erased by history. A recent article they promoted was from *The New Inquiry* called "The Lady Vanishes" on the invisible workforce that digitizes the books available to us to do introductory DH work using the Google Ngram.[29] Or how there is an entire underclass of Blacks and Latinos working for poverty wages in Silicon Val-

---

26 Amy E. Earhart, "Can Information Be Unfettered? Race and the New Digital Humanities Canon," in *Debates in the Digital Humanities,* 309–18, at 313.

27 See *Digitizing 'Chinese Englishmen,'* http://chineseenglishmen.adelinekoh. org/decolonizing-the-archive/ and *Mukurtu,* http://mukurtu.org/.

28 *#TransformDH,* http://transformdh.org/.

29 Shawn Wen, "The Lady Vanishes," *The New Inquiry,* 11 November 2014.

ley, who design the computers and smart phones we use.[30] Or the content moderators who live invisibly overseas.[31] We love our tools, but we often don't want to confront the labor issues inherent in what went into making them. Or the "crowdsourcing" some DH projects have come to rely on, as articulated by Martin Eve (but he is by far not the only one): "whenever we want a job done and somebody volunteers to do it, that person must, in the vast majority of cases, surely be supporting themselves through some other form of paid employment. What, though, about people who need jobs but find their livelihoods undercut by others volunteering to do work for free because it looks like a game? Perhaps you don't care about this and think that the labour "market" should fix this. I think I do care and don't believe in a market for labour that would make this work, evidenced by rises in under- and un-employment and top-loading of wealth."[32]

Adjuncts are excluded from most DH work because of their precarious positions, while graduate students often do much of the heavy lifting on the building of projects then postdoctoral positions that place unreasonable demands on the newest digital super-humanist. Or, the super-humanist can't find a position at all because they didn't quite get to be super enough: know too much code and not enough traditional scholarship, you are qualified only for what are known as "coding monkey" jobs, where you code someone else's project, at low wages at that. Don't know enough code and you might be left in the thankless DH postdocs that expect you to do all the things. As put by Miriam Posner, "I've been frank, as you may know, about what I think of taking someone fresh out of grad school, giving her a temporary gig, and expecting her to be the sole torchbearer for some amorphous DH initiative. In brief, it's a bad idea, for a lot

---

30 Sam Biddle, "The Non-White Workers of Silicon Valley Get Screwed Every Day," *Valleywag,* 26 August 2014.

31 Adrian Chen, "The Laborers Who Keep Dick Pics and Beheadings Out of Your Facebook Feed," *Wired,* 23 October 2014.

32 Martin Eve, "Universities, DH, 'the crowd', and Labor That Looks Like a Game," blog post, 12 November 2014.

of different reasons. It's not fair to the person you're hiring, who will spend her entire tenure trying desperately to impress you at this impossible task so she can keep her job."[33]

Increasingly, though, DH positions, beyond the postdoc, are non-tenure track positions. They are, for all of their fancy titles, contingent. I think here of people like Brian Croxall, who ran and won a spot within the MLA executive council on contingency and alt-academic issues, but also Bethany Nowviskie, a now-tenured professor at University of Virginia, who addresses contingency within DH and beyond in a 2014 blog post: "But we have to address the downsides of a culture of abundance: of conspicuous consumption; the increasing carbon footprint of DH; the increasing adjunctification and contingency not only of our teaching faculty but of knowledge workers of all stripes, including software developers and librarians."[34] We are beginning to notice, to speak out and speak up, and this, too, for me, is DH work.

Another question around DH work is what is considered "scholarship" — particularly around more public-facing DH work and around teaching. Mark Sample poses the question in this way: "When does service become scholarship? When does anything — service, teaching, editing, mentoring, coding — become scholarship? My answer is simply this: a creative or intellectual act becomes scholarship when it is public and circulates in a community of peers that evaluates and builds upon it."[35] This, to me, is where the most exciting work is being done in DH. But I think it reflects the areas where those who felt (or who really are) cut off from the traditional DH centers have looked to grow their own work and make space for themselves. The most interesting tools that are being currently developed focus on sharing, community-building, a more public-facing DH, and

33 Miriam Posner, "Here and There: Creating DH Community," blog post, 18 September 2014.

34 Bethany Nowviskie, "Speculative Computing and the Centers to Come," blog post, 15 November 2014.

35 Mark Sample, "When Does Service Become Scholarship?" *samplereality* (blog), 8 February 2013.

enhancing the student educational experience. And because of technology, it can cross over into the realm of scholarship more easily and more visibly because there are so many more opportunities for engagement with our peers, who then take it and build from it.

What are we trying to build?

No matter how we posit the space that DH occupies, many of these debates elide the affective labor that is required. Should the building of communities be as, if not more, important than the tools they produce? And do the community-building tools represent a way to dismantle traditional hierarchical power structures, including the Center?

But to start, how do we build community? Lynne Siemens, in the article "The Balance between On-line and In-person Interactions: Methods for the Development of Digital Humanities Collaboration," points to the importance of balancing the different kinds of contact and communication in order for a collaborative DH project to work.[36] It isn't until the end of the essay that she mentions the importance of developing trust to ensure a successful collaboration using "a variety of channels." Certainly, getting to know your collaborators through Twitter and Facebook are ways to develop trust through a more, for lack of a better word, intimate relationship. Those banalities that are often pointed to as being why social media should be avoided are those very things that help build relationships and trust between individuals. Working within open-access channels can also disrupt the hierarchical structures that have dominated higher education.

While once upon a time, academics saw each other once or twice a year at conferences, we can now more easily grow these connections and relationships into more intimate (and fruit-

---

36 Lynne Siemens, "The Balance between On-Line and In-Person Interactions: Methods for the Development of Digital Humanities Collaboration," *Digital Studies/Le champ numérique* 2, no. 1 (2010).

ful) relationships that benefit us. We "share" our triumphs, our bad days, our Pandora stations (replacing the mixed tape), our ideas, our favorite shows, our favorite tools. We are not "intimate" with everyone on our timelines or our "friends" list, and that is where email, the direct message, and Skype/Google chat can play a role. Much like writers and artists once intimated in letters to close friends that mixed the professional and the personal, we can begin by talking about a project at hand and end with a conversation about our lives at that moment. This is not a waste of time, but a way to grow trust, allowing for intimacy, and then from there, community.

What does community mean in the 21st century, and how does it specifically relate to digital humanities? We typically talk about "networks" now when we look at the relationships today because perhaps community is too utopian a concept for our particular moment. In their introduction to a special issue of *Digital Humanities Quarterly,* "Theorizing Connectivity: Modernism and the Network Narrative," Wesley Beal and Stacy Lavin talk about how some Modernists used what we now consider a network to "perform an important mediation of the period's impulses towards totalization and dispersal, unity and fragmentation that typify the period's tensions in, for example, the U.S.'s changing demographic makeup."[37] As traditional forms of community dissolved, the network as we now know came in to replace it. Now, networking can also symbolize the professionalization that has taken over higher education, where trust and distrust are replaced by elements such as cost-benefit analyses to judge the amount of time to invest in relationships. Communities, I think, are more necessary than ever in academia.

In the article "Community and Consumption: The Transformation of Social Space Online," Mark McGuire offers a number of definitions of community, but that tend to include "a group of people; social interaction; shared cultural practices; common

---

37 Wesley Beal and Stacy Lavin, "Theorizing Connectivity: Modernism and the Network Narrative," *Digital Humanities Quarterly* 5, no. 2 (2011).

geographic space; and a sense of belonging."[38] Today, most communities within DH would qualify as "communities of interest, which do not depend on shared space." Howard Rheingold
identifies the important element of intimacy in these virtual (or
semi-virtual, as the case may be) communities because these
"social aggregations [...] emerge from the Net when enough
people carry on those public discussions long enough, with sufficient human feeling, to form webs of personal relationships in
cyberspace."[39] Without "important and meaningful" communication, a virtual community cannot exist.

But these "virtual communities" have existed since before the
web. My experiences with the research "collaboratory" EMiC
(Editing Modernism in Canada) has reminded me of the period of literary growth that we all studied: the Modernist movement in Canada that was nurtured in large part by epistolary
dialogues carried out between many different members.[40] While
something like the Montreal Movement started in and around
McGill, it continued through letters and other forms of communication and collaboration from a distance once the members
dispersed because of jobs, family, or other reasons (including
seeking better opportunities in New York or the ex-pat experience in Paris). Frank Scott, quoted in Patricia Godbout's book
on "sociabilité interculturelle," laments this new distance between member of his artistic community:

> Yet, I still do not like seeing the rare good talent of this coun-
> try go elsewhere. Why, I wonder? Am I too nationalist? I
> think not; I think selfishness is at the bottom of it. I don't in-
> tend to leave myself (though I have been away several times)
> so I want more good people around me. Especially poets.[41]

---

38 Mark Maguire, "Community and Consumption: The Transformation of So-
    cial Space Online," *Digital Studies/Le champ numérique,* no. 10 (2005).

39 Howard Rheingold, *The Virtual Community: Homesteading on the Electron-
    ic Frontier,* rev. edn. (Cambridge: The MIT Press, 2000), xx.

40 *Editing Modernism in Canada,* http://editingmodernism.ca/.

41 Patricia Godbout, *Traduction littéraire et sociabilité interculturelle au Cana-
    da, 1950–1960* (Ottawa: University of Ottawa Press, 2004), 6.

Given the job market the way it is for academics, we can wish for more people who share our interest in DH and (in the case of EMiC) Modernist poetry and poets in Canada to live and work within our own institutions (or city, if we are so lucky), but the reality is that we go where the jobs are, traveling across the country, or (as in my case) leaving the country altogether. But we can still form these communities despite geography.

DH centers represent communities that have developed more normally, sharing both geography and common interests. But again, these physical centers can increasingly reach beyond their physical location; I'm thinking in particular of a place like HASTAC centered in Michigan (now Arizona), but reaching hundreds of DH enthusiasts through its extended community. Again, these kinds of communities have long existed, created and nurtured by aspiring artists and intellectuals. In his book, *When Canadian Literature Moved to New York,* Nick Mount describes the communities, both official and unofficial, that grew in New York to help support Canadian writers:

> Here, in New York, Canadian writers formed their country's first professional literary communities. At these gathering places they shared setbacks and successes, read and discussed each other's work, exchanged literary gossip, and argued new literary trends. They helped each other into print, passing on tips about copy-hungry editors and warning about those slow with a paycheque. Most directly, they published each other's work in the magazines they edited. [...] And, crucially for some Canadian writers, they extended the reach of these communities beyond New York, using their positions and connections to promote and publish the work of friends still at home and in other literary centers.[42]

DH Centers and "collaboratories" (I think ProfHacker and Grad-Hacker are two great examples of DH communities as well) grew

---

42 Nick Mount, *When Canadian Literature Moved to New York* (Toronto: University of Toronto Press, 2005), 15.

as an antidote to the potentially isolating and dehumanizing elements of academia. We do all of those things that Mount describes; we help each other, we publish each other, we give feedback to each other, we even, when we can, hire each other. That we happen to live dispersed around the world, meeting infrequently, makes no difference as we are able to maintain our bounds through the virtual communities we contribute to, participate in, and share. These communities, using a mixture of both open source and commercial digital tools, also help to challenge the profit-driven motives of both the modern university and modern society. As put by McGuire,

> The conversion of the Internet, which began as a public service project, into a collection of privately owned online communities, repeats this transformation in electronic space. In both cases, the image of what has been replaced is all that remains. This façade mitigates the loss and conceals the revised function.[43]

These companies (such as Amazon in McGuire's study) create "the illusion of community." Questions of "ownership and control" are some of the central issues that concern DH practitioners and theorists, but also in how we choose to use such tools mindfully and critically (practices that we also often use in our pedagogical approaches when teaching with/about these same tools).

The communities within the Digital Humanities also work to counter what Alan Liu describes in his book *The Laws of Cool* (paraphrased eloquently here by Andrew Prescott),

> how modern computing is an instrument of that managerial impulse which seeks to make knowledge work as mechanical and controlled as work on a production line. Liu reminds

---

43 See Alan Liu, *The Laws of Cool: Knowledge Work and the Culture of Information* (Chicago: University of Chicago Press, 2004) and Andrew Prescott, "An Electric Current of Imagination: What the Digital Humanities Are and What They Might Become," lecture, King's College London, 25 January 2012.

us how the aesthetics and language of computing, with its excitement about the latest "cool" medium, are a refuge from the grim reality of a cubicle in an open plan office on an industrial scale. In the end Liu sees the digital humanities as an escape from the tyranny of the cool.[44]

I have never, ever been cool in my life, so this seems like the perfect fit for me and I think many of us. But the "managerial" concerns (and financial limitations) of the modern university encroach on this ethos, forcing what McGann describes as "a haphazard, inefficient, and often jerry-built arrangement of intramural instruments, free-standing centers, labs, enterprises, and institutes, or special digital set up outside the traditional departmental structure of the university."[45] Is this, from an academic's (or alt-academic's) perspective, really a bad thing? I read his description of the situation of DH (in this particular case, the UK), I see a number of different kinds of communities that developed organically to meet the needs of the members, rather than the members bending their needs according to the limitations placed on them by the structure of the modern university. We see some of these limitations in the US when many DH programs are housed in English departments when the interests and services are much more than English (and specifically) literary interests. Here, some good microhistory would help us understand these communities, all seemingly anomalous, but clearly (if we can judge by McGann's tone) in need of a face.

These aren't the only challenges that DH communities face as they try to resist and yet thrive in a traditional academic environment. Susan Brown, director of the Orlando Project, as well as the Canadian Writers Research Collaboratory, has worked extensively on not only ways to encourage and facilitate collaboration within the research communities, but also how to

---

44 Prescott, "An Electric Current of Imagination."

45 Jerome McGann, *A New Republic of Letters: Memory and Scholarship in the Age of Digital Reproduction* (Cambridge: Harvard University Press, 2014), 131. Cited in Prescott, "An Electric Current of Imagination."

measure said contributions by individual academics within the community. This is a direct response to the pressures of modern academia to measure in the name of accountability and ultimately hiring and promotion decisions. While I don't disparage accountability, I do have an issue with artificial ways of "measuring" productivity that don't originate organically from the community. This also isn't to criticize the work that others are doing; I only wish to illustrate the tensions that have arisen between DH communities as they integrate themselves more fully within the traditional structures and expectations of academia.

Even working collaboratively is a challenge for most academics. Michael Best, in his article "'A Marvelous Convenient Place': Collaborations in the Electronic Text," outlines how academics, even academics working in DH, tend to approach collaboration, the "familiar pattern of most Humanities scholars, working on their own, meeting occasionally at conferences, but otherwise self-directed. In a project for which there conventions of publication have been well established, there is no particular reason for scholars to do otherwise than to collaborate with the books on their desks, but in the still new electronic medium, where there are as yet no strictly defined principles for the presentation of the text, collaboration between scholar and interface designer is vital if the potential of the medium is to be realized."[46] For the first time here, we've heard mention of someone outside of what we would traditionally consider the humanities: the interface designer. This is an even bigger challenge for digital humanists on both sides of the traditional divide: how does a humanist talk to a programmer and vice-versa? Indeed, the relationship between programmer/developer has usually been once sided: the developer develops something and we figure out how to use it (see, Word, Explorer, Microsoft in general). These kinds of collaborative DH communities further disrupt the traditional power structure and hierarchies both within the university and in society at large; again, we are expected to be consumers of

---

46 Michael Best, "'A Marvellous Convenient Place': Collaboration in the Electronic Text," *Digital Studies/Le champ numérique* 1, no. 1 (2009).

technology, not the makers of it unless we possess a certain almost magical skill. Open source software and projects challenge that traditional dichotomy, but how many humanists feel comfortable communicating their needs to the larger community which include "technical" people. We also need systems in place, systems that don't often exist within the university, to facilitate these kinds of collaborations.

But these aren't the only challenges that DH communities face. In fact, the nature of communities is that some are excluded. In his introductory essay for a special issue of Digital Studies, Brent Nelson outlines some of the large lines of the DH ethos: "The digital humanities are in the business of building bridges" and goes on to describe how "in the digital age community has indeed become scalable: even large groups can be made to feel that they occupy a common space in pursuit of common interests."[47] These bridges, to many, only seem to extend so far. In the recent (and, in some cases, already outdated!) collection, *Debates in the Digital Humanities,* two essays (and only two) deal explicitly with issues of race and none with the issue of class. In, "Why are the Digital Humanities So White?" Tara McPherson shows how the bridges between DH and Black Cultural Studies have not yet been adequately built, showing how important it is for those doing critical race studies to interrogate the foundations of computing and computational language.[48] Amy Earhart examines those elements that are still missing in the "open" ethos of DH, texts that originate from ethnic communities, often works and projects that once existed, but for lack of support, become lost.[49] She points to how "the canon" is still the primary impetus for funding and grants, leaving ethnic scholars on the outside of the funding circle necessary to sustain a vibrant DH community interested in working on large-scale projects.

---

47 Brent Nelson, "Introduction: Bridging Communities," *Digital Studies/Le champ numérique* 1, no. 3 (2009).

48 Tara McPherson, "Why are the Digital Humanities So White? or Thinking the Histories of Race and Computation," in *Debates in the Digital Humanities,* 139–60.

49 Earhart, "Can Information Be Unfettered?"

Take for example a debate that "exploded" on Twitter concerning the ethics of live-tweeting conference presentations (tweet away, I need all the exposure I can get). The discussion on Twitter was initiated by women of color, one of whom also identifies as queer. The debate was quickly trivialized (problematically by IHE, who I wrote for) and also dismissed by many members of the DH community as "been there, done that." This response angered Tressie McMillan Cottom and she took to her blog to vent her frustrations:

> Digital Humanities scholars said the debate was "dated." The tone of posts and articles and tweets was decidedly condescending. The line went something like this: are these silly people talking about something I wrote about once TWO YEARS AGO?! The exchange has been characterized as both trivial and humorous. To which I say: it must be nice to think power, privilege, privacy, status competition, and access are so damn funny. Except, wait. It's NOT funny when other people are talking about it…Is the new academic vanguard advocating for open access and dialogue or is it arguing to replace the existing elite with its own? It is a fair question, I think, but I will be sure to ask a white man to ask it so that I have a shot at an answer. Because the answer matters more to some of us than to others. Some can talk about discussions of privileged information with condescension. I cannot. I don't have that privilege. Who I am, in body and in ascribed status, is so entangled with these discussions of access and power that they cannot seem dated or humorous or inconsequential to me. And if there is no room in digital humanities or open access or the new academic model to realize that, then I'm not interested. I'd rather keep the master I do know than fight to legitimize one that I don't, if you please.[50]

---

50 Tressie McMillan Cottom, "Black, Female, Thinking Out Loud & #Twittergate," blog post, 5 October 2012.

This particular community clearly has not been bridged. I have heard similar critiques from those in other disciplines who have been doing what we now consider DH work for some time but have been excluded (particularly those who are in Rhetoric and Composition, traditionally the disrespected cousin in the English department). McMillan Cottom took to Twitter later to say that she is more interested in "the tools than the community. Anytime issues of race and tools are mentioned, I am reminded of Audre Lorde's essay: "The Master's Tools will never dismantle the master's house."[51] Have we reached the point where we have, perhaps because and through our increasing connections to traditional academic structures, simply re-created the same exclusionary structures within DH? Again, communities are necessarily exclusionary, and as observed by Roopika Risam, "some point in the future, the tribes would reform DH. The future is now."[52]

That tweet just quoted from Risam, and much of these discussions, are taken from three-plus years ago. I want to be optimistic about the potential of DH to be transformed, to be a site of resistance, to be a place where marginalized scholars come together and transform not just scholarship, but the institutions themselves. This paper is a combinations of two talks that I gave: one in 2012 at Western University and another in 2014 at the University of Cincinnati.[53] I've organized a panel at the MLA, Building Bridges in Digital Humanities, in 2013, as well as co-presented at DH 2013, "Digital Humanities: Egalitarian or the New Elite?"[54] This essay here is informed by what I wrote then, what was said during those panels, and the feedback I

51 Audre Lorde, "The Master's Tools Will Never Dismantle the Master's House," in *Sister Outsider: Essays and Speeches,* 110–14 (Berkeley: Crossing Press, 1984; reprint 2007).

52 Roopika Risam (@roopikarisam), Twitter post, 6 October 2012, 12:35 p.m.

53 Lee Skallerup Bessette, "Intimacy, Community, and Collectivity: Interdisciplinarity and Digital Humanities," Talk at Western University, 11 October 2012 and "Wither DH? New Tension, Directions, Evolutions in the Digital Humanities," Talk at the University of Cincinnati Libraries, 17 November 2014.

54 Lee Skallerup Bessette et al., "Expanding Access: Building Bridges within Digital Humanities," special session, MLA, 3 January 2013 and "Digital Hu-

received afterwards. But the time between then and now have not seen any great change in the literal color and tone within the digital humanities community. Recent controversies within the community saw the establishment of a "Working Group on Inclusivity."[55] It remains to be seen if this will lead to any real change. As Scott B. Weingart et al. point out, DH has not been the most inclusive of spaces, and it doesn't look like it has changed much.[56]

And so this essay collection appears, as do others, such as the forthcoming (as of this writing) *Feminist Debates in Digital Humanities,* edited by Jacque Wernimont and Elizabeth Losh, University of Minnesota Press. Are we publishing, speaking, writing for change or for tenure (or the hope of getting a tenure-track position)? This essay was a challenge to write because I myself also have no skin in the small-tent DH game anymore, so to speak; in an alt-academic, pedagogically focused position that doesn't require publishing. Not that I ever did, really. I came to DH naive and optimistic, and wanted to contribute to the discussion and process of making it, if not better, than something different and, dare I say it, aspirational, within the spaces of the current form of academia. DH will not whither and die, but it may not flourish as an alternative and aspirational model. The Center will remain and recreate.

---

manities: Egalitarian or the New Elite?" long paper, *Digital Humanities,* University of Nebraska-Lincoln, 17 July 2013.

55 "ADHO Establishes Working Group on Inclusivity," *Alliance of Digital Humanities Organizations,* http://adho.org/announcements/2013/adho-establishes-working-group-inclusivity.

56 Scott B. Weingart, Nickoal Eichmann, and Jeana Jorgensen, "Representation at Digital Humanities Conferences (2000–2015)," *the scottbot irregular* (blog), 22 March 2016 and Scott B. Weingart, "Acceptances to DH2016 (pt. 1)," *the scottbot irregular* (blog), 22 March 2016.

Bibliography

#TransformDH. http://transformdh.org/.

"ADHO Establishes Working Group on Inclusivity." *Alliance of Digital Humanities Organizations.* http://adho.org/announcements/2013/adho-establishes-working-group-inclusivity.

Beal, Wesley, and Stacy Lavin. "Theorizing Connectivity: Modernism and the Network Narrative." *Digital Humanities Quarterly* 5, no. 2 (2011). http://www.digitalhumanities.org/dhq/vol/5/2/000097/000097.html.

Best, Michael. "'A Marvellous Convenient Place': Collaboration in the Electronic Text." *Digital Studies/Le champ Numérique* 1, no. 1 (2009). https://www.digitalstudies.org/articles/10.16995/dscn.136/.

Biddle, Sam. "The Non-White Workers of Silicon Valley Get Screwed Every Day." *Valleywag.* 26 August 2014. http://valleywag.gawker.com/the-real-title-should-have-black-and-latino-instead-0-1627045918.

Bogost, Ian. "A New Department Marks the Rise of a Discipline: 'Computational Media.'" *Wired Campus* (blog). 13 October 2014. http://chronicle.com/blogs/wiredcampus/a-new-department-marks-the-rise-of-a-discipline-computational-media/54883.

Bourg, Chris. "Never Neutral: Libraries, Technology, and Inclusion." *Feral Librarian* (blog). 28 January 2015. https://chrisbourg.wordpress.com/2015/01/28/never-neutral-libraries-technology-and-inclusion/.

———. "The Once and Future Librarian." *Feral Librarian* (blog). 18 March 2015. https://chrisbourg.wordpress.com/2015/03/18/the-once-and-future-librarian/.

Center for Digital Research in the Humanities. University of Nebraska-Lincoln. http://cdrh.unl.edu/.

Center for Digital Humanities and Social Sciences. Michigan State University. http://www2.matrix.msu.edu/.

The Roy Rosenzweig Center for History and New Media. George Mason University. http://chnm.gmu.edu/.

Chen, Adrian. "The Laborers Who Keep Dick Pics and Beheadings Out of Your Facebook Feed." *Wired*. 23 October 2014. http://www.wired.com/2014/10/content-moderation/.

Cottom, Tressie McMillan. "Black, Female, Thinking Out Loud & #Twittergate." Blog post. 5 October 2012. https://tressiemc.com/2012/10/05/black-femalethinking-out-loud-twittergate/.

de jesus, nina. "Locating the Library in Institutional Oppression." *In the Library with the Lead Pipe* (blog). 24 September 2014. http://www.inthelibrarywiththeleadpipe.org/2014/locating-the-library-in-institutional-oppression/.

Earhart, Amy E. "Can Information Be Unfettered? Race and the New Digital Humanities Canon." In *Debates in the Digital Humanities,* edited by Matthew K. Gold, 309–18. Minneapolis: University of Minnesota Press, 2012.

Editing Modernism in Canada. http://editingmodernism.ca/.

Elk, Lauren Chief (@chiefelk). Twitter post. 25 March 2015, 12:23 p.m. https://twitter.com/chiefelk/status/580766904056303616.

Eve, Martin. "Universities, DH, 'the crowd', and Labor That Looks Like a Game." Blog post. 12 November 2014. https://www.martineve.com/2014/11/12/universities-dh-the-crowd-and-labour-that-looks-like-a-game/.

Global Outlook::Digital Humanities. http://www.globaloutlookdh.org/.

Godbout, Patricia. *Traduction littéraire et sociabilité interculturelle au Canada, 1950–1960.* Ottawa: University of Ottawa Press, 2004.

Golumbia, David. "Building and (Not) Using Tools in Digital Humanities." *uncomputing* (blog). 5 February 2013. http://www.uncomputing.org/?p=206.

———. "Definitions That Matter (Of 'Digital Humanities')." *uncomputing* (blog). 21 March 2013. http://www.uncomputing.org/?p=208.

The Institute of Advanced Technology in the Humanities. University of Virginia. http://www.iath.virginia.edu/.

Kim, Dorothy. "Social Media and Academic Surveillance: The Ethics of Digital Bodies." *Model View Culture.* 7 October 2014. https://modelviewculture.com/pieces/social-media-and-academic-surveillance-the-ethics-of-digital-bodies.

Kirschenbaum, Matthew G. "What Is Digital Humanities and What's It Doing in English Departments?" In *Debates in the Digital Humanities,* edited by Matthew K. Gold, 3–11. Minneapolis: University of Minnesota Press, 2012.

Koh, Adeline. "Introducing Digital Humanities Work to Undergraduates: An Overview." *Hybrid Pedagogy.* 14 August 2014. http://www.digitalpedagogylab.com/hybridped/introducing-digital-humanities-work-undergraduates-overview/.

———. *Digitizing "Chinese Englishmen": Representations of Race and Empire in the Nineteenth Century.* http://chineseenglishmen.adelinekoh.org/decolonizing-the-archive/.

Liu, Alan. *The Laws of Cool: Knowledge Work and the Culture of Information.* Chicago: University of Chicago Press, 2004.

Lorde, Audre. "The Master's Tools Will Never Dismantle the Master's House." In *Sister Outsider: Essays and Speeches,* 110–14. Berkeley: Crossing Press, 1984; reprint 2007.

Maguire, Mark. "Community and Consumption: The Transformation of Social Space Online." *Digital Studies/Le champ numérique,* no. 10 (2005). DOI: 10.16995/dscn.159.

Maryland Institute for Technology in the Humanities. University of Maryland. http://mith.umd.edu/. Accessed 29 June 2016.

McGann, Jerome. *A New Republic of Letters: Memory and Scholarship in the Age of Digital Reproduction.* Cambridge: Harvard University Press, 2014.

McPherson, Tara. "Why are the Digital Humanities So White? or Thinking the Histories of Race and Computation." In *Debates in the Digital Humanities,* edited by Matthew K. Gold, 139–60. Minneapolis: University of Minnesota Press, 2012.

Mount, Nick. *When Canadian Literature Moved to New York.* Toronto: University of Toronto Press, 2005.

Mukurtu. http://mukurtu.org/.

Murray, Heather. "Literary History as Microhistory." In *Home-Work: Postcolonialism, Pedagogy, and Canadian Literature.,* edited by Cynthia Sugars, 405–22. Ottawa: University of Ottawa Press, 2004.

Nelson, Brent. "Introduction: Bridging Communities." *Digital Studies/Le champ numérique* 1, no. 3 (2009). https://www.digitalstudies.org/articles/10.16995/dscn.97/.

Nowviskie, Bethany. "Asking for It." Blog post. 8 February 2014. http://nowviskie.org/2014/asking-for-it/.

———. "Speculative Computing and the Centers to Come." Blog post. 15 November 2014. http://nowviskie.org/2014/speculative-computing/.

Posner, Miriam. "Here and There: Creating DH Community." Blog post. 18 September 2014. http://miriamposner.com/blog/here-and-there-creating-dh-community/.

Prescott, Andrew. "An Electric Current of Imagination: What the Digital Humanities Are and What They Might Become." Lecture, King's College London, 25 January 2012. http://digitalriffs.blogspot.com/2012/01/electric-current-of-imagination-what.html.

Priego, Ernest. "Various Shades of Digital Literacy." HASTAC (blog). 22 October 2012. https://www.hastac.org/blogs/ernesto-priego/2012/10/22/various-shades-digital-literacy-new-digital-divides.

Pulizzi, James. "In the Near Future, Only Very Wealthy Colleges Will Have English Departments: Adapt (Not Publish) or Perish." *New Republic.* 8 June 2014. https://newrepublic.com/article/118025/advent-digital-humanities-will-make-english-departments-pointless.

Rheingold, Howard. *The Virtual Community: Homesteading on the Electronic Frontier.* Revised edition. Cambridge: The MIT-Press, 2000.

Risam, Roopika (@roopikarisam). Twitter post. 6 October 2012, 12:35 p.m. https://twitter.com/roopikarisam/status/254620824408035330?ref_src=twsrc%5Etfw.

Sample, Mark. "When Does Service Become Scholarship?" *samplereality* (blog). 8 February 2013. http://www. samplereality.com/2013/02/08/when-does-service-become-scholarship/.

Siemens, Lynne. "The Balance between On-Line and In-Person Interactions: Methods for the Development of Digital Humanities Collaboration." *Digital Studies/Le champ numérique* 2, no. 1 (2010). .

Skallerup Bessette, Lee. "Intimacy, Community, and Collectivity: Interdisciplinarity and Digital Humanities." Talk at Western University. 11 October 2012. Storify. https:// wakelet.com/wake/0e249301-2011-45ef-a0b7-df11f4eab6f8.

———— et al. "Expanding Access: Building Bridges within Digital Humanities." Special session. MLA. 03 January 2013. https://apps.mla.org/conv_listings_detail?prog_id=22&year=2013.

———— et al. "Digital Humanities: Egalitarian or the New Elite?" Long paper. *Digital Humanities.* University of Nebraska-Lincoln. 17 July 2013. http://dh2013.unl.edu/abstracts/ab-214.html.

————."Wither DH? New Tension, Directions, Evolutions in the Digital Humanities." Talk at the University of Cincinnati Libraries. 17 November 2014. *Storify.* https://wakelet.com/wake/d3cd5d38-c901-4fd9-a464-bcb2f1d8acbb.

Underwood, Ted. "Digital Humanities Might Never Be Evenly Distributed." *The Stone and the Shell* (blog). 7 September 2015. https://tedunderwood.com/2015/09/07/digital-humanities-might-never-be-evenly-distributed/.

Weingart, Scott B. "Acceptances to DH2016 (pt. 1)." t*he scottbot irregular* (blog). 22 March 2016. http://scottbot.net/acceptances-to-dh2016-pt-1/.

————, Nickoal Eichmann, and Jeana Jorgensen. "Representation at Digital Humanities Conferences (2000–2015)." *the scottbot irregular* (blog). 22 March 2016. http://scottbot.net/representation-at-digital-humanities-conferences-2000–2015/#note-41565–1.

Wen, Shawn. "The Lady Vanishes." *The New Inquiry.* 11 November 2014. http://thenewinquiry.com/essays/the-ladies-vanish/.

25

The Library Is Never Neutral

Chris Bourg

I want to acknowledge from the outset that this article has been heavily influenced by a number of people who have shared their work and their thoughts with me over the years. I've been privileged to learn from them, in some cases formally through their publications and in some cases through conversations on twitter or even in person. These aren't the only folks whose work and thinking have influenced me, but they are the key people I think of when I think of critical work on the intersections of libraries, technology, higher education, and social justice. These are their names — a mix of students, librarians, scholars, and technologists. Again, this is not a comprehensive list of the people whose work inspires me, but they are my top 7 right now on these topics.

- nina de jesus
- Myrna Morales
- Cecily Walker
- Bess Sadler
- Tressie McMillan Cottom
- Safiya Noble
- Andromeda Yelton

Let me also acknowledge that I'm well aware that the fact that I am a white woman working at an elite private US university that gives me access to a platform like this one to talk about issues of bias and exclusion in libraries and technology. But there are plenty of folks who have been and continue to talk about and write about these issues, with far more insight and eloquence than I can, but who don't get invitations like this for a variety of reasons. And the sad truth is that what I say, as an associate director at Stanford Libraries or as Director of MIT Libraries, often gets more attention than it deserves because of my title; while folks with less impressive titles and less privilege have been talking and thinking about some of these issues for longer than me and have insights that we all need to hear.

So next time you are looking for a speaker, please consider one of the names listed above.

A fundamental tenet that undergirds this article, and frankly undergirds much of the work I have done in and for libraries, is the simple assertion that libraries are not now nor have they ever been merely neutral repositories of information. In fact, I'm personally not sure "neutral" is really possible in any of our social institutions… I think of neutral as really nothing more than a gear in your car.

But what I mean when I say libraries are not neutral is not just that that libraries absorb and reflect the inequalities, biases, ethnocentrism, and power imbalances that exist throughout our host societies and (for those of us who work in academic libraries) within higher education.

I mean that libraries are not neutral in a more direct and active way.

For an exceptionally compelling take on libraries as not just not neutral, but as instruments themselves of institutional oppression, please read "Locating the Library in Institutional Oppression" by my friend and colleague nina de jesus.[1] nina argues that "Libraries as institutions were created not only for a specific

---

1   nina de jesus, "Locating the Library in Institutional Oppression," *In the Library with the Lead Pipe*, 24 September 2014.

ideological purpose, but for an ideology that is fundamentally oppressive in nature."[2] It is a bold argument, convincingly made; and I urge you to read it. As a bonus, the article itself is Open Access and nina elected to use only Open Access sources in writing it.

So I start with the premise that it isn't just that libraries aren't perfectly equitable or neutral because we live in a society that still suffers from racism, sexism, ableism, transphobia and other forms of bias and inequity; but libraries also fail to achieve any mythical state of neutrality because we contribute to bias and inequality in scholarship, and publishing, and information access.

Let me step back for a minute and own up to a few of my own biases — my library career thus far has been solely and squarely within large academic libraries; so my perspective, my examples, and my concerns come out of that experience and are likely most relevant to that sector of libraries. But, I hope we can have a conversation at the end of my talk about what the differences and similarities might be between the way these issues play out in large academic libraries and the way they play out in all kinds and sizes of libraries. I'm also definitely speaking from an American perspective, and I look forward to hearing where and how cultural differences intersect with the ideas I'll talk about.

OK — so libraries are not neutral because we exist within societies and systems that are not neutral. But above and beyond that, libraries also contribute to certain kinds of inequalities because of the way in which we exercise influence over the diversity (or lack thereof) of information we make available to our communities and the methods by which we provide access to that information.

I have a whole talk that I've given on how the collection development decisions we make impact not just how inclusive or not our own collections are, but also what kinds of books and authors and topics get published. The short version of that talk is that when we base our purchasing decisions on circulation and popularity, we eliminate a big part of the market for niche top-

---

2   Ibid.

ics and underrepresented authors. That is bad for libraries, bad for publishing, and bad for society. But that's another talk. This article is about library technologies.

But before we get into technology per se, I think a word about our classification systems is necessary, because the choices we make about how our technologies handle metadata and catalog records have consequences for how existing biases and exclusions get perpetuated from our traditional library systems into our new digital libraries.

Many of you are likely well aware of the biases present in library classification systems.

Hope Olson — one of the heroes of feminist and critical thinking in library science — has done considerable work on applying critical feminist approaches to knowledge organization to demonstrate the ways in which libraries exert control over how books and other scholarly items are organized and therefore how, when, and by whom they are discoverable.[3] Our classification schemes — whether Dewey Decimal or Library of Congress — are hierarchical, which leads to the marginalization of certain kinds of knowledge and certain topics by creating separate sub-classifications for topics such as "women and computers" or "black literature."

Let me give a couple of examples of the effects of this.

The power of library classification systems is such that a scholar browsing the shelves for books on military history is unlikely to encounter Randy Shilts's seminal work *Conduct Unbecoming: Gays & Lesbians in the US Military,* because that book has been given a call number corresponding to "Minorities, women, etc. in armed forces." In my own library at Stanford University, that means the definitive work on the history of gays and lesbians serving in the armed forces is literally shelved between *Secrets of a Gay Marine Porn Star* and *Military Trade* — a collection of stories by people with a passion for military men. Now I'm not

---

3   See, for example, Hope Olson, *The Power to Name: Locating the Limits of Subject Representation in Libraries* (Dordrecht: Kluwer Academic Publishers, 2002).

saying we shouldn't have books about gay military porn stars or about those who love men in uniform. I am saying that there is nothing neutral about the fact that the history of gay and lesbian service members is categorized alongside these titles, while the history of "ordinary soldiers" (that's from an actual book title) is shelved under "United States, History — Military."[4]

Another example is one I learned of from my friend and colleague Myrna Morales, and you can read about it in an article I co-authored with her and Em Claire Knowles. In that article, Myrna writes about her experience doing research for her undergraduate thesis on the Puerto Rican political activism that took place in NYC in the 1960s, with a special interest in the Young Lords Party.

Here is how Myrna described her experience:

> I first searched for the YLP with the subject heading "organizations," subheading "political organization," in the *Reader's Guide to Periodical Literature*. Here I found no mention of the YLP. I was surprised, as I had known the YLP to be a prominent political organization — one that addressed political disenfranchisement, government neglect, and poverty. A (twisted) gut feeling told me to look under the subject heading of "gangs." There it was — Young Lords Party. This experience changed my view of the library system, from one impervious to subjectivity and oppression to one that hid within the rhetoric of neutrality while continuing to uphold systemic injustices.[5]

I suspect that this kind of experience is all too common for people of color and other marginalized people who attempt to

---

4 More specifically, the Stanford University Libraries' call number for *Conduct Unbecoming* is UB418.G38 S55 1993, meanwhile *A People's History of the U.S. military: Ordinary Soldiers Reflect on Their Experience of War, from the American Revolution to Afghanistan* is found under E181.B535 2012.

5 Myrna Morales, Em Claire Knowles, and Chris Bourg, "Diversity, Social Justice, and the Future of Libraries," *Libraries and the Academy* 14, no. 3 (2014): 439–51, at 445.

use the resources we provide. I'll go so far as to wonder if these sorts of experiences aren't at least partially responsible for the incredibly low proportion of people of color who pursue careers in librarianship. So our traditional practices and technologies are not neutral, and without active intervention we end up with collections that lack diversity and we end up classifying and arranging our content in ways that further marginalizes works by and about people of color, queer people, indigenous peoples, and others who don't fit neatly into a classification system that sets the default as the western, white, straight, and male.

Of course, the promise of technology is that we no longer need rely on arcane cataloging rules and browsing real library stacks to discover and access relevant information. With the advent of online catalogs and search engines, books and other information items can occupy multiple "places" in a library or collection. But despite the democratizing promise of technology, our digital libraries are no more capable of neutrality than our traditional libraries; and the digital tools we build and provide are likely to reflect and perpetuate stereotypes, biases, and inequalities unless we engage in conscious acts of resistance.

Now when most people talk about bias in tech generally or in library technology, we talk about either the dismal demographics that show that white women and people of color are way underrepresented in technology, or we talk about the generally misogynistic and racist and homophobic culture of technology; or we talk about both demographics and culture and how they are mutually reinforcing. What we talk about less often is this notion that the technology itself is biased — often gendered and/ or racist, frequently ableist, and almost always developed with built in assumptions about binary gender categories.

For some folks, the idea that technologies themselves can be gendered, or can reflect racially based and/or other forms of bias is pretty abstract. So let me give a few examples. Most librarians will agree that commercial search engines are not "neutral" in the sense that commercial interests and promoted content can and do impact relevancy. Or, as my colleague Bess Sadler says, the idea of neutral relevance is an oxymoron.[6] Safiya Noble's work demonstrates how the non-neutrality of commercial search engines reinforce and perpetuate stereotypes, despite the fact that many assume the "algorithm" is neutral.[7]

What Noble's analysis of Google shows us is that Google's algorithm reinforces the sexualization of women, especially black and Latina women.[8] Because of Google's "neutral" reliance on popularity, page rank, and promoted content, the results for searches for information on black girls or Latina girls are dominated by links to pornography and other sexualized content. Noble suggests that users "Try Google searches on every variation you can think of for women's and girls' identities and you will see many of the ways in which commercial interests have subverted a diverse (or realistic) range of representations."[9] Search technologies are not neutral — just as basing collection development decisions on popularity ensures that our collections reflect existing biases and inequalities, so too does basing relevancy ranking within our search products on popularity ensure the same biases persist in an online environment.

But it isn't just search engines. In an article called "Teaching the Camera to see my skin," photographer Syreeta McFadden describes how color film and other photographic technologies were developed around trying to measure the image against white skin.[10] Because the default settings for everything from film stock to lighting to shutter speed were and are designed to best capture white faces; it is difficult to take photos of non-white faces that will be accurately rendered without performing

6 Bess Sadler and Chris Bourg, "Feminism and the Future of Library Discovery," *Code4Lib Journal* 28 (2015).

7 Safiya U. Noble, "Missed Connections: What Search Engines Say About Women," *Bitch Magazine* 54 (Spring 2012): 36–41. See also her article, "Google search: Hyper-Visibility As a Means of Rendering Black Women and Girls Invisible," *InVisible Culture* 19 (October 2013).

8 Ibid.

9 Noble, "Missed Connections," 39.

10 Syreeta McFadden, "Teaching the Camera To See My Skin," *Buzzfeed News*, 2 April 2014.

post-image adjustments that sacrifice the sharpness and glossy polish that is readily apparent in photos of white faces.[11]

Finally, in an example of a technology that betrays its lack of neutrality by what it ignores, Apple's recently released health app allows users to track a seemingly endless array of health and fitness related information on their iPhone. But strangely, Apple's health app did not include a feature for tracking menstrual cycles — an important piece of health data for a huge percentage of the population. As one critic noted, Apple insists that all iPhone uses have an app to track Stock prices — you can't delete that one from your phone — but fails to provide an option for tracking menstrual cycles in its "comprehensive" health tracking application.[12]

I hope these examples demonstrate that technology does not exist as neutral artifacts and tools that might sometimes get used in oppressive and exclusionary ways. Rather, technology itself has baked-in biases that perpetuate existing inequalities and exclusions, and that reinforce stereotypes. So how do we intervene, how do we engage in acts of resistance to create more inclusive, less biased technologies? Note that I don't think we can make completely neutral technologies … but I do think we can do better. One way we might do better is simply by being aware and by asking the questions that the great black feminist thinkers taught us to ask:

Who is missing?

Whose experience is being centered?

Many, many folks argued — rather convincingly to my mind — that the dearth of women working at Apple may have contributed to the company's ability to overlook the need for menstrual cycle tracking in its health app. So we might also work on recruiting and retaining more white women and people of color into library technology teams and jobs. There is much good work being done on trying to increase the diversity of the pipe-

---

11 Ibid.

12 Rose Eveleth, "How Self-Tracking Apps Exclude Women," *The Atlantic*, 15 December 2014.

line of people coming into technology — Black Girls Code and the Ada Initiative are examples of excellent work of this type.

I also think the adoption of strong codes of conduct at conferences and other library and technology events make professional development opportunities more welcoming and potentially safer for all — and I think those are important steps in the right direction. But in the end, one of the biggest issues we need to address if we truly want a more diverse set of people developing the technologies we use is the existence of a prevailing stereotype about who the typical tech worker is.

I want to turn now to some research on how stereotypes about who does technology, and who is good at it, affect how interested different kinds of people are in pursuing technology related fields of study, how well people expect they will perform at tech tasks, and how well people already working in tech feel they fit in, and how likely they are to stay in tech fields.

First a definition — stereotypes are widely shared cultural beliefs about categories of people and social roles. The insidious thing about stereotypes is that even if we personally don't subscribe to a particular stereotype, just knowing that a stereotype exists can affect our behavior.[13]

Second, a caution — much of this research focuses on gender, to the exclusion of intersecting social identities such as race, sexuality, or gender identity. In fact, in many studies of gender and attitudes towards or experiences in technology, results are reported according to a binary concept of gender; and rarely broken down by any other intersecting categories.[14] Clearly

13 See Cecilia L. Ridgeway and Chris Bourg, "Gender as Status: An Expectation States Approach," in *Psychology of Gender*, 2nd ed., ed. by Alice Eagly, Anne Beall, and Robert Sternberg, 217–41 (New York: Guilford, 2004). See also Cecilia L. Ridgeway, *Framed by Gender: How Gender Inequality Persists in the Modern World* (New York: Oxford University Press, 2011).

14 See David Beede, Tiffany Julian, David Langdon, George McKittrick, Beethika Khan, and Mark Doms, "Women in STEM: A Gender Gap to Innovation," *U.S. Department of Commerce, Economics and Statistics Administration Report,* 3 August 2011; Catherine Hill, Christianne Corbett, and Andresse St. Rose, "Why So Few? Women in Science, Technology, Engineering, and Mathematics," *American Association of University Women Re-*

more research is needed to capture the full range of experiences that marginalized people, especially those with multiple marginalized identities, have with and in technology.

That said, there is a huge body of research documenting the effect of negative stereotypes about women's math and science abilities.[15] These kinds of stereotypes lead to discriminatory decision making that obstructs women's entry into and advancement in science and technology jobs. Moreover, negative stereotypes about women and math affects women's own self-assessment of their skill level, interest, and suitability for science and technology jobs.

In a study of men and women working in Silicon Valley technology firms, Stanford sociologists Alison Wynn and Shelley Correll looked at the impact of how well tech workers felt they matched the cultural traits of a successful tech worker on a number of outcomes.[16]

First they developed a composite scale based on how tech employees, men and women, described successful tech workers. The stereotype that emerged was masculine, obsessive, assertive, cool, geeky, young, and working long hours.

Their data show that women tech workers are significantly less likely than their male counterparts to view themselves as fitting the cultural image of a successful tech worker. While that may not be a surprising finding, their research goes on to show that the sense of not fitting the cultural image has consequences.[17]

_____

port (American Association of University Women, 2010); Caroline Simard, Andrea Davies Henderson, Shannon K. Gilmartin, Londa Schiebinger, and Telle Whitney, Climbing the Technical Ladder: Obstacles and Solutions for Mid-Level Women in Technology (Anita Borg Institute for Women and Technology and the Michelle R. Clayman Institute for Gender Research at Stanford University, 2007).

15 Brian A. Nosek, Mahzarin R. Banaji, and Anthony G. Greenwald, "Math = Male, Me = Female, Therefore Math ≠ Me," *Journal of Personality and Social Psychology* 83, no. 1 (2002): 44–59.

16 Alyson T. Wynn and Shelley J. Correll, "Gendered Perceptions of Cultural and Skill Alignment in Technology Companies," *Social Sciences* 6, no.2 (2017), 45.

17 Ibid.

Because women are less likely to feel they fit the image of a successful tech worker, they are less likely to identify with the tech field, more likely to consider leaving the tech field for another career, and less likely to report positive treatment from their supervisors.

The bottom line is that cultural fit matters — not just in the pipeline, as women decide whether to major in STEM fields or to pursue tech jobs — but also among women who are currently working in technology. In other words, stereotypes about tech work and tech workers continue to hinder women even after they have entered tech careers. If we want to ensure that our technologies are built by diverse and inclusive groups of people, we have to find ways to break down the stereotypes and cultural images associated with tech work.

How do we do that?

If we want to look to success stories, Carnegie Mellon University is a good example. At Carnegie Mellon they increased the percentage of women majoring in computer science from 7% in 1995 to 42% in 2000 by explicitly trying to change the cultural image of computer scientists. Faculty were encouraged to discuss multiple ways to be a computer scientist and to emphasize the real world applications of computer science and how computer science connects to other disciplines. They also offered computer science classes that explicitly stated that no prerequisites in math or computer science were required.[18]

For libraries, we can talk about multiple ways to be a library technologist, and we can emphasize the value of a wide variety of skills in working on library tech projects — metadata skills, user experience skills, design skills. We can provide staff with opportunities to gain tech skills in low-threat environments and

---

18 Lenore Blum, "Women in Computer Science: The Carnegie Mellon Experience," 27 January 2001, http://www.cs.cmu.edu/~lblum/PAPERS/women_in_computer_science.pdf. See also "Press Release: Women Comprise 40 Percent of Computer Science Majors Among Carnegie Mellon's Incoming First-Year Class," *Carnegie Mellon University,* 5 June 2014 and "School of Computer Science Courses," *Carnegie Mellon University,* http://coursecatalog.web.cmu.edu/schoolofcomputerscience/courses/.

in environments where white women and people of color are less likely to feel culturally alienated.

RailsBridge workshops and AdaCamps seem like good fits here, and I'd like to see more library administrators encouraging staff from across their org's to attend such training.[19] At Stanford, my colleagues Bess Sadler and Cathy Aster started basic tech training workshops for women on the digital libraries' staff who were doing tech work like scanning, but who didn't see themselves as tech workers. Providing the opportunity to learn and ask questions, in a safe environment away from their supervisors and male co-workers gave these women skills and confidence that enhanced their work and the work of their groups.

Another simple way we can make progress within our own organizations is to pay attention to the physical markers of culture.

In a fascinating experimental study, psychologist Sapna Cheryan and colleagues found that women who enter a computer science environment that is decorated with objects stereotypically associated with the field — such as Star Trek posters — are less likely to consider pursuing computer science than women who enter a computer science environment with non-stereotypical objects — such as nature or travel posters. These results held even when the proportion of women in the environment was equal across the two differently decorated settings.[20]

We need to pay attention to the computer labs and maker spaces in our libraries, and we need to pay attention to physical work environments our technical staff work in. By simply ensuring that these environments aren't plastered with images and objects associated with the stereotypes about "tech guys," we will remove one of the impediments to women's sense of cultural fit.

So let me try to sum up here.

---

19 See the homepages for *Railsbridge,* http://railsbridge.org/ and *Ada Initiative,* https://adainitiative.org/.

20 See Sapna Cheryan et al., "The Stereotypical Computer Scientist: Gendered Media Representations as a Barrier to Inclusion for Women," *Sex Roles* 69, no. 1 (2013): 58–71.

I've argued that like libraries, technology is never neutral. I've offered examples from search engines to photography to Apple's health tracking app.

I've talked about how the pervasive stereotypes about who does tech work limit women's participation in tech fields, through both supply and demand side mechanisms.

The stereotypes about tech workers also contain assumptions about race and sexuality in the US context, in that the stereotypical tech guy is white (or Asian) and straight. Sadly, there is significantly less research on the effect of those stereotypes on black and Latino men and women and queer people who are also vastly underrepresented in technology work.

Let me offer some parting thoughts on how we might make progress.

We need to think and we need to do.

We need to think about the technology we use in our libraries, and ask where and how it falls short of being inclusive. Whose experiences and preferences are privileged in the user design? Whose experiences are marginalized? Then we need to do what we can to push for more inclusive technology experiences. We likewise need to be transparent with our patrons about how the technology works and where and how the biases built into that technology might affect their experience. The folks who do work in critical information literacy provide great models for this.

We should think about how libraries and library staff reinforces stereotypes about technology and technology work. Subtle changes can make a difference. We should drop the term "tech guy" from our vocabulary and we should ditch the Star Trek posters. I'd like to see more libraries provide training and multiple paths for staff to develop tech skills and to become involved in technology projects. We need to pay attention to the demographics and to the culture — and remember that they are mutually reinforcing.

We also need to remember that we aren't striving for neutral, and we aren't aiming for perfectly equitable and inclusive technology.

While neutral technologies are not possible — or necessarily desirable — I believe that an awareness of the ways in which technology embodies and perpetuates existing biases and inequalities will help us make changes that move us towards more inclusive and equitable technologies.

Work in the digital humanities frequently relies on library expertise and resources, but often in ways that are surprisingly uncritical. It seems to me that a truly disruptive digital humanities must shine a critical lens on the library itself — recognizing and interrogating the biases built into our collections; our staffing, expertise, and credentialing; our tools and technologies; and our systems for organizing and providing access to information.

Bibliography

Ada Initiative. https://adainitiative.org.

Beede, David, Tiffany Julian, David Langdon, George McKittrick, Beethika Khan, and Mark Doms. "Women in STEM: A Gender Gap to Innovation." *U.S. Department of Commerce, Economics and Statistics Administration Report.* 3 August 2011. http://www.esa.doc.gov/reports/women-stem-gender-gap-innovation.

Blum, Lenore. "Women in Computer Science: The Carnegie Mellon Experience." 27 January 2001. http://www.cs.cmu.edu/~lblum/PAPERS/women_in_computer_science.pdf.

Cheryan, Sapna, et al., "The Stereotypical Computer Scientist: Gendered Media Representations as a Barrier to Inclusion for Women." *Sex Roles* 69, no. 1 (2013): 58–71.

Eveleth, Rose. "How Self-Tracking Apps Exclude Women." *The Atlantic.* 15 December 2014. http://www.theatlantic.com/technology/archive/2014/12/how-self-tracking-apps-exclude-women/383673/.

Hill, Catherine, Christianne Corbett, and Andresse St. Rose. "Why So Few? Women in Science, Technology, Engineering, and Mathematics." *American Association of University Women Report.* American Association of University Women, 2010, http://www.aauw.org/resource/why-so-few-women-in-science-technology-engineering-mathematics.

de jesus, nina. "Locating the Library in Institutional Oppression." *In the Library with the Lead Pipe.* 24 September 2014. http://www.inthelibrarywiththeleadpipe.org/2014/locating-the-library-in-institutional-oppression/.

McFadden, Syreeta. "Teaching the Camera To See My Skin." *Buzzfeed News.* 2 April 2014. https://www.buzzfeed.com/syreetamcfadden/teaching-the-camera-to-see-my-skin?utm_term=.pc6oX7jnwO#.xdZoNm6GqP.

Morales, Myrna, Em Claire Knowles, and Chris Bourg. "Diversity, Social Justice, and the Future of Libraries." *Libraries and the Academy* 14, no. 3 (2014): 439–51. DOI: 10.1353/pla.2014.0017.

Noble, Safiya U. "Missed Connections: What Search Engines Say About Women." *Bitch Magazine* 54 (Spring 2012): 36–41.
———. "Google search: Hyper-Visibility As a Means of Rendering Black Women and Girls Invisible." *InVisible Culture* 19 (October 2013). http://ivc.lib.rochester.edu/google-search-hyper-visibility-as-a-means-of-rendering-black-women-and-girls-invisible/.

Nosek, Brian A., Mahzarin R. Banaji, and Anthony G. Greenwald. "Math = Male, Me = Female, Therefore Math ≠ Me." *Journal of Personality and Social Psychology* 83, no. 1 (2002): 44–59. http://www.ncbi.nlm.nih.gov/pubmed/12088131.

Olson, Hope. *The Power to Name: Locating the Limits of Subject Representation in Libraries.* Dordrecht: Kluwer Academic Publishers, 2002.

"Press Release: Women Comprise 40 Percent of Computer Science Majors Among Carnegie Mellon's Incoming First-Year Class." *Carnegie Mellon University.* 5 June 2014. http://www.cmu.edu/news/stories/archives/2014/june/june5_womenincomputerscience.html.

Railsbridge. http://railsbridge.org.

Ridgeway, Cecilia L., and Chris Bourg. "Gender as Status: An Expectation States Approach." In *Psychology of Gender,* 2nd edition, edited by Alice Eagly, Anne Beall, and Robert Sternberg, 217–41. New York: Guilford, 2004.
———. *Framed by Gender: How Gender Inequality Persists in the Modern World.* New York: Oxford University Press, 2011.

Sadler, Bess, and Chris Bourg. "Feminism and the Future of Library Discovery." *Code4Lib Journal* 28 (2015). http://journal.code4lib.org/articles/10425.

"School of Computer Science Courses." *Carnegie Mellon University.* http://coursecatalog.web.cmu.edu/schoolofcomputerscience/courses/.

Simard, Caroline, Andrea Davies Henderson, Shannon K. Gilmartin, Londa Schiebinger, and Telle Whitney. *Climbing the Technical Ladder: Obstacles and Solutions for Mid-Level Women in Technology.* Anita Borg Institute for Women and Technology and the Michelle R. Clayman Institute for Gender Research at Stanford University, 2007. https://www.gedcouncil.org/publications/climbing-technical-ladder-obstacles-and-solutions-mid-level-women-technology.

Wynn, Alyson T. and Shelley J. Correll, "Gendered Perceptions of Cultural and Skill nAlignment in Technology Companies," *Social Sciences* 6, no. 2 (2017), 45. DOI: 10.3390/socsci6020045.

26

After the Digital Humanities, or, a Postscript

Fiona Barnett

As a meditation on the theme of "disrupting" the digital humanities, I offer five moments of disruption for consideration:

1

At an MLA 2015 panel, #QueerOS: Queerness as Operating System, my fellow panelist Jacob Gaboury gave an amazing paper on "Compiling a Queer Computation."[1] He examined the possibility and purpose of a queer computation, and how it might be compiled — or how it might be uncomputable in some fundamental sense. As some of the examples, he offered a number of esolangs, or esoteric programming languages, including one of the most infamous, Brainfuck. Esolangs vary in purpose, but they all tend to work in that they can be used to write a program that actually compiles and runs, though usually for non-productive or humorous reasons.

---

[1] Session details can be found at "Queer OS: Queerness as Operating System," *Modern Language Association,* https://apps.mla.org/conv_listings_detail?prog_id=162&year=2015.

The disruption of an esolang isn't that it fundamentally won't compile, but that it doesn't operate according to the principles of mainstream languages which tend to privilege flexibility and productivity. The disruption isn't one of a stoppage, but a slow-down, a pointed diversion, a deliberate detour through opaque rules and oftentimes humorous commands.

2

While we were tweeting our #QueerOS panel, I was following the tweets from another DH panel happening at the same time. Folks in that room were tweeting about emulators. Meanwhile, folks in our room were tweeting as Zach Blas touched on the aesthetic and political appeal of Dildotectonics (as originally imagined by Beatriz Preciado).[2] Emulation often presumes an exact imitation or duplication of a system, whereas dildotecton-ics seems to presume a superfluous simulation, one which is sensitive to the slippage between mimesis and mimicry.

It's not that dildotectonics disrupt the history of emulators, so much as it showcases the limits of their presumed status as the definitive computational method of duplication and imita-tion. Sometimes disruptions point to the impossibility of the very system they are interrupting, and sometimes they are ex-posing just how pervasive and insidious that system might be.

3

As part of the ongoing Ferguson protests and social actions, groups have been staging "Black Brunches" as a way to point out the "people who have money and privilege have the leisure to brunch."[3]

The protests had already disrupted city streets, highway ramps, shopping malls, airport terminals and courthous-

---

2   See Beatriz Preciado, *kontrasexuelles manifest* (Berlin: b_books, 2003).

3   Justin Wm. Moyer and Nick Kirkpatrick, "#BlackBrunchNYC Protests 'White' N.Y. Breakfast Spots," *The Washington Post*, 5 January 2015.

es — and now weekend brunches. These disruptions are an especially ingenious method for pointing out that power operates not only in the arcane rules of our legal system, but in our bodies, spaces, communities and quotidian moments. The disruption is not intended to "ban brunch" as one commenter proclaimed, but to make visible and tangible the ways through which power is pervasive and invasive at a fundamental level, even at the breakfast table. Disrupting comfort is affective, performative, declarative and community-building. The intervention of the black brunches brings to mind the question of how to respond when you are "disrupted" — do you continue trying to eat your pancakes in peace? Or try to find out how to engage with the disruptors? Or file a complaint? Or even wonder if the disruption was not about you as an individual, but about a systemic structure of power and community?

4

A recent *New York Times* article, Among the Disrupted, begins with the treatise, "Amid the bacchanal of disruption, let us pause to honor the disrupted."[4] Here, the disrupted entity appears to be the human body and our social world before it was transformed by the tyranny of pervasive technology. What has been disrupted is a way of life, an ethos, an end to our humanism and an introduction to our posthumanity. The article declares that the concept of the human has been disrupted by the posthuman, a disruption, I might add, that apparently is not very welcome.

5

At a recent workshop I attended, the developers were showcasing a new mapping tool. They promised that the tool would not only be able to intake diverse data points, but also output many other formats which would instantly revolutionize how we interpreted the data. This claim to rationalism has a history, or

---

4   Leon Wieseltier, "Among the Disrupted," *New York Times,* 7 January 2015.

perhaps many histories, and doesn't always make visible the ethics of the proliferation of data.

We know that data doesn't determine its own outcome, and the desire for evidentiary proof isn't always enough. We saw this happen this past December, when a grand jury came back with the verdict not to indict the officer who killed Eric Garner, even though the video evidence offered exactly the kind of data called for by proponents of body cameras to prevent police brutality.

Evidence doesn't always offer its own interpretation. Evidence isn't rational. Evidence doesn't come with its own decryption code, or encryption code, or guide to interpretation. Data doesn't determine the way in which it can be mobilized, deciphered, or deployed. More evidence doesn't always make better evidence. The promise of data alone to disrupt our status quo completely negates the ways in which data and its interpretation have a history. This isn't to say we don't need new data, better data, or more ways of providing and interpreting and comparing data; but along with that we must foreground the questions we are asking of the data throughout its natural life.

In some ways, the Digital Humanities has been understood (or has staged itself) as "disrupting" the humanities, offering a new way to think and interact with other scholars, with materials, with objects and texts and disciplines. So does disrupting the digital humanities signal a return to... the humanities? Or is DH incommensurate with the sort of data we want it to compile and contain?

Bibliography

Moyer, Justin Wm., and Nick Kirkpatrick. "#BlackBrunchNYC Protests 'White' N.Y. Breakfast Spots." *The Washington Post.* 5 January 2015. https://www.washingtonpost.com/news/morning-mix/wp/2015/01/05/blackbrunchnyc-brings-police-brutality-protests-to-white-breakfast-spots/.

Preciado, Beatriz. *kontrasexuelles manifest.* Berlin: b_books, 2003.

"Queer OS: Queerness as Operating System." *Modern Language Association.* https://apps.mla.org/conv_listings_detail?prog_id=162&year=2015.

Wieseltier, Leon. "Among the Disrupted." *New York Times.* 7 January 2015. http://www.nytimes.com/2015/01/18/books/review/among-the-disrupted.html.

How to #DecolonizeDH: Actionable Steps for an Antifascist DH

Dorothy Kim

A number of groups have written manifestos to the digital humanities especially in relation to DH's issues with openness, race, disability, LGBTQ, feminist, and other kinds of non-normative bodies in the field.[1] There have been critical discussions in #transformdh, HASTAC, femtechnet and groups have been writing such manifestos and working to change DH. What I want to do here, then, is outline a set of practical steps to #decolonizedh, to make it less white, to begin working on an antifascist DH.

#DisruptDH

It is not a secret that DH as a field is a bastion of white masculinity. Its demographics are as terrible as what has been regularly criticized in tech circles in regards to the diversity of places like

---

1   In fact, you can often see these manifestos tweeted in entirety at the @DH-Manifesto_Bot and their site, *The Digital Manifesto Archive,* https://www.digitalmanifesto.net.

Twitter, Google, etc.[2] DH also has a problem acknowledging that women are prime movers and shakers in the field to the point that we have seen in Deb Verhoeven's talk at a recent major DH conference in Australia.[3] Colleagues in a feminist tech group had to create a "binder" of digital humanities women in order to point out to so many DH committees and organizations that there are women who can keynote and be on panels in a range of areas and subjects.[4] Jacqueline Wernimont writes:

> There are no more excuses for having an all-male panel, an all-male editorial board, an all-male DH qualifying exams reading list, an all male anything. [...] There are no more excuses. You know we are here and that we do damn fine work. Going forward, all-male panels can only be construed as a choice, not an issue of ignorance. We have been busy building the communities we want to see within DH, and now we've taken time from our reaching, our teaching, our lives to pull together information for you — now it's your turn to do your part.[5]

The fact that Femtechnet, and specifically Jacqueline Wernimont, had to create a "binder" in itself is telling. More recently, there has been furious conversation at the #DH2016 conference on the hashtag #DHDiversity about precisely the difficulties of inclusiveness in the field.

The discussions of the tech industry's "diversity" problems have moved away from just discussing inclusiveness, to address the interwoven prongs of white supremacy/white nationalist/ neo-nazi groups and their symbiotic relationship with the tech

2 Riley H., "Your Half-Assed Diversity Initiatives Aren't Going to Cut It in 2016," *Model View Culture,* 14 December 2016.

3 Deb Verhoeven, "Has Anyone Seen a Woman?" talk, Digital Humanities Conference, Sydney, Australia, 2 July 2015.

4 See Jacqueline Wernimont, "No More Excuses," blog post, 19 September 2015, https://jwernimont.wordpress.com/2015/09/19/no-more-excuses/ and "Build a Better Panel: Women in DH," blog post, 19 September 2015.

5 Wernimont, "No More Excuses."

industry and academia.[6] There can be no discussion about #De-colonizeDH or making DH less white without a discussion about resisting white supremacy/white nationalist/neo-nazi ideologies as well as bodies and organizations in our classrooms, campuses, epistemologies, field histories, and methodologies.

DH itself is a contested field and its definition is a work in progress. Various DH volumes have defined the digital humanities as DH1 or DH2[7]: DH1 being the kind of digital humanities work that builds things (like tools and archiving projects) and requires a knowledge of coding. DH2 is seen as the work that criticizes DH1 particularly in relation to gender, race, disability, access/class, sexuality.[8] There has been an ongoing argument of defining DH as only just what fits in the rubric of DH1. This particular move is bullshit and intellectually narrow, closing up the possibilities of inclusiveness, let alone finding ways to resist the weaponization of the digital in fascist regimes.

Such a move — one that completely imagines DH and any kind of archive or tool-building as a neutral act — participates in a rhetoric of heteropatriarchal, ableist, white supremacy. I say this because, in fact, data, algorithms, databases, even metadata,

---

6 See Matthew Rozsa, "Tech-bros and White Supremacists: A Union Based in Paranoia and Power," *Salon,* 6 October 2017; Joseph Bernstein, "Alt-white: How the Breitbart Machine Laundered Racist Hate," *BuzzFeed News,* 5 October 2017; David Lewis, "We Snuck into Seattle's Super Secret White Nationalist Convention," *The Stranger,* 4 October 2017; and Phillip Tracy, "March on Google Tries to Distance Itself from Alt-right White Supremacy," *The Daily Dot,* 15 August 2017.

7 See Melissa Terras, Julianne Nyhan, and Edward Vanhoutte, eds., *Defining the Digital Humanities: A Reader* (New York: Routledge, 2013); Stephen Ramsay, "DH Types One and Two," blog post, 3 May 2013; Matthew K. Gold, "The Digital Humanities Moment," in *Debates in the Digital Humanities,* ed. Matthew K. Gold, ix–xvi (Minneapolis: University of Minnesota Press, 2012); Ramsay, "On Building," blog post, 11 January 2011; and Ramsay, "Who's In and Who's Out," blog post, 8 January 2011.

8 Adeline Koh's article, "Niceness, Building, and Opening the Genealogy of the Digital Humanities: Beyond the Social Contract of the Digital Humanities," *differences: A Journal of Feminist Cultural Studies* 25, no. 1 (2014): 93–106, explains the complexities of this split as one of hack vs. yack. Her article also explains the complexities of imagining these rubrics as separate spheres.

all of these, are never neutral. And usually, the bodies attached to the data, algorithms, databases, etc. who are most harmed are communities of color or those with intersectional identities. In fact, the data, algorithms, databases, metadata, tools, archives, and methods of the digital humanities have become violently weaponized to attack, abuse, deport, assault, mob, silence, spy, imprison, and kill targeted non-white, non-cisgendered, non-male, non-Christian, and differently abled bodies.

Currently, DH is not a safe or comfortable space for most scholars who are not white, cisgendered, able, Christian, and upper middle-class males. And considering the current rise of late-fascism and the involvement of the "alt-tech" sector inter-secting with white academic supremacy to help create the nexus of what is called the "alt-right," I believe we have to move be-yond a discussion of diversity and inclusiveness to move into discussing DH justice and equity.[9] The digital world and aligned with that, academic digital humanities, helped shape and cre-ate the rise of the "alt-right" who have robustly used all digital means to weaponize the tools, methods, results, data, and struc-tures of this field to violently harm specific minority groups. Why, at this point, would anyone see these methods as benign? Why would the general public see them as benign? And why would we, as scholars, imagine them so? We must begin to dis-cuss how to decolonize DH. We must discuss how to deliberately create structures and frames for an antifascist DH — to deliber-ately dismantle the methodologies, epistemologies, structures, data, databases, tools, archives, code that have created a world in which technology is now used to consistently bludgeon its most vulnerable denizens.

In fact, I was spurred to write this piece on how to decolo-nize DH because of what Annemarie Perez says in her piece in this collection, "Lowriding through the Digital Humanities." In it, she offers an account of her experience in DH sessions over many years at MLA:

---

9   Dafina-Lazarus Stewart, "Language of Appeasement," *Inside Higher Ed,* 30 March 2017.

I'm taking a long time telling this. It is because the memory is painful. The panels and workshops I attended were a shock. Not only because the work was so exciting, especially, for me, the pedagogy, the mapping and time lining and other amazing projects. But because even at MLA, even at a literature conference, I had never experienced a stronger sense of being racially/ethnically other. The rooms, crowded to bursting were visibly, notably white spaces. This was a bit jarring, but what was even more so was that no one was taking about this. No one was asking where the brown people were. The absence of racialized bodies was un-noted.[10]

If nothing else, this volume and its pieces consistently mark and note the paucity of racialized bodies and other non-normative bodies and voices in DH.

In addition, almost all of the pieces in this volume, precisely grapple with this whiteness and lack of inclusiveness. From Mongrel Coalition Against Gringpo's "mongrel cliff notes" to Eunsong Kim's "The Politics of Visibility" which explores Reina Gossett's point that "visibility is a pillar of criminalization, not a tenant of liberation" and also Grace Hong's argument that "visibility is a rupture, an impossible articulation" for women of color.[11] Or Annemarie Perez's aforementioned "Lowriding through the Digital Humanities," where she writes an autoethnographic counternarrative of the spaces of the digital humanities. And in this way, these counternarratives are the central node of this collection. My piece then is a call to arms, a listicle, a checklist, a guide to how to make these non-normative bodies the center of what is the digital humanities. It is also a call to resist, dismantle, disarm, and find other ways to push back against

---

10 Annemarie Perez, "Lowriding through the Digital Humanities," *Disrupting the Digital Humanities: Digital Edition,* 6 January 2016.

11 "We Cannot Live Without Our Lives: A Conversation on Anti-blackness, Trans Resistance and Prison Abolition," forum, University of California, San Diego, 4 November 2014. See also Grace Kyungwon Hong, *The Ruptures of American Capital: Women of Color Feminism and the Culture of Immigrant Labor* (Minneapolis: University of Minnesota Press, 2006), xxviii.

the worldwide encroachment and centering of white heteropatriarchy and the "alt-right." We must center the idea that "methodology needs to be decolonised. The process of its decolonisation is an ethical, ontological and political exercise rather than simply one of approach and ways of producing knowledge."[12]

What I think many people hoped for the digital humanities was that technological access to a larger public would mean that communities of color, LGBTQ communities, differently abled bodies would finally get a chance to have their narratives told and their archives curated. But in fact, what has happened is that the digital humanities, particularly the kind of DH that builds projects and applies for government and foundation funding, has mostly reified and made more extreme the inequities we have seen in scholarship in regards to what gets discussed, whose narratives get published, what communities have a voice. Digital technologies (social media, surveillance, algorithms, databases, etc.) are now being used to deliberately silence and harm the most vulnerable voices. Thus, though we can read the announcements for the NEH grants and see big projects related to major figures of the American and English canon — Shakespeare, Melville, etc. — projects related to communities of color, narratives of disability, the lives and archives of the LBGTQ communities are rare to the point that their inclusion feels like a form of tokenization.

In the US political climate, I wonder about the ability of government agencies to break from the nazi/fascist ideologies of the Trump White House administration. We have already seen the dismantling of the EPA, the State Department, the Education Department, why do we imagine the NEH or NSF will not also turn into an arm of the fascist state? Culture is important to the ideologies of fascist administrations, but it's a culture that is shaped and messaged to back a white supremacist/white nationalist worldview. Before the recent rise in fascist governments, amazing projects like the *South Asian American Digital*

---

12   Sabelo Ndlovu-Gatsheni, "Decolonising Research Methodology Must Include Undoing Its Dirty History," *The Conversation*, 26 September 2017.

Archive ran on shoe strings and were powered by the uncompensated labor, dedication and hopes of committed librarians, scholars, and students who want to illuminate the narratives of their communities and tell polyvocal histories.[13] Now I wonder whether these vulnerable projects will be mobbed by the "alt-right." One significant bright spot is that I finally see a move by these granting agencies in regards to current issues and the racial climate in the US. In particular, I was delighted to see the librarian at Washington University St. Louis — who did such an amazing job archiving the Twitter feed of #Ferguson and #BlackLivesMatter — become the Librarian of the Carter Presidential library. Her work has garnered major funding to find a way to archive and curate social media.[14] Yet, now that bright spot also has to address the fact that Black Live Matter has been discussed by the FBI terrorism unit as "the black identity extremists" and a "threat."[15] Will these projects, especially government-funded ones, be used as a form of surveillance and information gathering against Black Lives Matter? Or will they be defunded without comment, as is the case with the UC Berkeley Black Panther documentary project whose $98,000 National Park Service grant was yanked without comment after a complaint from the local police union?[16]

At another MLA 2016 session ("Repair and Reparations in Digital Public Spaces" organized by Adeline Koh and Annemarie Perez) on archives for communities of color, several scholars pointed out the complexities of creating narratives for communities who usually do not get their own histories and are not part

13 *South Asian American Digital Archive,* https://www.saada.org/.

14 See "Meredith Evans Named New Director of Jimmy Carter Library," *Atlanta Georgia News,* 16 November 2016 and Ed Summers, "Introducing Documenting the Now," blog post, *Maryland Institute for Technology in the Humanities,* 16 February 2016.

15 Sam Levin, "FBI Terrorism Unit Says 'Black Identity Extremists' Pose a Violent Threat," *The Guardian,* 7 October 2017.

16 Angela Helm, "Government Yanks $98,000 for UC Berkeley Black Panther Project after Complaints from Police Union," *The Root,* 30 October 2017.

of the mainstream Western canon.[17] These communities often do not trust forms of institutional, government, and other kinds of "purported" philanthropy and especially any form of digital marking or counting. They have logical reasons for this fear. As we have seen in the use of DACA data to potentially deport the Dreamers in the US, handing over information to government agencies comes with extreme and often violent risks.[18]

This is precisely why things like Mukurtu were created to give consent and control to these native and first nation communities, who would not otherwise trust government agencies or universities (i.e., the military and university industrial complex) to "archive" them and their memories/documents/histories.[19] Sometimes they have also hacked corporate open-access, online platforms to create depositories and archives that are not permanent but rather ephemeral nodes.[20] For example, several Native American communities have used Facebook and its photo library as a way to crowd-source and collect the pictures in people's personal archives. This is then accessible to these communities and can hold some sort of shape and be used but remain an impermanent archive. DH has failed to embrace ephemerality and must acknowledge itself as at an "incubator stage."[21]

For scholars of color and other intersectional scholars, it already takes so much labor — political, social, and cultural capital — to work ethically and with respect for various communities. Then to juggle that and the politics, funding cycles, and whiteness of DH, these scholars are already doing double or triple the work, and usually on shoe string budgets with donated labor. And now they have to work in increasingly hostile, viru-

---

17 Adeline Koh, "#MLA 2016 Proposal: Repair and Reparations in Digital Public Spaces," blog post, 20 March 2015.

18 Ted Hesson, "Dreamers Fear Deportations from DACA Data," *Politico*, 5 September 2017.

19 Ibid. Siobhan Senier explained this explicitly in her talk for this panel. See also *Mukurtu*, http://mukurtu.org.

20 See Koh, "Repair and Reparations," specifically Senier's talk again.

21 See Johanna Drucker, *Graphesis: Visual Forms of Knowledge Production* (Cambridge: Harvard University Press, 2014).

lently white supremacist spaces in which they are under direct surveillance. This will deplete any scholar of color, making it increasingly difficult for them to live healthily.[22]

What practical steps, then, must we take to make DH less white, less heternormative, less cisgendered, and less ableist? What steps are required to begin to create an antifascist and antiracist DH? Our colleagues in STEM, whose funding, academic lines, and research are even more heavily dependent on government and foundational grants, have already organized an antifascist resistance with a list of actionable steps in SAFE (Scientists against a Fascist Establishment).[23] Before this, Chanda Prescod-Weinstein had been using her Medium platform to begin and add to a Decolonize Science Reading List.[24]

The steps here (and the continued outlining of them) should begin with undergraduates. In fact, the students on our campuses see the political stakes and issues of our work more clearly and with more political urgency than we probably do — as they have been protesting consistently to decolonize the university.[25] We do have a problem with the pipeline. There are so few scholars of color and non-normative bodies within academia, it is going to take more direct and explicit efforts to get more variant bodies into these spaces and it needs to start early. However, creating a robust pipeline is only one set of labors on one end of the spectrum. The other set requires a systematic shift in thinking, decolonizing our practices, addressing systematic structural problems in our cultures, hiring, and promotion. These include issues related to how few scholars of color get hired and how

---

22 For an excellent discussion on how academia depletes its scholars of color, see Sara Ahmed, "Feeling Depleted?" *feministkilljoys* (blog), 17 November 2013.

23 Chanda Prescod-Weinstein, "SAFE Actions: Proposed Actions from Scientists Against a Fascist Establishment," *Medium,* 1 February 2017.

24 Chanda Prescod-Weinstein, "Decolonising Science Reading List: It's the End of Science as You Know It," *Medium,* 25 April 2015.

25 See Krista Mahr, "South Africa's Student Protests Are Part of a Much Bigger Struggle," *Washington Post,* 23 September 2016 and Nick Roll, "Taking a Knee on Campus," *Inside Higher Ed,* 27 September 2017.

bias works.[26] And this goes beyond implicit bias, but must include discussions of deliberate white supremacist/white nationalist bias that targets certain marginal groups. I think it would be utterly naïve and some form of academic white innocence to imagine that just as the "alt-right" has infiltrated tech, "liberal" journalism, the entertainment industry, it has also not infiltrated academia and particularly the academic area in the humanities most connected to digital technology.[27] As Deb Verhoeven explained in her Australian talk, it also requires that white men step down from taking up the room. These efforts must be conscious, planned, and consistently set as a priority. Simply put, good intentions without planning, structure, and labor to make changes is not going to shift any of these demographics or the cultures in which they breed.

This then is a way to answer the manifestos, to come up with a practical guide, a listicle of how to #decolonizeDH.

A practical guide to decolonize DH (polyvocal, multitudes) and to make it less white (+cishetero, ableist, and male) and to begin shaping an antifascist DH

1. **The myth of "if you build it, they will come" must stop.** DH needs to make concerted efforts to bring scholars of color and scholars working on non-traditional areas, minority communities, different perspectives into the mix by incentivizing their presence in DH, with targeted and directed (1) Mentorship; (2) Seed money for projects; (3) Resources (both technical and in terms of working through ways to move projects along); (4) Money for conferencing and networking. DH must ask the questions pointed out by Dafina-Lazarus Stewart: "Equity responds: 'Who is trying to get in the room but can't? Whose presence in the room is under constant threat

---

26 Stefanie K. Johnson, David R. Hekman, and Elsa T. Chan, "If There's Only One Woman in Your Candidate Pool, There's Statistically No Chance She'll Be Hired," *Harvard Business Review,* 26 April 2016.

27 Bernstein, "Alt-white."

of erasure?' [...] Equity responds, 'What conditions have we created that maintain certain groups as the perpetual majority here?' [...] Justice challenges, 'Whose safety is being sacrificed and minimized to allow others to be comfortable maintaining dehumanizing views?'"[28]

2. **DH needs to stop being defensive about its whiteness** (particularly its insistence on compiling a list of "projects" about communities of color). That is not the point. The bodies at the margins are being violently attacked. If you are not digging in to help, fight back, and do the work against white supremacy, you are wasting our time.

3. **DH must stop ignoring critical race theory and postcolonial/ decolonial theory.** DH needs to stop pretending critical race and postcolonial/decolonial theory have not been discussed, theorized, prototyped, and implemented by scholars of color for decades. This begins with #inclusive citation, #inclusive panels, #inclusive syllabi, #inclusive grants, etc. The computational methods used by DH are implicated and complicit in helping organize the Holocaust, creating and running the algorithms that help targeted racist information cascades, deporting DACA and refugees from the US. Academics are not innocent nor have we ever been in our history. DH needs to ask how to dismantle and decolonize its standard histories, epistemologies, and methodologies.

4. **DH must find a way to break the "only lonely" system.** Stewart continues: "Equity answers, 'What are people experiencing on campus that they don't feel safe when isolated and separated from others like themselves?'"[29] This is a question that DH should ask and constantly work to change. This means creating inclusive DH cohorts so that non-white, non-cishetero, non-male groups do not feel alienated while trying to work in DH arenas.

5. **DH must have separate funds for inclusive projects.** DH must earmark separate money for projects related to and run

---

28 Stewart, "Language of Appeasement."
29 Ibid.

by communities of color, faculty of color, graduate students of color, etc. They must be separate and specifically geared to expand this range of work, give credit, give funding, give resource help.

6. **DH must stop writing narratives that attempt to wipe from memory the existence of entire fields.** DH must stop ignoring or coopting new media, computers and writing, digital rhetoric, and digital pedagogy as integral and central fields in its history, praxis, and future.

7. **DH must stop excessively citing white men.** DH must, as Sara Ahmed discusses, stop creating conference and panel structures that replicate white genealogies. In other words, stop citing only white men. Stop having only #manels. Stop having collections about the digital humanities almost entirely made up of white men.

8. **DH must decolonize its conferences and panels.** Likewise, DH must decolonize the biggest conferences in the field and start to apportion out panels and presence by a different standard of inclusiveness. Organizing committees must find participants and panelists that represent the larger populations of their worlds. Follow the models laid out by the decolonizing science colleagues who have collaborated with indigenous groups to discuss the epistemologies, structures, and methodologies of the field. DH must decolonize to make itself a "tool of resistance." But this can only be done with the most marginal groups centered and leading these conversations.[30]

9. **DH methods must not be only about tools.** Digital methods classes must stop being just about tools. They must include a balance of discussing critical issues like race, gender, disability, multimodality, sexuality, etc. Otherwise, graduate training works to further replicate the same frames that have made DH so white and so male and so focused on tool-building. All DH classes need to be decolonized. If they are not, they are upholding white supremacist heteropatriarchy.

---

30 Sophie Duncan, "Zapatistas Reimagine Science as Tool of Resistance," *Free Radicals,* 5 April 2017.

10. DH **must fund scholarships for training scholars of color.** DH training needs to directly give scholarships and particularly try to assemble groups to help potential scholars of color learn new skills but also these groups can allow people to talk to each other about some of the issues they see at stake and potentially find other collaborators.

11. DH **must have further resources for scholars of color.** Help DH scholars of color meet with publishers and funding officers to talk about both DH funding applications and how to thread their research work in multiple directions to get full credit in promotion and tenure. Scholars with diverse bodies are not going to do DH work unless they know they can get credit.

12. DH **must prioritize HBCUs, community colleges, and minority-serving institutions.** The schools in the trenches of inclusive higher education are not places like Yale or even the SLAC where I've taught, Vassar. They are community colleges, HBCUs, and minority-serving institutions. There must be separate and ear-marked funds, training, mentorship for faculty at these institutions and a pool of resources and money for them to create projects with their students.

13. DH **must train and include adjuncts.** As the statistics have shown that over 70% of the US professoriate are adjuncts, we must address how to find funds, support, infrastructure, and help to bring in the largest population of faculty in the country. We must train and include adjuncts. We must make sure they are on our panels, collections, grant proposals.

14. DH **must use the Rooney Rule.** The *Washington Post* article had a great suggestion on how to diversify Traditionally White Universities by using what Football has used, the Rooney Rule. Namely the requirement that at least 1 minority must be interviewed for every senior position. Facebook has decided to begin using this rubric. What would happen if the Rooney Rule were used in the digital humanities for every position, every major grant, every major conference keynote and panel?

As the *Washington Post* recently pointed out:

> African Americans make up 13 percent of the U.S. population
> and 15 percent of the enrolled student population at America's
> colleges, but only 5.5 percent of all full-time faculty are black.
> Back in 2007, when the black faculty rate was 5.4 percent, the
> *Journal of Blacks in Higher Education* predicted black faculty
> rates would reach parity with the percentage of blacks in the
> United States in about 140 years. Long time coming. Unfor-
> tunately, between 2009 and 2011, black faculty rates actually
> slipped back a little. So, that original prediction might be off
> by a generation or two […]. Remarkably, 96 percent of black
> tenured faculty are at HBCUs (even though HBCUs comprise
> only 3 percent of the nation's 3000 colleges and universities).
> If HBCUs disappeared, so would most of the nation's black
> academics […]. We had a 43 percent increase in the number
> of black PhDs between 2000 and 2010, but during that time
> black faculty appointments at TWIs increased only 1.3 per-
> cent. This is not a crisis of supply.[31]

There are people of color who work in various DH areas who
have Ph.Ds. Are the DH jobs going to minority candidates in the
field? As DH programs tend not to be established at HBCUs, what
exactly then does this say about the demographics of the digital
humanities? Or does this require that our students begin to protest DH centers, DH programs, DH organizations with signs calling to decolonize the field and calling for #BlackLivesMatter in
academia before DH begins to make concerted efforts to change?

I am of course giving DH the benefit of the doubt here. Its power brokers could respond like so many universities have already
responded by demonstrating further white fragility and defensiveness. As I said earlier, that is not the point. Isn't the point to

---

31 See Valerie Strauss, "The Academy Awards Isn't Alone with Its Color Problem, Look at Higher Education," *Washington Post*, 29 January 2016 and "It's
2015, Where Are All the Black College Faculty?" *Washington Post*, 12 November 2015.

make the digital humanities matter to communities of color, LG-BTQIAA, international, global south, disability communities, etc. The largest social and political phenomena of the last six years have implicated the digital (#arabspring, #solidarityisforwhite-women, #TrayvonMartin, #NotYourAsianSidekick, #Ferguson, #BlackLivesMatter) and the public. When is DH going to tell these narratives as the center of their methodological praxis?

The final goal should be about the benchmarks of equity and justice. Stewart explains:

> Equity celebrates reductions in harm, revisions to abusive systems and increases in supports for people's life chances as reported by those who have been targeted. [...] Justice celebrates getting rid of practices and policies that were having disparate impacts on minoritized groups."[32]

This volume has been entirely about speaking.

> *If we don't speak, if we don't prompt one another to speak, then yes, we are left with silence. Where has that gotten us so far?*
> — Audrey Watters[33]

> *Your silence will not protect you.*
> — Audre Lorde[34]

---

32 Stewart, "Language of Appeasement."
33 Audrey Watters, "On Silence," blog post, 16 August 2014.
34 Audre Lorde, "The Transformation of Silence into Language and Action," in *Sister Outsider: Essays and Speeches by Audre Lorde,* 40–44 (Berkeley: Crossing Press, 1984; reprint 2007), 41; cited in Watters's "On Silence."

Bibliography

Ahmed, Sara. "Feeling Depleted?" *feministkilljoys* (blog). 17 November 2013. https://feministkilljoys.com/2013/11/17/ feeling-depleted/.

Bernstein, Joseph. "Alt-white: How the Breitbart Machine Laundered Racist Hate." *BuzzFeed News.* 5 October 2017. https://www.buzzfeed.com/josephbernstein/heres-how-breitbart-and-milo-smuggled-white-nationalism.

The Digital Manifesto Archive. https://www.digitalmanifesto. net.

Drucker, Johana. *Graphesis: Visual Forms of Knowledge Production.* Cambridge: Harvard University Press, 2014.

Gold, Matthew K. "The Digital Humanities Moment." In *Debates in the Digital Humanities,* edited by Matthew K. Gold, ix–xvi. Minneapolis: University of Minnesota Press, 2012.

H., Riley. "Your Half-Assed Diversity Initiatives Aren't Going to Cut It in 2016." *Model View Culture.* 14 December 2016. https://modelviewculture.com/pieces/your-half-assed-diversity-initiatives-arent-going-to-cut-it-in-2016.

Helm, Angela. "Government Yanks $98,000 for UC Berkeley Black Panther Project after Complaints from Police Union." *The Root.* 30 October 2017. https://www.theroot. com/national-park-service-yanks-98–000-for-black-panther-p-1819968960.

Hesson, Ted. "Dreamers Fear Deportations from DACA Data." *Politico.* 5 September 2017. http://www.politico. com/story/2017/09/05/dreamers-fear-deportation-immigrants-242351.

Hong, Grace Kyungwon. *The Ruptures of American Capital: Women of Color Feminism and the Culture of Immigrant Labor.* Minneapolis: University of Minnesota Press, 2006.

Johnson, Stefanie K., David R. Hekman, and Elsa T. Chan. "If There's Only One Woman in Your Candidate Pool, There's Statistically No Chance She'll Be Hired." *Harvard Business Review.* 26 April 2016. https://hbr.org/2016/04/

if-theres-only-one-woman-in-your-candidate-pool-theres-statistically-no-chance-shell-be-hired.

Koh, Adeline. "Niceness, Building, and Opening the Genealogy of the Digital Humanities: Beyond the Social Contract of the Digital Humanities." *differences: A Journal of Feminist Cultural Studies* 25, no. 1 (2014): 93–106. DOI: 10.1215/10407391-2420015.

———. "#MLA 2016 Proposal: Repair and Reparations in Digital Public Spaces." Blog post. 20 March 2015. http://www.adelinekoh.org/blog/2015/03/30/mla-2016-proposal-repair-and-reparations-in-digital-public-spaces/.

Levin, Sam. "FBI Terrorism Unit Says 'Black Identity Extremists' Pose a Violent Threat." *The Guardian.* 7 October 2017. https://www.theguardian.com/us-news/2017/oct/06/fbi-black-identity-extremists-racial-profiling.

Lewis, David. "We Snuck into Seattle's Super Secret White Nationalist Convention." *The Stranger.* 4 October 2017. http://www.thestranger.com/news/2017/10/04/25451102/we-snuck-into-seattles-super-secret-white-nationalist-convention.

Lorde, Audre. "The Transformation of Silence into Language and Action." In *Sister Outsider: Essays and Speeches by Audre Lorde,* 40–44. Berkeley: Crossing Press, 1984; reprint 2007.

Mahr, Krista. "South Africa's Student Protests Are Part of a Much Bigger Struggle." *Washington Post.* 23 September 2016. https://www.washingtonpost.com/news/worldviews/wp/2016/09/23/south-africas-student-protests-are-part-of-a-much-bigger-struggle/.

"Meredith Evans Named New Director of Jimmy Carter Library." *Atlanta Georgia News.* 16 November 2016. http://www.ajc.com/news/lifestyles/meredith-evans-named-new-director-of-jimmy-carter-/npN2C/.

Mukurtu. http://mukurtu.org.

Ndlovu-Gatsheni, Sabelo. "Decolonising Research Methodology Must Include Undoing Its Dirty History." *The Conversation.* 26 September 2017. https://theconversation.

com/decolonising-research-methodology-must-include-undoing-its-dirty-history-83912.

Perez, Annemarie. "Lowriding through the Digital Humanities." *Disrupting the Digital Humanities: Digital Edition.* 6 January 2016. http://www.disruptingdh.com/lowriding-through-the-digital-humanities/.

Prescod-Weinstein, Chanda. "Decolonising Science Reading List: It's the End of Science as You Know It." *Medium.* 25 April 2015. https://medium.com/@chanda/decolonising-science-reading-list-339fb773d51f.

———. "SAFE Actions: Proposed Actions from Scientists Against a Fascist Establishment." *Medium.* 1 February 2017. https://medium.com/@chanda/safe-actions-980add3088e1.

Ramsay, Stephen. "Who's In and Who's Out." Blog post. 8 January 2011. http://stephenramsay.us/text/2011/01/08/whos-in-and-whos-out/.

———. "On Building." Blog post. 11 January 2011. http://stephenramsay.us/text/2011/01/11/on-building/.

———. "DH Types One and Two." Blog post. 3 May 2013. http://stephenramsay.us/2013/05/03/dh-one-and-two/.

Roll, Nick. "Taking a Knee on Campus." *Inside Higher Ed.* 27 September 2017. https://www.insidehighered.com/news/2017/09/27/inspired-kaepernick-and-nfl-professors-and-students-protest-field.

Rozsa, Matthew. "Tech-bros and White Supremacists: A Union Based in Paranoia and Power." *Salon.* 6 October 2017. https://www.salon.com/2017/10/06/tech-bros-and-white-supremacists-a-union-based-in-paranoia-and-power/.

South Asian American Digital Archive. https://www.saada.org/.

Stewart, Dafina-Lazarus. "Language of Appeasement." *Inside Higher Ed.* 30 March 2017. https://www.insidehighered.com/views/2017/03/30/colleges-need-language-shift-not-one-you-think-essay.

Strauss, Valerie. "It's 2015, Where Are All the Black College Faculty?" *Washington Post.* 12 November 2015. https://www.washingtonpost.com/news/answer-sheet/wp/2015/11/12/its-2015-where-are-all-the-black-college-faculty/.

———. "The Academy Awards Isn't Alone with Its Color Problem, Look at Higher Education." *Washington Post.* 29 January 2016. https://www.washingtonpost.com/news/answer- sheet/wp/2016/01/29/the-academy-awards-isnt-alone-with-its-color-problem-look-at-higher-education/.

Summers, Ed. "Introducing Documenting the Now." Blog post. *Maryland Institute for Technology in the Humanities.* 16 February 2016. http://mith.umd.edu/introducing-documenting-the-now/.

Terras, Melissa, Julianne Nyhan, and Edward Vanhoutte, eds. *Defining the Digital Humanities: A Reader.* New York: Routledge, 2013.

Tracy, Phillip. "March on Google Tries to Distance Itself from Alt-right White Supremacy." *The Daily Dot.* 15 August 2017. https://www.dailydot.com/debug/march-on-google-alt-right/.

Verhoeven, Deb. "Has Anyone Seen a Woman?" Talk, Digital Humanities Conference, Sydney, Australia, 2 July 2015. http://speakola.com/ideas/deb-verhoeven-has-anyone-seen-a-woman-2015.

Watters, Audre. "On Silence." Blog post. 16 August 2014. http://audreywatters.com/2014/08/16/on-silence.

"We Cannot Live Without Our Lives: A Conversation on Anti-blackness, Trans Resistance and Prison Abolition." Forum, University of California, San Diego, 04 November 2014. http://www.reinagossett.com/events/live-without-lives-conversation-anti-blackness-trans-resistance-prison-abolition-cece-mcdonald-janetta-johnson-eric-stanley/.

Wernimont, Jacqueline. "No More Excuses." Blog post. 19 September 2015. https://jwernimont.wordpress.com/2015/09/19/no-more-excuses/.

———. "Build a Better Panel: Women in DH." Blog post. 19 September 2015. https://jwernimont.com/2015/09/19/build-a-better-panel-women-in-dh/.

Contributors

Moya Bailey is an assistant professor in the Department of Cultures, Societies, and Global Studies and the program in Women's, Gender, and Sexuality Studies at Northeastern University. Her work focuses on marginalized groups' use of digital media to promote social justice as acts of self-affirmation and health promotion. She is interested in how race, gender, and sexuality are represented in media and medicine. She currently curates the #transformDH Tumblr initiative in Digital Humanities. She is also the digital alchemist for the Octavia E. Butler Legacy Network.

Maha Bali is Associate Professor of Practice at the Center for Learning and Teaching at the American University in Cairo. International Director of Digital Pedagogy Lab, Editor at the journal Hybrid Pedagogy, editorial board member of *Teaching in Higher Education, Online Learning Journal, Journal of Pedagogic Development* and *Learning, Media and Technology*. She is an open and connected educator and a writeaholic/learnaholic. She is a *Prof Hacker* blogger, DML *Central* blogger, and co-founder and co-director of virtuallyconnecting.org. She also blogs at http://blog.mahabali.me and tweets @bali_maha.

Fiona Barnett is director of HASTAC Scholars and a PhD candidate in the Literature Program and Women's Studies at Duke University. For the past five years, she has been the Director of HASTAC Scholars, an award-winning program for over 200 interdisciplinary graduate students around the world. She has overseen the community of emerging scholars and has developed dozens of highly viewed topical forums on topics such as race and queer theory in the digital age. She is a founding member of the #transformDH collective and continues to develop scholarly projects at the intersection of queer theory, race studies and the digital humanities. Her dissertation project, *Turning the Body Inside Out,* is a genealogy of the desire to see the inside of the body. In 2013 she was named a Future Leader of Higher Education by the Association of American Colleges & Universities.

Kathi Inman Berens, Assistant Professor of Book Publishing and Digital Humanities in Portland State University's English department, adjuncted for three years between full-time jobs. She aims to open avenues of access for adjuncts and other faculty at non-elite institutions to practice digital humanities. In addition to her research in DH, book publishing and electronic literature, Kathi specializes in digital pedagogy. She curated the keyword "Interface" for *Digital Pedagogy in the Humanities,* MLA's first open access publication. She pioneered a simultaneously embodied/virtual class at USC's Annenberg School of Communication, where she was a Research Fellow at the Annenberg Innovation Lab.

Lee Skallerup Bessette is a Learning Design Specialist at Georgetown University. She has published on issues around contingency, technology, digital pedagogy, as well as Canadian Literature.

Chris Bourg is the Director of Libraries at Massachusetts Institute of Technology (MIT), where she also has oversight of the MIT Press. Before assuming her role at MIT, Chris worked for 12 years in the Stanford University Libraries. Prior to that, she served for 10 years as an active duty US Army officer, including three years on the faculty at the United States Military Academy at West Point. She is currently the Chair of the Committee on Diversity and Inclusion of the Association of Research Libraries, and is on the Steering Committee of SocArXiv, a new open access platform for social science research. She is also a member of the Board of Directors for the Digital Public Library of America (DPLA), and a member of the Harvard Board of Overseers Committee to Visit the University Library.

Edmond Y. Chang is an Assistant Professor of English at Ohio University. He earned his PhD from the University of Washington and his dissertation is entitled "Technoqueer: Re/con/figuring Posthuman Narratives." His areas of interest include technoculture, gender and sexuality, video game studies, popular culture, and contemporary American literature. Recent publications include "Queergaming" in Queer Game Studies (University of Minnesota Press) and "A Game Chooses, A Player Obeys: *BioShock,* Posthumanism, and the Limits of Queerness" in Gaming Representation (Indiana University Press). He is currently working on his first book tentatively called *Queerness Cannot Be Designed: Digital Games and the Trouble with Technonormativity.*

Cathy N. Davidson is HASTAC's cofounding director (co-director, as of 2017). She is Distinguished Professor and Founding Director of The Futures Initiative at the Graduate Center, City University of New York. She served as an appointee, by President Obama, to the National Council of the Humanities from 2011 to 2017 and is on the Board of Directors of Mozilla. Her most recent book is, *The New Education: How To Revolutionize the University To Prepare Students for a World in Flux* (Basic Books, 2017).

Chris Friend is Assistant Professor of English at Saint Leo University, Director of *Hybrid Pedagogy,* and the producer of the *HybridPod* podcast. His scholarship explores the interactions among teaching, learning, and course delivery modes with a specific emphasis on critical pedagogy in digital spaces. He is particularly interested in how technologies of connection and communication influence writing-studies pedagogy. He tweets about all this @chris_friend and posts his digital work to http://chrisfriend.us.

Richard H. Godden is Assistant Professor of English at Louisiana State University, and has published in *postmedieval, Arthuriana,* and *New Medieval Literatures.* He is also co-author with Jonathan Hsy of "Analytical Survey: Encountering Disability in the Middle Ages." His current book project, *Material Subjects: An Ecology of Prosthesis in Medieval Literature and Culture,* focuses on the materiality of prosthesis in later medieval literature and culture, and he is also co-editor of the forthcoming essay collection, *Monstrosity, Disability, and the Posthuman in the Medieval and Early Modern World,* and of *The Open Access Companion to The Canterbury Tales.* He also works on the alliances between Disability Studies and the fields of Composition, Pedagogy, and Digital Humanities. His essay (co-written with Anne-Marie Womack) "Making Disability Part of the Conversation: Combatting Inaccessible Spaces and Logics" appeared in *Hybrid Pedagogy.*

Jonathan Hsy is Associate Professor of English at George Washington University, where he is also founding co-director of the GW Digital Humanities Institute. He specializes in medieval literature and culture with interests in translation theory, digital media, and disability studies. He is the author of *Trading Tongues: Merchants, Multilingualism, and Medieval Literature* (2013), co-director of *Global Chaucers,* and co-editor of Bloomsbury's *A Cultural History of Disability in the Middle Ages* (in progress). His current individual book project investigates life writing by medieval and postmedieval authors who self-identify as blind or deaf, and his publications on disability and digital media have appeared in *Accessus, Cambridge Companion to the Body in Literature, New Medieval Literatures* (co-authored essay with Richard H. Godden), *Journal of Literary and Cultural Disability Studies, postmedieval,* and PMLA. He blogs at *In The Middle,* a group medieval studies blog.

Spencer D.C. Keralis is a scholar of the past, present, and future of the book. He is a former Council on Library & Information Resources (CLIR) Fellow, and has held has held a Research Residency at the Queer Zine Archive Project, an Andrew W. Mellon Fellowship at the Library Company of Philadelphia, and a Legacy Fellowship at the American Antiquarian Society. His scholarship has appeared in *Texas Library Journal, Book History, American Periodicals,* the CLIR reports *The Problem of Data* (2012) and *Research Data Management: Principles, Practices, and Prospects* (2013), and is forthcoming in the Modern Language Association publication *Digital Pedagogy in the Humanities: Concepts, Models, and Experiments.* He is the Founding Director of Digital Frontiers, the largest and longest-running digital humanities conference in the southwest US. He holds a PhD in English and American Literature from New York University.

Dorothy Kim teaches Medieval Literature at Brandeis University. She was a 2013–2014 Fellow at the University of Michigan's Frankel Institute of Advanced Judaic Studies where she finished a monograph entitled *Jewish/Christian Entanglements: Ancrene Wisse and its Material Worlds* which is forthcoming from the University of Toronto press. She also has two books, *Digital Whiteness and Medieval Studies* and *Decolonize the Middle Ages,* forthcoming in 2018 with ArcPress/Western Michigan University Press. She is the co-project director in the NEH-funded Scholarly Editions and Translations project *The Archive of Early Middle English* that plans to create a 161 MSS database for medieval English manuscripts from 1100–1348 that include all items in Early Middle English. She is co-editing two collections in the Digital Humanities. The first collection, co-edited with Jesse Stommel (University of Mary Washington) on *Disrupting the Digital Humanities* (2018, punctum books). The second collection, co-edited with Adeline Koh on *Alternative Histories* *of the Digital Humanities* (forthcoming 2018, punctum books), examines the difficult histories of the digital humanities in relation to race, sexuality, gender, disability, fascism. She has co-written articles on "#GawkingatRapeCulture," "TwitterEthics," and written articles about "TwitterPanic" and "Social Media and Academic Surveillance" at Modelviewculture.com. She is the medieval editor for the Orland Project 2.0 and can be followed @dorothyk98. She was named by *Diverse: Issues in Higher Ed* 2015 Emerging Scholar under 40.

Eunsong Kim is an Assistant Professor in the English Department at Northeastern University. Her essays on literature, digital cultures, and art criticism have appeared and are forthcoming in: *Journal of Critical Library and Information Studies, Scapegoat, Lateral, The New Inquiry, Model View Culture,* AAWW's *The Margins,* and in the book anthologies, *Poetics of Social Engagement* and *Reading Modernism with Machines.* Her book, *gospel of regicide* was published in 2017 by Noemi press and she was the recipient of a 2015 Andy Warhol Foundation Arts Writers Grant for the arts forum *contemptorary,* which she co-edits.

Adeline Koh was an associate professor at Stockton University and an independent web designer. She is now the founder of a skincare product startup, Sabbatical Beauty. She's passionate about teaching, web design, technology, and the darker side behind tech: inequality and oppression. She writes a lot about gender, race, ethnicity, issues in higher education, digital pedagogy and the digital humanities.

Katarzyna Lecky is an Assistant Professor of English at Bucknell University. Her first book manuscript, *Pocket Empire: Portable Maps and Public Poetry, 1590–1649,* places Renaissance poetry into conversation with small-format cartography to show that Spenser, Jonson, Milton, and others banked their success as professional writers on portraying Britain as the property of the commonwealth rather than the Crown. She has published essays in *Exemplaria,* the *Journal of Medieval and Early Mod-*

ern Studies, Reformation, Studies in English Literature, Spenser Studies, and edited collections, and has earned fellowships and grants from the American Council of Learned Societies and the Andrew Mellon Foundation, the National Endowment for the Humanities, the Renaissance Society of America, and the Folger, Huntington, and Newberry Libraries, among others.

Mongrel Coalition Against Gringpo DECOLONIZE TO DIE | DECOLONIZE TO LIVE

Michelle Moravec is associate professor of history at Rosemont College in Philadelphia, PA. She is a member of the American Historical Association's Digital History Standards Committee and the Digital Projects editor at *Women and Social Movements.* Her writing on digital history has appeared in The Chronicle of Higher Education, Inside Higher Education, the University of Venus, and Women in Higher Education, and in a monthly column for the Mid-Atlantic Research Center for the Humanities.

Sean Michael Morris is Director of Digital Pedagogy Lab and Lead Instructional Designer in the Division of Teaching and Learning Technologies at University of Mary Washington. He is the former English Department chair for the Community Colleges of Colorado Online, and has been teaching digitally for 16 years. His research focuses on the nature of design, the practice of Critical Digital Pedagogy within and outside of institutions of education, and on pedagogy as a social justice practice. He identifies strongly with the post-digital and he learns, theorizes, and teaches from a contemplative perspective. You can follow him on Twitter @slamteacher, and his personal website can be found at seanmichaelmorris.com.

Angel Nieves is Associate Professor and Director of the American Studies Program at Hamilton College, Clinton, New York. He is currently Co-Directing Hamilton's Digital Humanities Initiative (DHi) funded by the Mellon Foundation. He is also Research Associate Professor in the Department of History at the

University of the Witwatersrand in Johannesburg, South Africa. His scholarly work and community-based activism critically engages with issues of memory, heritage preservation, gender and nationalism at the intersections of race and the built environment in cities across the Global South. His co-edited book *"We Shall Independent Be:" African American Place-Making and the Struggle to Claim Space in the U.S.* was published in 2008. He has published essays in numerous journals including *Journal of Planning History, Places Journal,* and *Safundi: The Journal of South African and American Studies,* and in several edited collections, including *Making Humanities Matter* (2017). His digital research and scholarship have been featured on MSNBC.com and in *Newsweek.*

Annemarie Perez is an Assistant Profesor of Interdisciplinary Studies at California State University Dominguez Hills. Her specialty is Latinx literature, with a focus on Chicana feminist writer-editors from 1965-to the present. Her interests include digital humanities and digital pedagogy work and its intersections and divisions with ethnic and cultural studies. She is currently writing a book on Latina feminist editorship using digital tools such as SCALAR to perform close readings across multiple editions and versions of early Latina articles and essays.

Laura Emily Sanders holds a PhD from the University of Southern California and is the interim associate division dean of Liberal Arts and Pre-College at the Portland Community College Southeast Campus. In this new role, she hopes to help address the many challenges adjuncts face by drawing upon over nine years of her experience teaching at community colleges in Oregon and California, in addition to serving as teaching learning center coordinator, online English faculty mentor, and community-based learning coordinator. Laura has taught composition and rhetoric at private research institutions, small liberal arts colleges, state universities, and community colleges since 1992. In recent years, she has served as co-editor of an accreditation self-study, interim grants officer, and academic department assessment coach. Combining her passions for professional development and social justice, Laura continues to seek the sweet spot between digital humanities and online community-based learning.

Liana Silva is a high school teacher, editor, and writer in Houston, TX. She is also the Managing Editor for the academic blog Sounding Out! She obtained her PhD from SUNY Binghamton. You can find her on Twitter at @liana_m_silva. Her website is www.lianamsilva.com and her blog is www.wordsaremygame. com

Bonnie Stewart is an educator and social media researcher fascinated by who we are when we're online. An instructor in the Faculty of Education at the University of Prince Edward Island, and Founder/Director of the media literacy initiative Antigonish 2.0, Bonnie is interested in the intersections of knowledge, technology, and identity. As Director of Edactive Technologies, Inc., Bonnie keynotes and consults about digital strategy, digital pedagogy, and community capacity-building around the world. Bonnie has been a teacher and facilitator for more than twenty years, and she enjoys few things more than a lively discussion. She does her best thinking out loud on Twitter as @bonstewart.

Shyam Sharma is Assistant Professor of Writing and Rhetoric at Stony Brook University (State University of New York). His research and teaching centers on writing in the disciplines, language policy, cross-cultural communication, and emerging media in writing studies and higher education.

Jesse Stommel is Executive Director of the Division of Teaching and Learning Technologies at University of Mary Washington. He is also Executive Director of *Hybrid Pedagogy: an open-access journal of learning, teaching, and technology* and Co-founder of Digital Pedagogy Lab. Jesse is a documentary filmmaker and teaches courses about digital pedagogy, film, and new media. Jesse experiments relentlessly with learning interfaces, both digital and analog, and works in his research and teaching to emphasize new forms of collaboration. He's got a rascal pup, Emily, two clever cats, Loki and Odin, and a badass daughter, Hazel. He's online at jessestommel.com and @Jessifer.

Matt Thomas was educated at the University of Southern California and the University of Iowa, where he got his PhD in American Studies. He currently lives, teaches, and writes in Iowa City, IA.

Audrey Watters is a writer who focuses on education technology — the relationship between politics, pedagogy, business, culture, and ed-tech. She has worked in the education field for over 15 years: teaching, researching, organizing, and project-managing. Although she was two chapters into her dissertation (on a topic completely unrelated to ed-tech), she decided to abandon academia, and she now happily fulfills the one job recommended to her by a junior high aptitude test: freelance writer. Audrey has written for *The Atlantic, Edutopia, MindShift, Inside Higher Ed, The School Library Journal, The Huffington Post,* and elsewhere across the web, in addition to her own blog *Hack Education.* She is the author of *The Monsters of Education Technology, The Revenge of the Monsters of Education Technology,* and *Claim Your Domain — And Own Your Online Presence.*

Robin Wharton holds a PhD and a JD from The University of Georgia, and she is currently a lecturer in English at Georgia State University, where she teaches courses in digital and multimodal rhetoric and composition, and designs professional development workshops related to digital pedagogy and scholarship for faculty and graduate students. In her classes, she engages students in public-oriented service learning and interdisciplinary digital humanities scholarship. Wharton is the technical development lead and co-editor of the *Hoccleve Archive* (hocclevearchive.org), a collection of resources for study of the life and works of fifteenth-century poet Thomas Hoccleve, and a contributing and special projects editor at *Hybrid*

Pedagogy. When she's not teaching, writing, or coding, she enjoys immersing herself in the adventures of motherhood, reading good books, and dancing."

Meg Worley is an associate professor of Writing & Rhetoric and Film & Media Studies at Colgate University. She is the editor of the *Ormulum Digital Critical Edition,* published by the *Archive of Early Middle English,* and has written articles on a variety of medieval and contemporary topics.

33434727R00286

Made in the USA
San Bernardino, CA
22 April 2019